An Introduction to Genetic Engineering

Fourth Edition

The fourth edition of this popular textbook retains its focus on the fundamental principles of gene manipulation, providing an accessible and broad-based introduction to the subject for beginning undergraduate students. It has been brought thoroughly up to date with new chapters on the story of DNA and genome editing, and new sections on bioethics, significant developments in sequencing technology and structural, functional and comparative genomics and proteomics, and the impact of transgenic plants. In addition to chapter summaries, learning objectives, concept maps, glossary and key word lists, the book now also features new concluding sections, further reading lists and websearch activities for each chapter to provide a comprehensive suite of learning resources to help students develop a flexible and critical approach to the study of genetic engineering.

Desmond S. T. Nicholl was Senior Lecturer in Biological Sciences, Head of Bioscience, Head of Quality Enhancement and Assistant Dean for Education at the University of the West of Scotland. As well as three previous editions of *An Introduction to Genetic Engineering*, he also authored *Cell and Molecular Biology* (Learning & Teaching Scotland, 2000).

'Genetic engineering represents a toolbox that all students within the basic and applied biology fields must get acquainted with. The fourth edition of *An Introduction to Genetic Engineering* is an excellent up-to-date version of a classic textbook. This ambitious book excellently balances the molecular biology knowledge required to grasp the comprehensive gene technology toolbox with a discussion of its impact on society.'

Per Amstrup Pedersen, University of Copenhagen

'As a biomedical engineering professor teaching an undergraduate Genetic Engineering course for close to 10 years, I use Dr Nicholl's *An Introduction to Genetic Engineering* as my go-to textbook. It is not one of those overly thick textbooks that overwhelm students. Its comprehensiveness captures readers' attention with succinct fundamental concepts that truly promote one's interest in exploring the wonder of many genetic engineering techniques and applications. To facilitate that further, the material provided at the end of each chapter encourages readers to expand their learning with relevant resources ... Many of my students become so interested that they pursue graduate degrees and have a career in this field. Dr Nicholl's textbook has a long-term influence on its readers.'

M. Ete Chan, State University of New York at Stony Brook

'Dr Nicholl's book covers all the basic material that one would expect from its title, but what particularly impressed me was how it isn't afraid to move into political and socio-economic arenas. In Chapter 16, for example, balanced arguments are presented for and against the development of transgenic organisms, and these don't always come out in favour of the science.'

Neil Crickmore, University of Sussex

An Introduction to Genetic Engineering

Fourth Edition

Desmond S. T. Nicholl

Shaftesbury Road, Cambridge CB2 8EA, United Kingdom

One Liberty Plaza, 20th Floor, New York, NY 10006, USA

477 Williamstown Road, Port Melbourne, VIC 3207, Australia

314–321, 3rd Floor, Plot 3, Splendor Forum, Jasola District Centre,
New Delhi – 110025, India

103 Penang Road, #05–06/07, Visioncrest Commercial, Singapore 238467

Cambridge University Press is part of Cambridge University Press & Assessment,
a department of the University of Cambridge.

We share the University's mission to contribute to society through the pursuit of
education, learning and research at the highest international levels of excellence.

www.cambridge.org
Information on this title: www.cambridge.org/highereducation/isbn/9781009180597

DOI: 10.1017/9781009180610

First published 1994
Second edition 2002
Third edition 2008
Fourth edition 2023

Printed in the United Kingdom by TJ Books Limited, Padstow, Cornwall, 2023

A catalogue record for this publication is available from the British Library.

A Cataloging-in-Publication data record for this book is available from the Library of Congress

ISBN 978-1-009-18059-7 Hardback
ISBN 978-1-009-18060-3 Paperback

Additional resources for this publication at www.cambridge.org/nicholl4

Contents

Detailed Contents

Part 3 | The Methodology of Gene Manipulation

Preface

Advances in genetics continue to be made at an ever increasing rate, which presents something of a dilemma when writing an introductory text on the subject. In the years since the third edition was published, many new applications of gene manipulation technology have been developed; genome sequencing has become available at bench-top scale and cost, and gene editing can be achieved using very modest laboratory infrastructure. Personal genome profiling is available from a range of companies, and genetic technology has played a major role in managing many aspects of the COVID-19 pandemic, from diagnostic testing to rapid development of safe and effective vaccines.

Information technology resources, coupled with the internet and World Wide Web, have been critical parts of all these developments, providing tools for the analysis of DNA sequences and instant sharing of data across the globe. At the same time, a level of mistrust has developed among some sections of society, largely driven by misinformation on social media channels, which has illustrated the power of the internet in a less positive way. It is against this background that some themes began to emerge for the fourth edition, reflecting the aim of encouraging students to use the excellent resources on the web, whilst retaining a level of critical assessment of the information. Aspects around the ethics of using genetic technology are perhaps now even more important than before, so these are discussed early in the text to enable the applications to be placed within an appreciation of the ethical framework.

Whilst aiming for a slight broadening in scope, I remain convinced that a basic technical introduction to the subject should be the major focus of the text. Thus, some of the original methods used in gene manipulation have been kept as examples of how the technology developed, even though some of these have become little used or even obsolete. From the educational point of view, this should help the reader cope with more advanced information about the subject, as a sound grasp of the basic principles is an important part of any introduction to genetic engineering. I have been gratified by the many positive comments about the third edition of the text, and I hope that this new edition continues to serve a useful purpose as part of the introductory literature on this fascinating subject.

This book is organised as four parts. *Part 1* (*Genetic Engineering in Context*; Chapters 1–3) sets the scene and brings the discussion of the ethical issues around DNA technology to the start of the book. *Part 2* (*The Basis of Genetic Engineering*; Chapters 4–6) provides an introduction to molecular biology and outlines the tools available to the genetic engineer, and *Part 3* (*The Methodology of Gene Manipulation*; Chapters 7–12) extends this theme further by examining how these tools enable sophisticated experiments and procedures to be carried out. Finally, in *Part 4* (*Genetic Engineering in Action*; Chapters 13–17), we look at the impact of DNA technology across a range of key areas.

In the fourth edition, I have expanded the range of features that should be useful as study aids where the text is used to support a particular academic course. In the book, there are *text boxes* sprinkled throughout the chapters. These highlight key points on the way through the text, and can be used as a means of summarising the content. At the start of each chapter, the *aims* of the chapter are presented, along with a *chapter summary* in the form of *learning objectives*. These have been written quite generally, so that an instructor can modify them to suit the level of detail required. A list of the *key words* in each chapter is also provided for reference. These are shown as bold in the text; terms in blue can also be found in the Glossary. A new addition to the end of each chapter is a *websearch* page that provides some structured web-based search exercises that help to set the chapter in context and act as a start point for further study using the resources available online. As in previous editions, a *concept map* has been generated for each chapter, showing how the main topics are linked. The concept maps provided here are essentially summaries of the chapters, and may be examined either before or after reading the chapter.

As this remains an introductory text, no in-text reference has been made to the primary (research) literature, but some suggestions for *further reading* are given at the end of each chapter. Most of these are available in open-access format or may be available through an institution's library subscription service. A *glossary* of terms used has also been provided.

A new development for the fourth edition is a set of *online resources* at www.cambridge.org/nicholl4. This provides access to a range of materials from the book (and additional information) that I hope will be useful in building a learning system to suit your preferred learning style. The resources have been provided in electronic format as a *study guide* to enable collation into a set of student-generated notes.

My thanks go to the anonymous (but appreciated) reviewers of the proposal and the early versions of the manuscript. Their comments and suggestions have made the book better; any errors of fact or interpretation of course remain my own responsibility. Special thanks to Megan Keirnan, Susan Francis, Helen Shannon and Rachel Norridge at Cambridge University Press, and to Joyce Cheung, for their cheerful advice, support, encouragement and patience, which helped bring the project to its conclusion.

My final and biggest thank you goes as ever to my wife Linda and to Charlotte, Thomas and Anna, who have grown up along with the various editions of 'IGE'. I dedicate this new edition to them.

Part I

Genetic Engineering in Context

Chapter 1 Summary

Learning Objectives

When you have completed this chapter, you will be able to:

- Define genetic engineering as it will be described in this book
- Outline the basic features of genetic engineering
- Describe the emergence of gene manipulation technology
- Explain the steps required to clone a gene
- Appreciate elements of the ethical debate surrounding genetic engineering
- Identify a range of internet-based resources related to DNA technology

Key Words

Genetic engineering, bioinformatics, gene manipulation, gene cloning, recombinant DNA (rDNA) technology, genetic modification, new genetics, DNA technology, molecular agriculture, genethics, Gregor Mendel, James Watson, Francis Crick, DNA ligase, type II restriction enzyme, plasmid, extrachromosomal element, replicon, clone, genetically modified organism (GMO), internet, World Wide Web, Tim Berners-Lee, uniform resource locator (URL), domain (*re.* URL), search engine, valid, reliable, peer review, Encyclopedia Britannica, Wikipedia, social media, misinformation, disinformation, suggested search term (SST), digital object identifier (DOI).

Chapter 1

Introduction

1.1 | What Is Genetic Engineering?

Making progress in any scientific discipline depends on continually developing techniques and methods to extend the range and sophistication of experiments that may be performed. Within the biosciences, this has been demonstrated in a spectacular way by the emergence and development of **genetic engineering**. In 2022, we marked the fiftieth anniversary of the creation of the first recombinant DNA molecules, an event that is often used to note the start of the recombinant DNA era of genetics. The five decades since 1972 have seen astonishing progress in the breadth and scope of the technology, and it is now routine practice to identify a specific DNA fragment from the genome of an organism, determine its base sequence and assess its function. The sequence might then be altered and replaced into the organism it came from, or a different organism, to achieve a particular goal. We have seen the expansion of the technology into the domain of 'big science' in the era of the Human Genome Project, and its return to the small-scale laboratory as new developments have appeared. Whole genomes can now be sequenced using a benchtop machine, and genome editing enables researchers to alter the genome of an organism with a high level of precision. All of this is now underpinned by the astonishing developments in bioinformatics, with sophisticated computational tools available to analyse almost unimaginable amounts of data that are generated on a daily basis.

The term genetic engineering is often thought to be rather emotive or even trivial, yet it is probably the label that most people would recognise. However, there are several other terms that can be used to describe the technology, including **gene manipulation**, gene cloning, recombinant DNA (rDNA) **technology** and **genetic modification**. You may also come across the term the '**new genetics**', although we are at a point where this is perhaps less useful than was the case previously. A more useful generic term that covers a wide range of techniques and applications is simply **DNA technology**. There are also legal definitions used in administering regulatory mechanisms in countries where genetic engineering is practised.

Several terms may be used to describe the technologies involved in manipulating genes.

The genetic material provides a rich resource in the form of information encoded by the sequence of bases in the DNA.

Although there are many diverse and complex techniques involved, the basic principles of genetic manipulation are reasonably simple. The premise on which the technology is based is that genetic information, encoded by DNA and arranged in the form of genes, is a *resource* that can be manipulated in various ways to achieve certain goals in both pure and applied science, medicine, biotechnology and agriculture. There are many areas in which genetic manipulation has made a significant impact, including:

- Basic research on gene structure and function
- Production of useful proteins by novel methods
- Generation of transgenic plants and animals
- Medical diagnosis and treatment
- Forensic analysis of crime scene samples
- Molecular anthropology and the study of evolution
- Genome analysis and genome editing

In later chapters, we will look at how DNA technology has contributed to these areas.

Gene cloning enables isolation and identification of individual genes.

The mainstay of genetic manipulation is the ability to isolate a single DNA sequence from the genome. This is the essence of gene cloning and can be considered as a series of four steps (Fig. 1.1). Successful completion of these steps provides the genetic engineer with a specific DNA sequence, which may then be used for a variety of purposes. A useful analogy is to consider gene cloning as a form of **molecular agriculture**, enabling the production of large amounts (in genetic engineering, this means nanograms or micrograms) of a particular DNA sequence. Although the basic cloning methodology has been extended (and in many cases replaced) by technologies such as the polymerase chain reaction, large-scale DNA sequencing and genome editing, this ability to isolate a particular gene sequence is

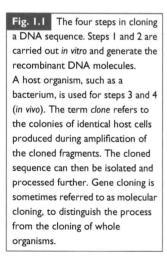

Fig. 1.1 The four steps in cloning a DNA sequence. Steps 1 and 2 are carried out *in vitro* and generate the recombinant DNA molecules. A host organism, such as a bacterium, is used for steps 3 and 4 (*in vivo*). The term *clone* refers to the colonies of identical host cells produced during amplification of the cloned fragments. The cloned sequence can then be isolated and processed further. Gene cloning is sometimes referred to as molecular cloning, to distinguish the process from the cloning of whole organisms.

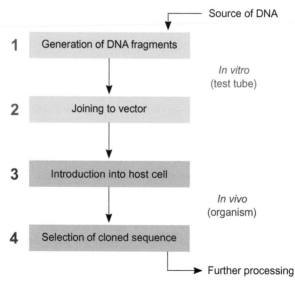

Source of DNA

1 Generation of DNA fragments

In vitro
(test tube)

2 Joining to vector

3 Introduction into host cell

In vivo
(organism)

4 Selection of cloned sequence

Further processing

still a major part of gene manipulation as carried out on a day-to-day basis in research laboratories worldwide.

One aspect of genetic engineering that has given cause for concern is the debate surrounding the potential applications of the technology. The term **genethics** is sometimes used to describe the ethical problems that exist in modern genetics, which are likely to increase in both number and complexity as genetic engineering technology becomes more sophisticated and implemented more widely. The use of transgenic plants and animals, investigation of the human genome, gene therapy, genome editing and many other topics are of concern not just to the scientist, but also to the population as a whole. Developments in genetically modified foods have provoked a well-documented public backlash against the technology in many countries. Additional developments in the cloning of organisms, and in areas such as *in vitro* fertilisation and xenotransplantation, raise further questions. Although not strictly part of gene manipulation technology, organismal cloning will be considered later in this book, as this is an area of much concern and can be considered as genetic engineering in its broadest sense. Research on embryonic stem cells, and the potential therapeutic benefits that this may bring, is another area of concern that is part of the general advance of genetics. We will look at some of these ethical aspects in more detail in Chapter 3.

> As well as technical and scientific challenges, modern genetics poses many moral and ethical questions.

1.2 | Laying the Foundations

Although the techniques used in gene manipulation began to appear in the 1970s, we should remember that development of these techniques depended on the knowledge and expertise provided by chemists, biochemists and microbial geneticists working in the earlier decades of the twentieth century. We can consider the development of genetics as falling into three main eras (Fig. 1.2). The science of genetics really began with the rediscovery of **Gregor Mendel**'s work at the start of the century, and the next 40 years or so saw the elucidation of the principles of inheritance and genetic mapping. Microbial genetics became established in the mid-1940s, and the role of DNA as the genetic material was confirmed. During this period, great advances were made in understanding the mechanisms of gene transfer between bacteria, and a broad knowledge base was established, from which later developments would emerge.

> Gregor Mendel is often considered the 'father' of genetics.

Determination of the structure of DNA by **James Watson** and **Francis Crick** in 1953 provided the stimulus for the development of genetics at the molecular level, and the next few years saw a period of intense activity and excitement as the main features of the gene and its expression were determined. This work culminated in the deciphering of the complete genetic code in 1966, and the stage was now set for the appearance of the new discoveries that would lead to the development of the early techniques in recombinant DNA technology.

> Watson and Crick's double helix is perhaps the most 'famous' and most easily recognised molecule in the world.

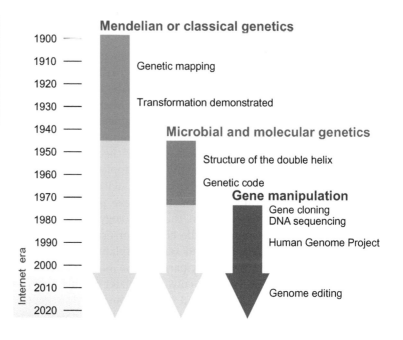

Fig. 1.2 The history of genetics since 1900. Three eras can be identified. Darker shaded areas represent the periods of major development in each branch of the subject, although advances continue to be made in all of these areas.

1.3 First Steps in DNA Cloning

By the end of the 1960s, most of the essential requirements for the emergence of gene technology were in place.

In the late-1960s, there was a sense of frustration among scientists working in the field of molecular biology. Research had developed to the point where progress was being hampered by technical constraints, as the elegant experiments that had helped to decipher the genetic code could not be extended to investigate the gene in more detail. However, a number of developments provided the necessary stimulus for gene manipulation to become a reality. In 1967, the enzyme **DNA ligase** was isolated. This enzyme can join two strands of DNA together, a prerequisite for the construction of recombinant molecules, and can be regarded as a sort of molecular glue. This was followed by the isolation of the first **type II restriction enzyme** in 1970, a major milestone in the development of genetic engineering. Restriction enzymes are essentially molecular scissors that cut DNA at precisely defined sequences. Such enzymes can be used to produce fragments of DNA that are suitable for joining to other fragments. Thus, by 1970, the basic tools required for the construction of recombinant DNA were available.

The first recombinant DNA molecules were generated at Stanford University in 1972, utilising the cleavage properties of restriction enzymes (scissors) and the ability of DNA ligase to join DNA strands together (glue). The importance of these first tentative experiments cannot be overstated. Scientists could now join different DNA molecules together and could link the DNA of one organism to that of a completely different organism. The methodology was extended in

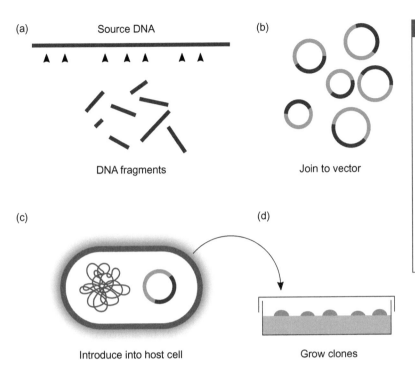

(a) Source DNA

DNA fragments

(b) Join to vector

(c) Introduce into host cell

(d) Grow clones

Fig. 1.3 Cloning DNA fragments. (a) The source DNA is isolated and fragmented into suitably sized pieces. This may be carried out mechanically or by using restriction enzymes. (b) The fragments are then joined to a carrier molecule or vector to produce recombinant DNA molecules. In this case, a plasmid vector is shown. (c) The recombinant DNA molecules are then introduced into a host cell (a bacterial cell in this example) for propagation as clones (d). These can be stored as a stable resource that can be used for further analysis.

1973 by joining DNA fragments to the **plasmid** pSC101, which is an **extrachromosomal element** derived from an antibiotic resistance plasmid originally isolated from the bacterium *Salmonella typhimurium*. These recombinant molecules behaved as **replicons**, *i.e.* they could replicate when introduced into *Escherichia coli* cells. Thus, by creating recombinant molecules *in vitro*, and placing the construct in a bacterial cell where it could replicate *in vivo*, specific fragments of DNA could be isolated from bacterial colonies that formed **clones** (colonies formed from a single cell, in which all cells are identical) when grown on agar plates. This development marked the emergence of the technology that became known as gene cloning (Fig. 1.3).

The discoveries of 1972 and 1973 triggered what is perhaps the biggest scientific revolution of all – the 'new genetics' era had arrived. The use of the new technology spread very quickly, and a sense of urgency and excitement prevailed. This was dampened somewhat by the realisation that the new technology could give rise to potentially harmful organisms with undesirable characteristics. It is to the credit of the biological community that measures were adopted to regulate the use of gene manipulation, and that progress in contentious areas was restricted until more information became available about the possible consequences of the inadvertent release of organisms containing recombinant DNA. However, the development of **genetically modified organisms (GMOs)**, particularly crop plants, has reopened the debate about the safety of these organisms and the consequences

The key to gene cloning is ensuring that the target sequence is replicated in a suitable host cell.

The development and use of GMOs pose some difficult ethical questions that may not arise in other areas such as gene cloning.

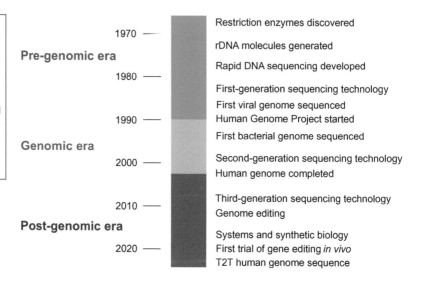

Fig. 1.4 Eras of genomics. Technological development in DNA science and computation in the 1970s and 1980s led to the project to sequence the human genome, which largely defined the genomic era. The knowledge gained has continued to drive the development of techniques and applications in the post-genomic era.

Pre-genomic era

1970 — Restriction enzymes discovered

rDNA molecules generated

Rapid DNA sequencing developed

1980 —

First-generation sequencing technology

First viral genome sequenced

1990 — Human Genome Project started

First bacterial genome sequenced

Genomic era

Second-generation sequencing technology

2000 —

Human genome completed

Third-generation sequencing technology

2010 — Genome editing

Post-genomic era

Systems and synthetic biology

2020 — First trial of gene editing *in vivo*

T2T human genome sequence

of releasing GMOs into the environment. In addition, many of the potential medical benefits of gene manipulation, genetics and cell biology pose ethical questions that may not be easy to answer. We will come across some of these issues later in this book.

DNA technology continued (and continues) to expand at pace, with a number of key techniques developed in the late-1970s and early-1980s that would enable a step-change in scale to be achieved, with the ambitious project to sequence the human genome being completed in 2003. Further developments have changed DNA sequencing significantly, to the extent that we can now usefully describe DNA technology as itself falling into three eras (Fig. 1.4). As we explore the topics in this book, it may be useful to keep this diagram in mind as a 'roadmap' to help us place the technology in context.

1.4 | Using the Web to Support Your Studies

Since the first edition of this book was published in 1994, the growth and development of DNA technology has been impressive, and often astonishing. In many ways, the parallel development of the **internet** and World Wide Web (www) is equally impressive, and the generation of students who may be using this book has grown up in a world that is completely immersed in the technology associated with 'the web'. Although information on how to access the internet is no longer needed, some of the guidance for using the web that was included in earlier editions remains appropriate.

The internet was conceived and developed by **Tim Berners-Lee** (now Sir Tim) in 1989 whilst working at CERN in Geneva. If you are not familiar with CERN and Berners-Lee, look these up now using your computer, laptop, tablet or smartphone! The term *internet* is used to describe the network of computers that together provide the means to publish and share information, whilst the *World Wide Web* is a more general description of the information that is published using the internet. However, the two terms are often used interchangeably, and the phrases 'surfing the net', 'surfing the web' and 'look it up online' have become part of modern-day language. We will use 'the web' in this book when referring to the World Wide Web.

A really positive feature of the web is the range of material that can be found with a few clicks of a mouse or swipes with a finger. The fluidity and dynamic nature of the web mean that a printed book (like this one) cannot possibly compete; so we will not try to. A subtle shift in emphasis is therefore required – this book is a resource that can help steer you through the confusion of the web, as well as providing a structured look at the subject matter. I have therefore assumed that you will be reading this text with pretty much instant and continuous access to the internet, and that you will be able to use the web to help gather, collate and interpret additional information that you find.

This book can be used as a guide to explore topics more broadly, and/or in more depth, using the web.

Finding websites is generally straightforward if you know where you are going. Each website has an 'address' known as a **uniform resource locator (URL)**. A URL generally begins with http://www. followed by the specific **domain** address. This may end with a country identifier such as .uk, .ca, .cn, .us, .au, *etc*. The most common domain ending is .com (where the term 'dot com' comes from), which is used by around 50 per cent of all websites. Many student textbooks have associated websites, as do research groups, university departments, companies, *etc*.

If you don't know the URL, one of the many **search engines** enables you to look for information using a range of terms, and it is astonishing what you can find. However, caution is needed: there is an awful lot of information on the web, and there is a lot of awful information there as well! It is very easy to get sidetracked and end up wasting a lot of time searching through sites that are of no value (but may be interesting nonetheless). As I write this, I have just typed in the search terms that I used to illustrate this section in the third edition of this book. The number of 'hits' for each of the various terms, in 2007 and 2021, is shown in Table 1.1.

A number of points can be made when we consider the data. Firstly, the number of retrieved items is often far too large to be useful, and thus more specific, defined or restricted search terms will generally produce fewer results. However, even with something like 'sheep cloning' or 'plasmid vector', there is still too much information to look through, so learning to use the search and filter facilities provided by various search engines and websites is well worth the effort. Secondly, it seems that there has been an orders-of-magnitude increase in the number of hits generated by all of these terms. At first glance, this may not seem implausible, given the increase in research and development that will have occurred in these areas since 2007. However, we do need to be a little cautious about inferring too much,

Table 1.1 | Comparison of the number of web search 'hits' obtained using some search terms in 2007 and 2021

Search term	'Hits' 2007	'Hits' 2021	Increase
genetics	5 564 000	578 000 000	104 ×
cloning	2 244 000	378 000 000	168 ×
cystic fibrosis	966 400	61 400 000	64 ×
DNA cloning	499 800	33 200 000	66 ×
sheep cloning	127 700	5 380 000	42 ×
plasmid vector	71 200	18 600 000	261 ×
homopolymer tailing	123	65 000	528 ×

Care is needed when interpreting information from the web. A critical approach is always helpful.

for reasons that are not immediately obvious from the data presented. Factors such as developments in web-searching techniques and the algorithms used will have a major impact on the returns, and it is unlikely that these variables were the same in 2007 and 2021. The search engine used also has an impact – for a particular search term, different variants return different numbers of hits, and even the same search engine can return different numbers of hits for the same term on different days. Thus, we can't be sure that the information is an accurate reflection of the actual number of items on a topic at any point in time. Whilst it does give some indication of the expansion of the science itself, its reporting across the web and its 'searchability', it is of limited value. In fact, the comparison presented here is neither **valid** nor **reliable**. These terms have specific meaning in scientific methodology, which we will explore further in Section 2.1.

The main point to be aware of when using the web is there is very little restriction on *what* gets published, and *by whom*. The **peer review** system that is used in science is not often applied to web-based information, so not all information may be accurate. The type of website can often give some indication as to how reliable the information may be. The last part of the URL usually indicates the sort of organisation, although this does vary from country to country. Examples are *.gov* for government sites, *.ac.uk* for UK academic sites (*.edu* used in some other countries) and *.com* or *.co* for commercial or company sites. The characteristics of these types of site can be described as follows:

- *Government sites* – usually accurate information, but this can have a particular emphasis. Political influences and 'spin' can sometimes make certain types of information highly suspect for some government-controlled sites. A knowledge of the national political, social and cultural situation may be useful in assessing the reliability of government-controlled websites.
- *Research institutes and universities* – very often the most useful sites for supporting academic study, as these offer a high level of specificity and detail. They tend to be managed by people who are familiar with publishing information *via* the peer review mechanism, and therefore are aware of the need to be accurate and unbiased.

- *Special interest groups/professional societies* – similar to research institutes and universities in terms of authoritative and accurate information. Can range from fairly broad to very narrowly specific coverage.
- *Company sites* – again can provide a wealth of information, but this is obviously presented to support company aims, ethos and policy in most cases. Many responsible companies take care to present a balanced view of any contentious issues, often including links to other sources of information.
- *Publishers* – range from book publishers to online journals. Often provide excellent resource material, but for some journal sites, access may require registration and payment of a fee for each issue or article downloaded.
- *Information 'gateway' sites* – many specialist research areas have associated sites that collate and distribute information about the latest developments. Often it is possible to register for email updates to be sent directly when they are published. As with journals, some of these sites require registration and subscription to access the full range of facilities. May be associated with a special interest group or professional body.
- *Pressure group sites* – can provide a lot of useful information, very often gathered to counteract some particular claim made by another party. Despite the obvious dangers of accepting the views presented, many of these sites do take care to try to present accurate information. Often there is a higher risk of confirmation bias (see Section 2.1) affecting content and opinion than in other types of website. Views can range from moderate to extreme.
- *Personal sites* – prepared by individuals, therefore likely to be the most variable in terms of content and viewpoint. Range from highly respected sites by well-known experts to the sort of site that is best avoided as far as serious study is concerned.

The difficulty in weeding out the less useful (or downright weird) sites is that anyone with a flair for design can produce a website that looks impressive and authoritative. Conversely, some extremely useful and respected sites can look a little dull if a good web designer has not been involved. However, as long as you remember that *critical evaluation is essential*, you should have few problems.

Online encyclopedias are very useful as starting points for researching a topic. Long-established versions such as **Encyclopedia Britannica** are now available online, with others set up specifically for the web environment (*e.g. encyclopedia.com*). The interactive encyclopedia **Wikipedia**, which was set up in 2001, is 'editable' by its contributors, and thus represents a novel publishing phenomenon. The theory is that it is effectively a self-regulating project that grows as more articles are added and refined. In general, the articles are written by authors who are expert in their fields and who take care to ensure accuracy and objectivity. This is strengthened by the post-publication 'peer-amended' editing process (as opposed to the peer-reviewed pre-publication approach of more traditional publishing). Wikipedia is therefore a useful resource but (as with any source) should be backed up by further validation of the content if serious study is undertaken on a particular topic.

There are, of course, less positive aspects of the web, so care is needed. In particular, the all-pervasive nature of **social media**, and the impact on how we receive, filter and process information, is perhaps one of our most critical challenges in dealing with online information. Whilst social media sites provide some tremendous benefits, they are not often reliable sources for authoritative and objective information. The spread of **misinformation, disinformation**, fake news stories, conspiracy theories and the like can have a significant destabilising influence across a wide spectrum of society, with international penetration achieved easily *via* the web. A recent example affecting the public perception of scientific information was the anti-vaccination reaction to the SARS-CoV-2 viral pandemic; the spread of mis- and disinformation around the vaccines developed to reduce the risk of serious illness from COVID-19 undoubtedly led to additional loss of life that could have been prevented.

Despite the challenges, part of the fun of using the web is that it *is* such a diverse and extensive medium, and it is constantly changing and expanding. This can present a problem when including URLs in a printed text, as website addresses may change over time as sites develop, appear and disappear. The 'site not found' message can be frustrating, so where URLs are included in the book, these tend to be long-established sites that are unlikely to change. In practice, it is now unusual to copy a URL manually into a web browser from a printed source, and often the best way to find the URLs in this text is to type in a search term related to the site and select the matching URL from the list that appears. Thus, in many cases where websites are suggested, I have included a **suggested search term (SST)** instead of a URL. If you are using this book as a module or course text, your tutor will be able to direct you to any specific websites that support your course. Finally, using a search engine is the most effective way of getting up-to-date information about what is available, and by finding a few relevant sites, you can quickly build up a set of links that take you into other related sites. Good surfing!

> Social media sites present particular challenges around how we receive and process information. Scientific topics, and scientists themselves, can be affected (often in a negative way) by what circulates on social media platforms.

1.5 | Conclusion: The Breadth and Scope of Genetic Engineering

In this introduction, we have looked at genetic engineering in the context of the history of genetics from 1900. The rediscovery of Mendel's work stimulated the study of patterns of inheritance in organisms such as the fruit fly. In parallel, microbiologists began to consider how bacteria might be useful as model systems for studying genetics, and in 1928, transformation of bacterial characteristics was demonstrated in one of the classic experiments in genetics.

Progress in genetics is often dependent on the technology available; both 'hard' technology (such as laboratory and computing equipment) and 'soft' technology (biological supplies such as enzymes) are important. The start of the *molecular genetics era* is often dated to 1953 when Watson and Crick determined that DNA was a *double helix*, using data from a number of sources, including the ratios of the DNA bases in various organisms and X-ray crystallography of DNA samples. Over the next 20 years, a series of elegant experiments showed how DNA functions as the genetic material.

In 1972, the availability of *type II restriction enzymes* and *DNA ligase* enabled the first rDNA molecules to be generated, and we entered the era of DNA technology. *Gene cloning* enabled the isolation of specific DNA sequences, and the structure of the gene was studied at the molecular level for the first time. *DNA sequencing* developed from a laborious and difficult process to a relatively straightforward one with the advent of rapid sequencing methods in 1977, and within a few years, automation and ambition had brought the previously unthinkable within reach – determining the sequence of the human genome. This astonishing feat was completed in 2003, although we will see in Chapters 6, 11, 12 and 13 that the term 'completed' is not really accurate and genome sequencing continues to expand and have a significant impact on a number of different disciplines.

In this chapter, we have touched on the ethical questions that DNA technology raises; we will examine some of these in more detail in Chapter 3, and also when we consider the applications of genetic engineering in Part 4. This aspect is a fundamental part of the deployment of any technology, but is particularly important when we are dealing with things that can have a direct impact on human healthcare, the diagnosis and treatment of disease, reproduction, nutrition and forensic analysis of samples from scenes-of-crime.

To complete the introduction, we looked at the impact of the internet and World Wide Web, without which much of modern genetics could not be managed efficiently. In some ways, the growth and development of internet-based technologies parallels that of DNA technology itself, and is a good example of how technological and scientific progress are inextricably linked.

Further Reading

Suggestions for further reading include books, commentary, technical notes, opinion articles, formal reviews and primary literature, appropriate to each chapter. Some are classic papers that helped define the subject; others are general personal reflections, and some have quite a challenging level of detail. Where possible, articles have been selected as open access, with the URL and digital object identifier (DOI) listed.

Berg, P. (2001). The Nobel Prize interview. Interview by J. Rose. URL [www
.nobelprize.org/prizes/chemistry/1980/berg/interview/].

Cohen, S. N. (2013). DNA cloning: a personal view after 40 years. *Proc. Natl. Acad. Sci. U. S. A.*, 110 (39), 15521–9. URL [www.pnas.org/doi/full/10.1073/pnas.1313397110]. DOI [https://doi.org/10.1073/pnas.1313397110].

Fredrickson, D. S. (1991). Asilomar and recombinant DNA: the end of the beginning. In: K. E. Hannah, editor. *Biomedical Politics*. National Academy Press, Washington, DC; pp. 258–98. URL [www.ncbi.nlm.nih.gov/books/NBK234217/].

Websearch

People

Four people who played a major role in gene manipulation becoming a reality – have a look at what they did. They are: *Werner Arber, Daniel Nathans* and *Hamilton Smith* (Nobel Prize 1978), and *Paul Berg* (Nobel Prize 1980).

Places

What has *Asilomar State Beach* in California got to do with genetic engineering?

Processes

The process of *publishing information* on the web is something that affects us all. How do we sift out what is accurate and truthful from what is not? Pick a topic that you are interested in (not necessarily biological) and look to see how it is presented by the types of website listed in Section 1.4. Evaluate critically.

Reflections

What topics in this chapter have you found most challenging? Look for resources that help to illustrate the key points.

Genetic engineering

is also known as
- gene cloning
- recombinant DNA technology
- molecular cloning
- genetic modification

arose from microbial and molecular genetics **and was first achieved** when recombinant DNA was generated **by** Paul Berg in 1972 at Stanford University

has applications in
- basic science
- biotechnology
- medicine

but raises some complex personal, societal, ethical and legal questions

requires four basic steps
- generation of DNA fragments
- joining to a vector or carrier molecule
- introduction into a host cell
- selection of desired sequence

for use in a range of applications

is an enabling technology

that involves cutting, modifying and joining DNA molecules

using various types of enzymes

such as
- DNA ligase
- type II restriction enzymes

DNA ligase **used for** (generation of DNA fragments steps)

type II restriction enzymes **can be used for**

with sub-disciplines

such as
- genomics
- proteomics
- bioinformatics
- DNA profiling
- forensics
- ancient DNA

Concept Map I

Chapter 2 Summary

Learning Objectives

When you have completed this chapter, you will be able to:

- Outline elements of the philosophy of science
- Describe the scope and ethos of the scientific method
- Identify the infrastructure and support requirements of research
- Discuss types of data and their evaluation
- Summarise key events, people and places in the history of DNA research

Key Words

Aristotle, William Whewell, Isaac Newton, Robert Hooke, Louis Pasteur, Albert Szent-Györgyi, Nobel Prize, Thomas Huxley, Charles Darwin, Jacob Bronowski, George Wald, iterative, François Jacob, descriptive, empirical, experimental, hypotheses, hypothesis-driven, hypothesis-free, genome-wide association studies (GWAS), classical, reductionist, Occam's razor, emergent properties, systems biology, deductive, inductive, consensus sequence, hypothetico-deductive, Karl Popper, falsifiable, objective, subjective, data, quantitative, qualitative, valid, reliable, accuracy, precision, variables, controls, negative controls, positive controls, independent variable, dependent variable, confounding variables, biological variation, uncertainty, error bars, mean, confidence limits, standard error, bias, unconscious bias, selection bias, confirmation bias, Maurice Wilkins, primary literature, secondary literature, tertiary literature, 'publish or perish', basic science, pure science, fundamental science, applied science, technology, paradigm shift, Thomas Kuhn, geocentric, heliocentric, Newtonian mechanics, quantum mechanics, analogue, digital, step-changes, paradigm nudges.

Chapter 2

The Story of DNA

In this chapter, we will look in a little more detail at how the science of DNA has evolved and developed over the past 160 years. Although this is a long time in terms of human lifespan, it is only a fraction of the history of science as a recorded discipline. The Ancient Greek philosopher and polymath **Aristotle**, who lived in the fourth century BCE, is often considered the first scientist (although he did not know the term, as it was first used by **William Whewell** in 1833). Aristotle was the first to study biology systematically, so in some ways, we can trace the history of natural sciences back some 2 400 years. As we saw in Chapter 1, DNA technology is a relatively young area of DNA science, with the 20 years from 1953 to 1973 a time of intense activity that answered the fundamental questions about DNA structure and function, and paved the way for the creation of rDNA and the discoveries that followed. The next 30 years led to the sequence of the human genome, and we are now firmly in the post-genomic era. So how did we get to this point?

It took only 50 years to progress from determining the structure of DNA to working out the sequence of the human genome.

2.1 | How Science Works

The phrase *standing on the shoulders of giants* has a long history, but perhaps its most famous iteration in science is by **Isaac Newton** (later Sir Isaac) who, in a 1676 letter to **Robert Hooke**, wrote 'If I have seen farther, it is by standing on the shoulders of giants.' All progress in science depends on evaluation and extension of the work of others, often in small incremental steps, and this is one of the most important elements of science across the range of disciplines. **Louis Pasteur** alluded to this when he stated that 'chance favours the prepared mind', and the biochemist **Albert Szent-Györgyi**, who was awarded the **Nobel Prize** in 1937, said that 'research is to see what everybody has seen and think what nobody has thought'. These two statements provide a good baseline for our look at science; knowing the work of others is vital if an unsolicited observation is to be recognised as something potentially significant, and creative thinking is required to push science forward.

Science as an extension of common sense is also a theme that can be helpful as a mental backstop when things get convoluted. **Thomas Huxley** (a great supporter of his contemporary **Charles Darwin**) stated that science is 'nothing but trained and organised common sense', and **Jacob Bronowski**'s 1951 classic book *The Common Sense of Science* is still relevant today. Finally, **George Wald**, in his 1967 Nobel Prize lecture, gave what is perhaps the most evocative description of science; it is worth quoting the opening paragraph here in full:

> I have often had cause to feel that my hands are cleverer than my head. That is a crude way of characterizing the dialectics of experimentation. When it is going well, it is like a quiet conversation with Nature. One asks a question and gets an answer, then one asks the next question and gets the next answer. An experiment is a device to make Nature speak intelligibly. After that, one only has to listen.
>
> *Source:* © The Nobel Foundation 1967.

> **Science can be considered as having a conversation with the natural world, using the tools of curiosity and common sense.**

Science is not clear-cut and tidy; as we will see later, it involves a complex and sometimes messy journey, often using many approaches. It is certainly not the straightforward process-defined linear sequence of steps that simple models portray. The 'conversation with Nature' phrase reflects this very well – a conversation will ebb and flow, go down unpredicted avenues, sometimes excite and sometimes not, and often be frustrating. Science is the same, and for this reason, the back-and-forth nature of scientific investigation is an example of an **iterative** approach. Another helpful concept of how science works was first proposed by the 1965 Nobel Prize winner **François Jacob** who distinguished 'day science' from 'night science'. He thought of day science as the formal, hypothesis-based approach carried out in the workplace by scientists who have a goal that can be defined, tested and evaluated. In contrast, night science is an unconstrained, free-thinking, creative approach that Jacob called 'a sort of workshop of the possible'. As with many aspects of science, these two approaches are not mutually exclusive, but mutually supportive.

> **Science is not often a simple linear process, but is a complex web of ideas, discussions, experiments and interactions.**

A few more terms are needed if we are to make sense of science as a process. Many are presented as pairs, with what might be considered opposite or mutually exclusive meanings: only a brief outline of the following terms is given here; a useful exercise would be to have a look at some more extensive definitions of these terms. Be prepared for some head-scratching and even confusion!

A **descriptive** approach is characterised by observing, recording and classifying, whilst an **empirical** or **experimental** approach is a form of investigative science that constructs hypotheses (*sing.* hypothesis) to try to explain and predict the behaviour of a system. However, hypothesis testing is not *necessarily* part of all experimental science, and the terms **hypothesis-driven** and **hypothesis-free** science can be used. With the advent of genome sequencing, hypothesis-free investigations can be carried out using techniques like genome-wide association studies (GWAS) to look for patterns in large data sets. This is one example of how modern biology has changed significantly over the past 20 years or so.

In genetics teaching, one perennial question is whether to begin with a **classical** or a **reductionist** approach. The classical route starts from the Mendelian principles of transmission genetics (what we might call the 'behaviour' of genes), whereas the reductionist approach would begin with the structure of DNA, genes and genomes. In practice, both approaches are valid and sensible. The same applies to classical/reductionist approaches to genetics research where the topic under investigation will determine which methodology is required. DNA sequence data are perhaps the ultimate reductionist output; studying population genetics needs a different approach and produces different data sets. Reductionist methodology is sometimes associated with the principle defined by Occam's razor, in which the simplest explanation for a given circumstance is preferred. However, there is a danger in taking a reductionist approach too far, as biological systems are hierarchical and demonstrate the concept of emergent properties, where complex systems are built from simpler components. The aim of **systems biology** is to integrate the elements of a system and try to make sense of how it works as a whole. The reductionist and systems biology approaches are sometimes seen as antithetical (as are classical, *cf.* reductionist genetics, and 'day' and 'night' science). This tension is not a particularly helpful view; both approaches are needed and should complement each other in reaching valid conclusions about a particular topic. An interesting, yet unsurprising, paradox is that the ultimate reductionist output (a genome sequence) is proving to be the essential baseline resource that is enabling integrative investigation of how the genome actually *works*.

> It is usually sensible to consider the simplest explanation for a scientific question; natural systems often (but not always) avoid generating redundant complexity.

Let's assume that we have now selected our topic and set the parameters within which our investigation will be conducted. What other factors should we be aware of? The form of reasoning that is used in science is important conceptually but may not always be so in practice, as the design of an investigative system (if done correctly) should have considered any potential conflict. Two forms that are most often cited are **deductive** and **inductive** reasoning. Deductive reasoning takes accepted premises and reaches (deduces) valid conclusions from these premises, whilst inductive reasoning takes a set of specific observations and reaches (induces) a generalised conclusion. A more digestible description is that deductive is sometimes called a *top-down* approach, moving from the *general* to the *particular*, whereas inductive is a *bottom-up* approach that moves from the *particular* to the *general*. In practice, it is sometimes hard to determine which parts of an investigation use which method, but many parts of experimental science are based on deductive reasoning, as this enables sequential refinement of an investigation using the question/answer cycle noted by Wald. However, inductive reasoning also has a role to play – the identification of a consensus sequence for a particular part of a gene is one example where looking at a range of specific sequences across a

> Both deductive and inductive reasoning may be needed when designing a scientific study and evaluating the outcome.

number of genes or organisms leads to a generalisation (consensus) that sequence X usually has base sequence Y, with some variation.

The combination of forming a hypothesis and testing predictions using a deductive approach is often called the hypothetico–deductive process, and is considered a central tenet of the scientific method. A hypothesis should be able to be tested repeatedly by different scientists; in the words of **Karl Popper**, it should be **falsifiable**. Trying to disprove an interpretation of a particular phenomenon is just as important in science as marshalling support; gathering evidence (supportive or otherwise) is the key part of the process. Related to this, you may come across the terms **objective** and **subjective**, particularly when noting that scientists must be objective and remove any trace of subjectivity when conducting research. This is undoubtedly desirable, but we again must remember that a hypothesis proposed by an individual will inevitably be set within his or her frame of reference, and so will, to some extent, be subjective. The ideal of completely objective practice is not always attainable in the experimental life sciences, and thus awareness of the possibility of subjectivity is important. In our final consideration of the basis of the scientific process, we will look at some of the important parts of actually doing science – what tools scientists use, how they behave at a personal level and how they communicate.

Science is based on evaluation of information, often described as **data**. In a technical sense, data are (the term is a plural!) made up of single pieces of information (a datum) that identifies, describes or measures a particular part of a system. Scientists also talk about **quantitative** and **qualitative** data sets, where things are *measured* to generate quantitative data, or *described* to generate qualitative data. Although as ever the demarcation can often be blurry, most experimental science will produce quantitative data for analysis. In gathering and evaluating quantitative data sets, there are several factors that need to be considered. Is the measurement **valid** (does it measure what it is trying to measure?) and **reliable** (does the means of measurement remain consistent in application?). The 'means of measurement' might be a simple device like a ruler, or a more complex machine or a multi-step process. Clearly a ruler has intrinsically a high level of reliability (although its use may generate unreliable measurements), whereas a machine may require routine and frequent servicing and calibration if it is to remain reliable. Related to validity and reliability, **accuracy** and **precision** are important in gathering data. Accuracy refers to how close a measurement is to the 'true' value, whilst precision is a reflection of how close repeated measurements are to each other, or how much variation there is in a set of measurements. Trying to identify and control validity, reliability, accuracy and precision is always an interesting exercise in experimental design.

Experimental science relies on gathering evidence to support or refute the hypothesis under investigation, and confirmation of outcomes by colleagues who replicate the experiments.

Most experimental science involves gathering quantitative data for evaluation.

pressure to publish widely and frequently to ensure that a visible and respected profile, or perhaps a perceived competitive advantage, is maintained. This can lead to a tendency to overpublish, from both scientists and publishers, and many consider this to be a self-sustaining and problematic development in science publishing. Although it is not feasible to count the number of scientific journals accurately, estimates set this at some 30 000 or so, with several million papers published annually. This is a good example of how something that seems simple (count the number of papers published) can essentially be impossible (think through how validity, reliability, accuracy and precision might come into play here). Even trying to keep up to date in a narrow specialism can be problematic, and would be impossible without IT-based access and search tools.

2.1.1 A Simple Model for the Scientific Method

There is a fairly standard model for the scientific process, variants of which can be found in innumerable books, science magazines and websites. Most are based on the hypothetico-deductive process, with some variation in the number of stages listed. One version is shown in Fig. 2.1. This shows a process with six defined stages, often carried out sequentially. The observation can be a simple visual event, or can arise from a sophisticated technical process involving complex equipment. Asking a question and constructing a hypothesis usually follow (the

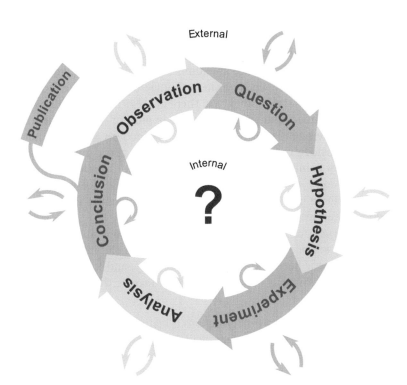

Fig. 2.1 A simple model for the scientific method. The process involves a series of steps beginning with an observation and progressing through to a conclusion, which may be negative or inconclusive and lead to amendment of the observation or restatement of the question/hypothesis. Not all stages may occur for all investigations, or they may be carried out in a different order. At all stages, internal and external influences (scientific literature, discussion, data processing) can be used to evaluate and inform. Eventually a consensus conclusion forms and leads to publication of the research.

Research requires continual internal and external questioning and discussion, covering a range of topics from day-to-day detail to high-level strategic planning.

'what if' question). Gathering data, either by observation or experiment, enables analysis and evaluation of the results, and leads to a conclusion. The conclusion may be negative or unhelpful, and require a rethink of the question or the methodology. At all stages, there will be processing of information in the scientist's own mind (internal) and interaction with other scientists and the scientific literature (external), much iterative thinking and talking, often a lot of repetition, problem-solving (why did my restriction enzyme not work?) and mini-hypothesis testing (if I add Y, instead of Z, will my restriction enzyme now work?). This 'day science' raises more questions, which are sometimes clear and well defined, but which can be rather hazy and speculative. This is usually where the creative and inspirational parts of 'night science' come to the fore (although this may, of course, occur in the middle of the afternoon whilst waiting for a reaction to finish!). And so the 'conversation with Nature' continues.

This fairly simple model is not incorrect, and it does show, to some extent, how the scientific process can be defined. It does not, however, describe the reality of how science works at the discipline or subject level. There has been an active debate about the failings of this type of model for almost as long as DNA has been studied, with many more sophisticated versions being proposed. We could look at some of these, but instead I want to broaden the view a little and consider the framework within which scientists work, with its attendant opportunities and caveats.

2.1.2 A More Realistic Model for How Science Works

The lines between **basic science** (often called **pure** or **fundamental science**), **applied science** and **technology** are often blurry, and to maintain this division has become almost redundant. An example is genome science; this depends on the technological developments that produced sequencing machines, and there are legitimate questions as to whether sequencing a gene or genome is pure or applied science. There are some scientists for whom these distinctions are important, and who consider that 'pure' science should remain untainted by application or commercialisation. The reality is that in practice, these definitions don't matter, and any organisation that does scientific research may be involved in all three areas concurrently in an interdependent and interdisciplinary setting.

The demarcation of basic science, applied science and technology is often imprecise and may actually be an unhelpful concept.

To construct our more realistic model, we will separate the *process* of doing the science (as outlined in Fig. 2.1) from the infrastructure and environment needed to support the activity, and look at the life science sector primarily. There are several factors that we need to consider:

Several factors influence how the scientific process is supported and enabled.

- Type of institution (university, research institute, company)
- Funding source(s) and recurrent availability/stability
- Location and links to other facilities and services

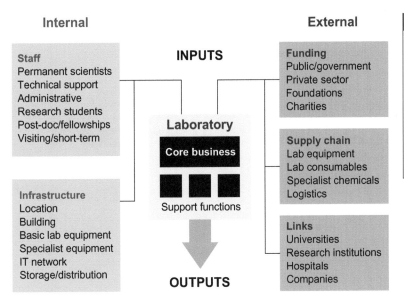

Fig. 2.2 Factors impacting on a research laboratory. Laboratories can vary in size and may be located within a larger organisational unit. Core business is where the scientific methodology and process happen. A range of internal and external factors provide inputs that need to be managed. Outputs may be publications, data, services or products.

- Staffing – permanent, short-term contract, students, technical, support
- Supply chain for equipment and consumables
- Distribution network for products (if needed)
- Ideological, cultural, social and political landscape (if relevant)

These factors apply across the sector, albeit at different levels. A small research group in a university will operate on a completely different scale than a multinational biotechnology company, but the basic requirements are broadly similar. Our more realistic model is shown in Fig. 2.2.

2.2 | DNA: A Biographical Timeline

The story of DNA may be unique in terms of the biological systems studied and the people involved, but it reflects how science works and has parallels in many other disciplines. Early discoveries helped set the scene for more detailed investigations, often involving experiments that were relatively simple in a technical sense but were elegant in their design and enabled Wald's 'conversation with Nature' in a fairly direct way. As more knowledge was gained, the techniques had to evolve to enable this to continue, and thus molecular biology has a co-dependence on technology that is clearly illustrated by a number of the developments since the early-1970s. To tell the story of DNA, we will look at some of the key discoveries on its biographical timeline, shown in Fig. 2.3.

Increasingly sophisticated experiments and procedures require the development of new technologies.

Mendel investigates transmission of characters in peas — **1865**

1869 — Meischer discovers DNA

1879 Mitosis described by Flemming

1900 — Correns, de Vries and von Tschermak rediscover Mendel's experiments

1902 Garrod studies the inheritance of alkaptonuria

1909 The word 'gene' is coined by Johannsen

1911 — Morgan establishes the 'fly room' and works out that genes are carried on chromosomes

Sturtevant constructs the first genetic linkage map — **1913**

1919 The word 'biotechnology' is used by Ereky

Griffith demonstrates bacterial transformation — **1928**

1941 — Beadle and Tatum establish the 'one gene one enzyme' hypothesis

Avery, MacLeod and McCarty show that DNA is the 'transforming principle' — **1944**

Hershey and Chase confirm that DNA is the genetic material — **1952** — Franklin takes the famous 'photo 51' of the X-ray diffraction pattern of the B-form of DNA

1953 — Watson and Crick determine the structure of the Double Helix

1955 Tjio confirms human chromosome number — **1955** — Kornberg isolates DNA polymerase
1956 Ingram establishes the cause of sickle cell disease

Meselson and Stahl show that DNA replication is semi-conservative — **1958**

1959 Lejeune shows Down syndrome is caused by trisomy-21

1961 Guthrie develops test for phenylketonuria in newborn babies — **1961** ⎡ Jacob and Monod show how the *lac* operon is regulated
⎣ Brenner, Jacob and Meselson find that mRNA carries the genetic code from DNA in protein synthesis

Nironborg, Holloy and Khorana docipher the genetic code — **1966**

1967 — Weiss and Richardson purify DNA ligase

1968 Arber, Nathans and Smith discover restriction enzymes

1970 — Smith and Nathans isolate type II restriction enzymes

1971 Cetus, the first biotechnology company, established

1972 — Berg generates the first rDNA

1973 Asilomar I conference on rDNA risks

Jaenish and Mintz develop transgenic mice — **1974**

1975 Asilomar II conference and moratorium
1976 The biotechnology company Genentech is formed

Web activity: find out what the key DNA-related discoveries were in the year that you were born.

Roberts and Sharp discover split genes — **1977** — Rapid DNA sequencing developed by Sanger and Gilbert

1982 GenBank database is established — **1982** — Birnster and Palmiter use rat growth hormone gene to create 'supermouse'
1982 Genentech rDNA insulin approved

Fig. 2.3 DNA timeline. *Source:* Background image by Robert Brook / Corbis / Getty Images.

1983 Huntington disease marker mapped to cs4
1983 Transgenic tobacco plant is generated
1983 — Mullis develops the polymerase chain reaction

Jeffreys develops DNA profiling — **1985**
1985 Discovery of zinc finger nucleases
1986 rDNA vaccine approved for hepatitis B
1986 Positional cloning of chronic granulomatous gene
1989 The cystic fibrosis *CFTR* gene is identified

Human Genome Project begins — **1990** — First human gene therapy treatment for SCID

Calgene's *Flavr Savr* GM tomato goes on sale — **1994**

1995 *H. influenzae* genome sequenced
1996 *S. cerevisiae* genome sequenced
1996 — Dolly the sheep cloned
1998 *C. elegans* genome sequenced
2000 *Drosophila* and *Arabidopsis* genomes sequenced

2001 — Human genome first draft

Mouse is first mammalian genome to be sequenced — **2002**
2002 *Plasmodium* genome sequenced

2003 Glo-Fish GM pet fish on sale
2003 — Human Genome Project completed

2004 Rat genome high-quality sequence completed

2005 Chimpanzee genome sequenced — **2005** — Rice genome sequenced
New DNA sequencing technologies developed —
2007

2008 — 1000 Genomes project launched

First comprehensive analysis of
cancer genomes published
2009

2010 Neanderthal genome sequence published
2010 — Venter creates first synthetic life form

2011 Discovery of TALENs
Doudna and Charpentier use CRISPR to edit genomes — **2012** — 100000 Genomes project launched

2014 — Esvelt proposes using CRISPR
to spread traits in a *gene drive*

A major controversy develops as a scientist claims
to have edited the genome of newborn twins — **2018**

The CRISPR baby scientists are convicted and
jailed for carrying out illegal medical practice — **2019** — Prime editing makes single-strand-cut editing a possibility

2020 — DNA technology plays a major role in managing the
SARS-CoV-2 pandemic by developing clinical
diagnostic tests, whole-genome sequencing for
variant analysis and vaccines
2021 CRISPR-edited tomato goes on sale — **2021**

A genetically modified pig heart is transplanted
into a patient; he dies after 2 months with pig
cytomegalovirus a possible cause — **2022** — T2T completes a gapless reference human genome

2.3 | People, Places and Progress: Paradigm Shifts or Step-Changes?

From time to time, there is a significant development in thinking that leads to a paradigm shift and changes the entire landscape of a subject. The phrase was used by the physicist and philosopher **Thomas Kuhn** to distinguish between what he saw as two types of science – 'normal' science (the day-to-day business of science, set within an agreed conceptual and technological framework) and scientific *revolutions* that change the way the subject is studied, usually in response to new thinking or findings that do not seem to make sense in the accepted framework. Examples include the move from the **geocentric** (Ptolemaic) to the **heliocentric** (Copernican) view of the solar system, from **Newtonian mechanics** to **quantum mechanics** and from **analogue** to **digital** electronics.

Although broadly accepted as a useful term, and subjected to extensive formal discourse among philosophers of science, the concept can present some problems. It can be difficult to assess if something represents a true paradigm shift or if it is just an extension of the existing methodology and thinking. Currently the *systems biology* approach, where multidisciplinary approaches are used to explore the vast amount of information available through, and derived from, genomic sequencing, can be considered as a newly emerging paradigm. Others take the view that the basic conceptual framework of systems biology (*i.e.* a holistic, rather than a reductionist, approach) has been around for thousands of years; this time, it is just the context and data sets that are different.

Rather less contentious than the identification of paradigm shifts, and much more frequent, are developments or discoveries that have a significant impact on the subject. These **step-changes** may be isolated insights by an individual, the culmination of a period of collaboration or an unforeseen development that sets off a new avenue of research. They fit well with the Pasteur and Szent-Györgyi quotations that we noted earlier in Section 2.1. Some step changes are of particular significance and almost fall into the paradigm shift category; we might borrow a phrase from behavioural science here and, perhaps slightly tongue-in-cheek, call these '**paradigm nudges**'. These concepts are not merely semantics; with limited funding for science, it is often prudent not to go against the current flow when applying for project grants, which can make it more difficult to fund so-called 'blue skies' research from which as yet unknown benefits might emerge.

The rather esoteric concepts of paradigms and step-changes can have real impacts on the funding of research.

Learned societies, institutions and professional bodies have procedures for the recognition of important contributions to science by individual scientists. These often involve awarding prizes, fellowships or honorary degrees, on an annual or occasional basis. The most well-known prize in science is the Nobel Prize, with the two categories relevant to life science research being for *Physiology or Medicine* and *Chemistry*. A useful way to collate the step-changes made in DNA-related fields is to look at the Nobel laureates in these areas, as shown in Table 2.1. From this list, where peer recognition has identified what

Table 2.1 | Milestones in cell and molecular biology recognised by the award of the Nobel Prize

Year	Prize	Recipient(s)	Awarded for studies on
1933	P/M	Thomas H. Morgan	Role of the chromosome in heredity
1935	P/M	Hans Spemann	The organiser effect in embryology
1958	C	Frederick Sanger	Primary structure of proteins
	P/M	Joshua Lederberg	Genetic recombination in bacteria
		George W. Beadle	Gene action
		Edward L. Tatum	
1959	P/M	Arthur Kornberg	Synthesis of DNA and RNA
		Severo Ochoa	
1962	C	John C. Kendrew	Three-dimensional structure of globular proteins
		Max F. Perutz	
	P/M	Francis H. C. Crick	Three-dimensional structure of DNA (the double helix)
		James D. Watson	
		Maurice H. F. Wilkins	
1965	P/M	François Jacob	Genetic control of enzyme and virus synthesis
		André M. Lwoff	
		Jacques L. Monod	
1968	P/M	Robert W. Holley	Elucidation of the genetic code and its role in protein synthesis
		H. Gobind Khorana	
		Marshall W. Nirenberg	
1969	P/M	Max Delbrück	Structure and replication mechanism of viruses
		Alfred D. Hershey	
		Salvador E. Luria	
1972	C	Christian B. Anfinsen	Three-dimensional structure and folding of ribonuclease
		Stanford Moore	Relationship between the chemical structure and active site of
		William H. Stein	ribonuclease
1975	P/M	David Baltimore	Reverse transcriptase and tumour viruses
		Renato Dulbecco	
		Howard M. Temin	
1978	P/M	Werner Arber	Restriction enzymes and their application to problems of
		Daniel Nathans	molecular genetics
		Hamilton O. Smith	
1980	C	Paul Berg	Recombinant DNA technology
		Walter Gilbert	DNA sequencing
		Frederick Sanger	
1982	C	Aaron Klug	Structure of nucleic acid/protein complexes
1983	P/M	Barbara McClintock	Mobile genetic elements
1984	P/M	Niels K. Jerne	Antibody formation
		Georges Köhler	Monoclonal antibodies
		César Milstein	

Table 2.1 | (cont.)

Year	Prize	Recipient(s)	Awarded for studies on
1989	C	Sidney Altman Thomas R. Cech	Catalytic RNA
	P/M	J. Michael Bishop Harold Varmus	Genes involved in malignancy
1993	C	Kary B. Mullis Michael Smith	The polymerase chain reaction Site-directed mutagenesis
	P/M	Richard J. Roberts Phillip A. Sharp	Split genes and RNA processing
1995	P/M	Edward B. Lewis Christiane Nüsslein- Volhard Eric F. Wieschaus	Genetic control of early embryonic development
1997	C	Paul D. Boyer John E. Walker Jens C. Skou	Synthesis of ATP Sodium/potassium ATPase
2001	P/M	Leland H. Hartwell R. Timothy Hunt Sir Paul M. Nurse	Key regulators of the cell cycle
2002	P/M	Sydney Brenner H. Robert Horvitz John E. Sulston	Genetic regulation of organ development and programmed cell death
2003	C	Peter Agre Roderick MacKinnon	Channels in cell membranes
2004	C	Aaron Ciechanover Avram Hershko Irwin Rose	Ubiquitin-mediated protein degradation
2006	C P/M	Roger D. Kornberg Andrew Z. Fire Craig C. Mello	Molecular basis of eukaryotic transcription RNA interference
2007	P/M	Mario R. Capecchi Sir Martin J. Evans Oliver Smithies	Principles for introducing specific gene modifications in mice
2008	C	Martin Chalfie Osamu Shimomura Roger Y. Tsien	Discovery and development of the green fluorescent protein (GFP)
2009	C	Venkatraman Ramakrishnan Thomas A. Steitz Ada E. Yonath	Studies of the structure and function of the ribosome
2009	P/M	Elizabeth H. Blackburn Carol W. Greider Jack W. Szostak	How chromosomes are protected by telomeres and the enzyme telomerase

Table 2.1 | *(cont.)*

Year	Prize	Recipient(s)	Awarded for studies on
2010	P/M	Robert G. Edwards	Development of *in vitro* fertilization
2012	C	Robert K. Lefkowitz Brian K. Kobilka	Studies of G-protein-coupled receptors
	P/M	Sir John B. Gurdon Shinya Yamanaka	Mature cells can be reprogrammed to become pluripotent
2015	C	Tomas Lindahl Paul Modrich Aziz Sancar	Mechanistic studies of DNA repair
2018	C	Frances H. Arnold George P. Smith Sir Gregory P. Winter	Directed evolution of enzymes Phage display of peptides and antibodies
	P/M	James P. Allison Tasuku Honjo	Cancer therapy by inhibition of negative immune regulation
2020	C	Emmanuelle Charpentier Jennifer A. Doudna	Development of a method for genome editing

Note: 'C' and 'P/M' refer to Nobel Prizes in Chemistry and Physiology or Medicine, respectively. Note also that Frederick Sanger is part of a very select group of Nobel laureates, having been awarded *two* Nobel Prizes for his work on proteins (1958) and DNA sequencing (1980). Information on past and current Nobel Prize winners can be found on the Nobel Prize website [www.nobelprize.org], from which this information was collated.
Source: Reproduced with permission.

are considered the most significant contributions to a particular discipline, we can identify a few of the discoveries on our DNA timeline as key step-changes (see Fig. 2.3), highlighted entries.

As well as people, *places* are important in science. We have seen that communication, collaboration and critical discussion are key parts of how a discipline progresses, and it is no surprise that many well-known laboratories around the world have shaped the field of DNA science since the early-1900s. As a laboratory develops a reputation for expertise in a particular area of research, more scientists are drawn to work there, either as permanent staff or as visiting researchers. This contributes to a vibrant research community, from which new discoveries emerge. A well-established laboratory with a good reputation is able to attract funding to support personnel and infrastructure, and so the cycle continues. In many ways, the evolution of DNA research laboratories reflects the science itself, with small-scale laboratories making many of the key early discoveries; the advent of large-scale DNA sequencing required scale-up in infrastructure and 'big science' came to biology for the first time. In the post-genomic era, large-scale research institutes with multidisciplinary approaches to a range of problems have become established. Some laboratories that have played a major part in the story of DNA are shown in Table 2.2. It is worth looking at the websites to explore the history of the laboratory and its contribution to the story.

Many world-famous laboratories and institutions have helped drive the science of molecular biology by providing a stimulating and well-funded workplace.

Table 2.2 | Some important laboratories and institutions

Laboratory	Location/comment	Suggested search term (SST)
T. H. Morgan's 'fly room'	Columbia University, New York City, early-1900s	t h morgan the fly room
The Cavendish Laboratory	Cambridge, UK. Physics laboratory where structure of DNA was discovered	cavendish laboratory structure of DNA history of the cavendish
Cold Spring Harbor Laboratory	Cold Spring Harbor, New York. Founded in 1890	cshl
Chinese Academy of Sciences	Beijing and multiple other sites. Largest research organisation in the world, multidisciplinary	chinese academy sciences
Wellcome Sanger Institute	Hinxton, Cambridge, UK. Established in 1992 to sequence the human genome	sanger
Broad Institute	Cambridge, Massachusetts. USA. Massachusetts Institute of Technology (MIT)/Harvard joint multidisciplinary initiative, launched in 2004	broad institute
Francis Crick Institute	London, UK. Multi-partnership collaborative approach to biomedical science	crick institute
European Molecular Biology Laboratory (EMBL)	Multi-site international European research organisation for life sciences	embl

Note: This is a small selection of organisations that have had, and continue to have, a major impact on life science and biomedical research. Links to other institutions can be found in the websites of each of the institutions listed in the table; you can find these easily using the SSTs.

2.4 | Conclusion: The Scientific Landscape

Our final task in this chapter is to integrate some of the ideas we have looked at so far – the 'how, where and who' of science. The process of doing science involves a number of elements, which we might think of as sitting in three domains:

- Intellectual
- Infrastructure
- Impact

The intellectual landscape is essentially made up of scientists and how they work. We looked at this in Section 2.1 when we considered the basis of the *scientific method*, and how some eminent scientists thought about the process of doing science. Communication, in discussion and through publication of results, is one of the key parts of how science works, and Wald's description of science as a *'conversation with Nature'* is one of the most useful.

The simple model of the scientific method we considered in Section 2.1.1 (shown in Fig. 2.1) is useful to some extent, but this is too constraining and we broadened this into the area of science infrastructure with the more realistic model shown in Fig. 2.2. The impact of the work that scientists do is also critical, as this will, to a large extent, determine how funding is allocated, particularly where very large expenditure is required to support a laboratory, institution or project. In Chapter 3, we will look at the impact of scientific progress when we consider the ethical dimension that is an integral part of the process.

Many of the aspects of how science works can be illustrated by looking at the history of DNA, which is outlined in diagrammatic form in Fig. 2.3. It may be useful to keep this in mind when we look at Parts 2, 3 and 4 of the book, to give a sense of perspective and illustrate Newton's reference to '*standing on the shoulders of giants*' (Section 2.1). We continued this theme into Section 2.3 when thinking about some of the people and places that have contributed to overall progress in DNA-related science.

The human story of DNA has always been of interest to a wider audience. Aspects of the science have always attracted media coverage, and factually based documentary programmes are made at regular intervals. These can be a very useful source of information, and help to set a topic in the context of its technology, benefits and risks. More fanciful depictions of DNA-related topics have also appeared in many films – just remember to treat these as entertainment rather than fact! Somewhere in the middle, there are factually based dramatisations of parts of the story. Three of these that are worth looking for are the films *Life Story* (1987) and *The Fly Room* (2014), and the stage play *Photograph 51* (2015). These present a human dimension to the protagonists in each case, and would be a useful (and entertaining) way to place the topics covered in this chapter in the historical framework of the DNA story.

Further Reading

Bronowski, J. (1951). *The Common Sense of Science*. A number of different editions and imprints are available from different publishers.

Goldacre, B. (2009). *Bad Science*. Harper Perennial, London. ISBN 978-0-00-728487-0.

International Science Council (2021). Opening the record of science: making scholarly publishing work for science in the digital era. International Science Council, Paris. URL [https://council.science/publications/sci-pub-report1/]. DOI [http://doi.org/10.24948/2021.01].

Levine, A. J. (2019). Seventy years and two paradigm shifts: the changing faces of biology. Institute for Advanced Study. Letter. 2019. URL [www.ias.edu/ideas/seventy-years-and-two-paradigm-shifts-changing-faces-biology].

Nurse, P. (2021). *What Is Life?: Understand Biology in Five Steps*. David Fickling Books, Oxford. ISBN 978-1-78845-142-0.

Schrödinger, E. (1944). *What Is Life?* A number of different editions and imprints are available from different publishers.

Websearch

People

Although James Watson, Francis Crick and Maurice Wilkins shared the 1962 Nobel Prize for their work on DNA, *Rosalind Franklin* is another scientist who will always be associated with the discovery. Have a look at her story; try to put yourself back in the early-1950s to try to get a sense of the issues she faced in her working life.

Places

The are 37 places called *Cambridge*, in six different countries. Arguably, the best known are those in the United Kingdom (the one in Cambridgeshire, not the other one) and the United States (the one in Massachusetts, not the other 24). Laboratories in these two cities have had a tremendous impact on the development of molecular biology and genetics. Find out what you can about the contributions made by Cambridge-based scientists, and construct a concept map of links between the topics studied and the scientists involved.

Processes

In this chapter, we looked at how science works, with the hypothetico-deductive method being one important part of the landscape. Explore the concepts of *hypothesis-driven* and *hypothesis-free* approaches in a little more depth, and note a few examples of each.

Reflections

What topics in this chapter have you found most challenging? Look for resources that help to illustrate the key points.

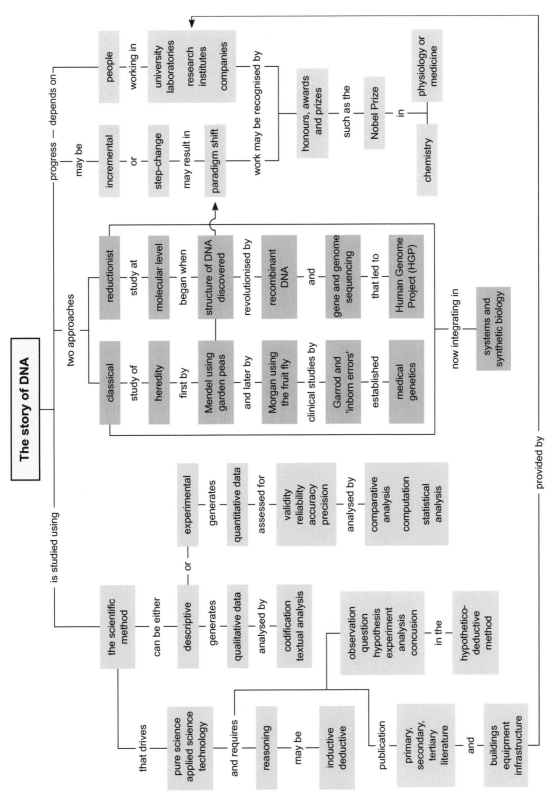

The story of DNA

is studied using

the scientific method

can be either

descriptive or **experimental**

descriptive generates
qualitative data
analysed by
codification
textual analysis

experimental generates
quantitative data
assessed for
validity
reliability
accuracy
precision
analysed by
comparative analysis
computation
statistical analysis

observation
question
hypothesis
experiment
analysis
concusion
in the
hypothetico-deductive method

that drives
pure science
applied science
technology

and requires
reasoning
may be
inductive
deductive

publication
primary, secondary, tertiary literature
and
buildings
equipment
infrastructure

progress — depends on
people working in
university laboratories
research institutes
companies

may be
incremental or *step-change*
may result in
paradigm shift

work may be recognised by
honours, awards and prizes
such as the
Nobel Prize
in
physiology or medicine
chemistry

two approaches

classical / **reductionist**

classical study of *heredity*
first by
Mendel using garden peas
and later by
Morgan using the fruit fly
clinical studies by
Garrod and 'inborn errors'
established
medical genetics

reductionist study at *molecular level*
began when
structure of DNA discovered
revolutionised by
recombinant DNA
and
gene and genome sequencing
that led to
Human Genome Project (HGP)

now integrating in
systems and synthetic biology

provided by

Concept Map 2

Chapter 3 Summary

Learning Objectives

When you have completed this chapter, you will be able to:

- Define the terms 'ethics' and 'bioethics'
- Outline the ethical framework within which DNA technology is placed
- Describe key ethical issues in genetics
- Appraise current and emerging ethical dilemmas in genetic engineering

Key Words

Ethics, moral philosophy, ethical legal and social implications (ELSI), code of practice, common good, golden rule, ethical framework, bioethics, genethics, absolute, relative, moral code, consequences, consequentialism, *primum non nocere*, Hippocratic oath, Auguste Chomel, uncertainty, Charles Darwin, Gregor Mendel, Sir Francis Galton, eugenics, selective breeding, positive eugenics, negative eugenics, cultural more, infanticide, assisted reproductive technologies (ARTs), genetic screening, pre-implantation diagnosis, genome sequencing, genome editing, informed consent, genetically modified organism (GMO), World Health Organization (WHO), over-medicalisation.

Chapter 3

Brave New World or Genetic Nightmare?

To conclude Part 1 of this book, we will consider the impact of genetic engineering in its widest sense by considering the ethical issues that accompany the use of DNA technology. In comparative terms, relatively few people work directly in genetics-related science and technology, but we will all have to cope with the consequences of gene-based research and its applications. Informed and vigorous debate is the only way that the developments in DNA technology can become established, accepted and extended. By placing the ethical issues firmly 'front and centre', I hope that you will be able to evaluate the technical details discussed in later chapters in the context of your own ethical framework.

3.1 | What Is Ethics?

Our first ethical dilemma is whether we say what *is* **ethics** (*i.e.* ethics as a singular) or what *are* ethics (*i.e.* ethics as a plural). In fact, both uses are correct, depending on the context. The singular form tends to be used when we refer to ethics as a discipline (in a technical sense, ethics is a branch of philosophy and is called **moral philosophy**). The plural form is more often used to describe the set of moral principles that are used to frame behaviour. As this outline of ethics is not a formal treatise of the subject, I hope the usage of the term will not present too many issues of interpretation. It is certainly worth looking into the subject in more detail than we can consider here, to provide a better appreciation of the complexities of this fascinating area.

Ethics is a branch of philosophy and a formal discipline in its own right.

In bioscience, the acronym ELSI (**ethical, legal and social implications**) is often used to refer to consideration of the impact of discoveries and their applications. Ethics is concerned with determining the best course of action in a particular circumstance. The ethics pertaining to a particular area, such as a business or profession, are usually set out as a consensual agreement among a group of interested parties, often called a **code of practice**. Some high-level principles that will be useful as we look at ethics in science include:

Where the benefits are clear, ethical concerns tend to be dealt with more easily than when there is some doubt as to the need for, or the effect of, a particular process or product.

- The concept of 'right' or 'wrong' as applied to actions or decisions
- Our individual (and/or collective) actions or decisions have consequences
- Ethical considerations involve judgement of value
- Ethical and legal frameworks may be aligned but are conceptually different
- Actions or decisions should follow the principles of the **common good** (what is best for the collective rather than the individual)
- Actions or decisions should be aligned with the **golden rule** (do to others what you would want them to do to you)

Not many people would object to this set of principles as a sound basis for the development of an **ethical framework**. However, achieving a workable set of ethics guidelines, rules and requirements for the range of disciplines in biological and health-related sciences is not a trivial task. The term bioethics is used to describe this area, with **genethics** (as described in Chapter 1) sometimes used to refer specifically to genetics-related ethical issues.

3.1.1 The Ethical Framework
The tenets, divisions and subdivisions of ethics as a discipline are complex and can sometimes be confusing to the non-expert. They can often appear as opposing points of view, each with its own values and justifications. Although we will not consider these in detail, we do need to look at some of the key elements to make sense of the overall ethical framework within which bioethics sits. I would suggest that you have a look at these elements in more detail – some suggestions are given in the *Websearch* section at the end of the chapter.

Our first question is whether we consider ethics as **absolute** or **relative**. This is an area that continues to generate significant debate. *Absolutists* favour the view that any ethical issue is either right or wrong, at all times and in all circumstances, whilst *relativists* see ethical issues as being placed in the context of current thinking and cultural norms. Often, though not exclusively, absolutists tend to be influenced by religious views and the perceived existence of an absolute **moral code**, and/or the concept of 'natural law'. Relativists may, in principle, agree with an absolute view of an issue, but will also consider the additional implications and may add a qualification. Most of us will agree that it is wrong to kill another human being; an absolutist would see this as applying in all circumstances, whilst a relativist would consider that there may be times when this can be justified (self-defence when life is at risk is often used as an example).

Two other aspects of the ethical framework are worth noting briefly. The **consequences** of an action (or inaction) form a major part of any evaluation of any medical or scientific ethics discussion (there is, in fact, a formal branch of ethics called **consequentialism**). This is perhaps the easiest ethical concept to grasp – the 'what if' question. There is also the interface between the ethical and legal frameworks in any given situation. Consider the topic of tax evasion/avoidance.

Ethical dilemmas very rarely have straightforward, clearly defined solutions.

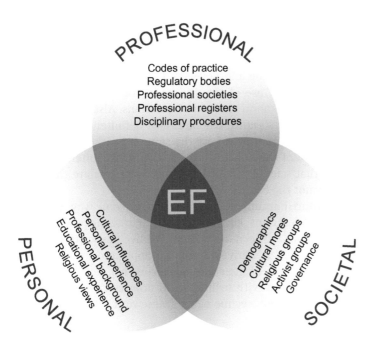

PROFESSIONAL
Codes of practice
Regulatory bodies
Professional societies
Professional registers
Disciplinary procedures

EF

PERSONAL
Cultural influences
Personal experience
Professional background
Educational experience
Religious views

SOCIETAL
Demographics
Cultural mores
Religious groups
Activist groups
Governance

Fig. 3.1 Factors influencing the development of an ethical framework. Factors come from three domains (personal, professional and societal), each of which has a subset of influencing factors. All of these may contribute to the development of the overall ethical framework (EF).

Tax *evasion* is illegal in most countries, but tax *avoidance* may not be. Thus, we may have a conflict between what is legal and what is ethical. Using an approved tax-free account to encourage saving is fine, but what if we begin to *exploit* existing rules and loopholes to avoid paying tax? Although this may be legal in terms of the letter of the law, we begin to get into that grey area that may be considered unethical and against the spirit of the legislation. On occasion, biological ethics may present a similar conflict, usually where a newly developed procedure may present a clear benefit from an ethics perspective but may not yet be legally permitted (it is less common for something to remain legal but be unethical, as legal frameworks are often amended in light of new knowledge or revised interpretation of existing practice).

Factors that inform the development of an ethical framework are shown in Fig. 3.1. In practical terms, we can think of scientific ethics as being a *relative* form of ethics, as many of the issues raised come from new discoveries that have, by definition, to be evaluated in the social and cultural context of the time. Consideration of the *consequences* of an action forms a major part of an ethical framework, which will also be supported by the development of the *legal system* to address new discoveries and procedures. How the framework is constructed for any given situation depends on the social, professional and personal norms that are prevalent at the time.

3.1.2 Is Science Ethically and Morally Neutral?

It is often said that science *per se* is neither 'good' nor 'bad', and that it is therefore ethically and morally neutral. Whilst this may be true of

The question of the moral and ethical neutrality of science is not easily answered, with many different aspects to be considered.

science as a *process*, it is the developments and applications that arise from the scientific process that pose the ethical questions. The example that is often quoted is the development of the atomic bomb – the science was interesting and novel, and of itself ethically neutral, but the application (*i.e.* use of the devices in conflict) posed a completely different set of moral and ethical questions. Also, science is, of course, carried out by *scientists*, who are most definitely not ethically and morally neutral, as they reflect the same breadth and range of opinion as the rest of the human race. An assumption often made by the layman is that scientists and the scientific process are the same thing, which is not the case.

Despite the purist argument that science is, in some way, immune from ethical considerations, I believe that to separate the process from its applications is an artificial distinction. We live in societies shaped by technology, which is derived from the application of scientific discoveries. The advent of genetic engineering has not altered the ethical issues associated with technological progress, but has simply placed them in a different technological framework.

3.1.3 The Scope of Bioethics

The scope of bioethics is closely aligned with the very broad and diverse range of subjects and disciplines in the biological/medical area. High-profile issues, such as abortion, animal rights, euthanasia, human cloning and environmental concerns, are perhaps the best known, but (depending on the level of definition) around 100 discrete areas of bioethics are studied actively. Although some of these are quite esoteric, there is still a wide range of topics that are important in bioethics. The *Nuffield Council on Bioethics* lists some 42 areas of current interest in five broad categories, of which we could consider 16 as being influenced, directly or indirectly, by the use of DNA technology. These are shown in Table 3.1.

All aspects of modern medical and genetic science have an ethical dimension.

A good foundation for bioethics is the phrase '*primum non nocere*', which translates to 'first do no harm'. This is often considered as part of the **Hippocratic oath** that medical professionals take. Although this exact phrase is not actually in the original version of the oath (but a very similar statement is), it does seem to date from the seventeenth century as part of the teaching philosophy of **Auguste Chomel**, a Parisian professor of medical pathology. It is quoted widely and serves well as a basic tenet for bioethics as well as medicine.

Organisations associated with the study, evaluation and promotion of bioethics are many and varied, and their online presence reflects the range that we considered in Chapter 1 (Section 1.4). This means that a particular organisation's view of a topic will not necessarily be neutral. As with other aspects of web-based information, care must be taken to ensure that any source used is accurate and authoritative, and that information is interpreted in the context of the type of organisation and its stated mission or aims. The field of bioethics can, not unexpectedly, be polarised and populated by individuals and groups with strongly held views, and thus it is important to

Table 3.1 | Areas of current interest in bioethics

Area of interest	Topic
Beginning of life	Genome editing and human reproduction Mitochondrial DNA disorders Non-invasive prenatal testing Whole genome sequencing of babies
Health and society	Genetics and genomics Personalised healthcare Screening
Animals, food and environment	Genome editing and farmed animals GM crops Xenotransplantation
Research ethics	Culture of scientific research Embryo and stem cell research Research in developing countries
Data and technology	Biological and health data Emerging biotechnologies Forensic use of bioinformation

Source: The Nuffield Council on Bioethics [www.nuffieldbioethics.org]. Reproduced with permission.

Table 3.2 | Some organisations in the field of bioethics

Organisation	Suggested search term (SST)
UNESCO International Bioethics Committee	unesco bioethics
Global Forum on Bioethics in Research	global forum bioethics
Council of Europe Bioethics	coe bioethics
International Association of Bioethics	international association bioethics
Nuffield Council on Bioethics	nuffield bioethics
NIH Department of Bioethics	nih bioethics
The Hastings Centre	hastings bioethics
Kennedy Institute of Ethics	kennedy bioethics

Note: UNESCO = United Nations Educational, Scientific and Cultural Organization; NIH = National Institutes of Health (USA). Websites can be found using the SSTs.

realise this when evaluating a particular issue. Some well-established organisations that provide a useful introduction to bioethics are shown in Table 3.2. In addition, many universities have departments of bioethics, and most countries have national frameworks and/or organisations that deal with bioethical issues.

3.2 | Elements of the Ethics Debate

In this section, we will highlight some of the ethical issues related to genetic engineering in a broad context. Later, chiefly in Part 4, we will consider some of the ethical issues raised by the development and use of the technology when we look at the applications of genetic engineering. In the ethics landscape, there are two major constituencies involved – the *scientists* who carry out the research and development work, and the *society* within which this is done. We will look at the role that each of these two groups has to play in the debate.

3.2.1   The Role of the Scientist

The scientist, and others who work in the applied industries that arise from the basic research and development, is often seen as the expert in the ethics debate. However, two points should be noted here. Firstly, as already indicated, scientists are of course part of society, with all the attendant opportunities and responsibilities that this brings. Often scientists will disagree on the scientific and/or ethical interpretation of a particular issue, which may of course be an entirely justifiable position. This can sometimes be misinterpreted by the public, who may expect 'the right answer' and whose trust in science can sometimes be undermined when **uncertainty** is part of the issue. This was a particular problem during the COVID-19 pandemic, with scientists rightly debating the issues (*e.g.* the efficacy of wearing face coverings) in public, but the public perception being clouded by both valid uncertainties around the issue and also the impact of mis- and disinformation that we considered in Chapter 1.

Secondly, scientists may be expert in their particular fields, but may not be expert in formal ethics or moral philosophy. Thus, the position of the scientist is not necessarily a simple one. All that we can ask of our scientific community is that its members carry out their work carefully, honestly and with a sense of professional and moral responsibility. Most scientists are also keen to consider the impact of their work in a wider context, often engaging with other experts in complementary fields to ensure full and informed debate.

Unfortunately, as in all walks of life, sometimes things in science do not go according to plan. Often progress is slow and can follow unproductive lines of investigation, but this is of course an inherent part of the scientific method and is accepted by all who are involved (although the general public sometimes sets expectations rather too high and expects scientists to be magicians). However, much more serious issues arise when scientists are found to have been falsifying data or otherwise compromising the integrity of the discipline. Thankfully these instances are rare, but they do happen and cause significant damage to the field when they come to light. The one positive aspect of this area of science is that the scientific community is very vigilant, and when cases do appear, they are dealt with

> Uncertainty in science means that individual scientists often disagree about the interpretation of the scale, consequences and impact of a particular issue.

> Scientists, individually and collectively, have a major role to play in presenting the arguments around the ethical and moral aspects of modern genetics.

> The scientific community expects a high standard of honesty, integrity and professional morality from its members.

appropriately. You can explore this aspect a little more by looking at the topics suggested in the *Websearch* section for this chapter.

3.2.2 The Role of Society

If scientists have a duty to act responsibly, the wider community also carries some of this responsibility, as well as the right to accurate and balanced information. Society is a complex mixture of many different types of people, groups and businesses, and presents many different points of view. Some of these can be extreme and often antagonistic, and care must be taken to ensure that a forceful stance taken by a minority of people does not compromise the level of serious debate that is needed to ensure that appropriate decisions are taken by those who regulate and govern (there may of course be a problem when a forceful stance is taken by governments, particularly when this may not align with the science of the situation). From a societal perspective, the three tenets of the *common good*, the *golden rule* and *first do no harm* are arguably more important than in an individual context; ensuring that these are central to the debate is not always an easy task.

Balancing the need to ensure open dialogue, there is also a need to guard against alienating various groups whose members may have genuine concerns. There is too often a tendency for a debate to become polarised, with the attendant risk of oversimplification of the issues. Given that we are now functioning in the 'global village', where information is available worldwide through various types of media, issues are immediately placed in this global dimension, and thus attract a much wider range of responses than would have previously been the case. The pervasive nature of social media often has a more significant impact on societal thinking than more reliable and authoritative sources, and countering this will be a significant part of managing the ethical landscape in the future.

Modern communication mechanisms mean that information placed in the public domain is immediately available worldwide, which has changed the dynamics of interactive debate.

Despite the challenges around ensuring that debate is based on accurate and unbiased information, it is important to accept that *all* members of democratic societies have a role to play in determining what policies and procedures are put in place to monitor, regulate and advise on the use of gene manipulation and its many applications. It is to the credit of those involved that, in most cases, full and frank discussion can enable consensus to be reached. Not everyone will be necessarily in agreement with the decisions made in this way, but most people accept that this is the most suitable way to proceed.

3.2.3 Current Issues in Bioethics

The areas of current interest shown in Table 3.1 are fairly high-level, and each of these has its own set of sub-topics, disciplines, techniques and ethical questions. We will examine some of these in more detail in later chapters when we look at the technical aspects of the topic. In this section, we will consider a few of the current issues that will continue to challenge our thinking on ethics in the future; to set the scene, we will begin by going back to the middle of the nineteenth century.

Charles Darwin published his seminal work *On the Origin of Species* (short form of the title) in 1859. At around the same period, **Gregor Mendel** was carrying out his experiments with peas that would lead (eventually in the early-1900s) to the realisation that his work, announced first in 1865, had established the principles of transmission genetics. Darwin and Mendel would rightly become recognised as perhaps the two major contributors to the development of biological sciences as a rigorous investigative discipline. Another significant contributor to genetics at this time was **Francis Galton** (later Sir Francis), who was a half-cousin of Darwin. Despite many important and positive contributions to the genetic thinking of the time, Galton is best known for coining the term **eugenics** in 1883, and establishing an area of study that would become a notorious example of the misuse of science.

Eugenics (and the eugenic movements that arose from its assumptions and predictions) advocates 'improving' the human population by influencing reproduction, essentially selective breeding applied to humans. Although aspects of eugenics have been practised for thousands of years, the modern eugenics movement spread from the United Kingdom across Europe, North America and Australia in the early decades of the twentieth century. People with desirable characteristics were encouraged to have more children (so-called **positive eugenics**), and those with undesirable characteristics discouraged by preventing marriage, segregation or using forced sterilisation (**negative eugenics**). I hope that your ethical alarm is now sounding loudly – the use of terms such as *improving*, *selective breeding* and *positive* and *negative* impacts on reproduction should raise immediate concerns. Many characteristics, traits or conditions were thought to be potential targets for eugenic intervention (*e.g.* physical disability, criminality, poverty, mental illness and even things like the desire to go to sea). Even allowing for the prevalent **cultural mores** of the time, it is difficult to see how these practices could be implemented and condoned by societies that were considered civilised.

The principles of eugenics were used to justify the events that occurred in Nazi Germany in the 1930s, with those citizens deemed 'unfit' facing sterilisation, segregation and later mass murder. This progressive dehumanisation of target groups, and the Nazi desire for Aryan racial purity, led of course to the Holocaust.

As the 1930s progressed, eugenics began to be discredited from both scientific and sociological perspectives and, by the end of World War II, had been largely abandoned as an identifiable government-endorsed policy in most countries. Whilst its demise was undoubtedly hastened by the events in Europe, and is to be welcomed, we are sometimes slow to learn lessons. The practice of involuntary sterilisation was implemented on a large scale in Peru from 1996 to 2000, and between 2005 and 2013, some 39 female prison inmates in California were sterilised without the proper consent being obtained.

The one-child policy that operated in China from 1980 to 2016 could also be considered a eugenics-related policy. Although the aim was to restrict population growth, and the policy was notionally non-invasive, it did lead to incidences of forced sterilisation and abortion, and abandonment (or even **infanticide**) of newborn female children.

Interest in eugenics and its validity has re-emerged in the twenty-first century with the development of new procedures. The increased availability and uptake of assisted reproductive technologies (ARTs), **genetic screening** techniques such as **pre-implantation diagnosis** and the potential impact of **genome sequencing** and genome editing all raise ethical issues that some consider should be aligned with, and evaluated against, the eugenics practices of the early-1900s. Others hold the view that we must disassociate ourselves from eugenics entirely, and see modern genetic practices as being different in that they involve individual choice and consent, rather than state-sponsored forced intervention, and are based on more robust science and medical practice than on the eugenics of the past. The debate will continue.

> Does an appreciation of the history of eugenics have a part to play in evaluating current and future applications of DNA technology?

The principle of **informed consent** is a critical part of research ethics when dealing with human participants and when a medical, genetic or clinical procedure is involved. When this is violated, we can have issues such as the non-consensual sterilisation discussed in this section. However, the principle is not only applicable to individuals, but also to groups of people. The deliberate release of a genetically modified organism (GMO) into an ecosystem is one example, where establishing informed consent from all members of the community that might be affected by the release can be difficult. Recently, updated guidance from the **World Health Organization (WHO)** (2021) on the use of genetically modified mosquitoes in controlling malaria is a good example of the range of points that need to be addressed when releasing GMOs in an interventionist way. Similar issues, but in a different context, apply to growing genetically modified crops. We will consider these areas in more detail in Chapter 16.

> The issue of genetically modified foods is an emotive matter for many people and continues to be an area of controversy.

A final point to consider brings us back to the overall scope and breadth of bioethics. Most of the ethical issues that we now face in genetics and medicine are associated with sophisticated, and thus often relatively expensive, applications. Genome analysis, ARTs, genetic diagnosis in embryos, the possibility of editing an individual's genome, gene therapy and other aspects of personalised genetics-based medicine are all examples of current or emerging technologies. Most people would, on balancing the risks and benefits, consider these as positive developments (when used within a sound ethical and regulatory framework). However, there is a valid point of view that these developments can represent a form of over-medicalisation, leading to disparity of access to the treatment or diagnostic. This disparity is usually resource-dependent, with 'rich' (people or countries) more able to access expensive technology than 'poor' (people or countries). A related issue is that the *under*-medicalisation and treatment of disease in relatively poor countries is often a result of a lack of basic infrastructure (clean water, accommodation and basic healthcare provision) as much as it is a lack of expensive medicines or treatments. This presents a different ethical issue – more a question of inequitable resource distribution than of too much intervention. A potential dilemma in the future will arise as technologies become more affordable and more accessible, and we need to ensure that we do not necessarily try to replicate practice from previous experience just because the technology has a wider geographical or socio-economic reach.

Inequality of access remains a problem in ensuring that the benefits of modern science are distributed widely and equitably across the global socio-economic spectrum.

3.3 | Conclusion: Has Frankenstein's Monster Escaped from Pandora's Box?

The rather awful question that forms the title for this final section illustrates how two phrases that would not usually exist in one sentence can be joined together. This of course is exactly the essence of gene manipulation – splicing pieces of DNA together to generate *recombinant* molecules that would not exist in nature. In Parts 2, 3 and 4 of this book, we will be considering the techniques and processes that are used in gene manipulation, from basic principles up to advanced techniques. I hope that this brief treatment of *bioethics* will help you to think about the ethical issues around genetic engineering, particularly when we look at its applications and potential in Part 4.

You should be prepared to play your part in the debates that are yet to take place, so that appropriate and due consideration is given to any new developments that have the potential to impact on all our lives.

As with any branch of human activity, the responsibility for using genetic technology lies with those who discover, adapt, implement and regulate it. However, the pressures that exist when commercial development of genetic engineering is undertaken can sometimes change the balance of responsibility. Most scientists ply their trade with honesty and integrity, and would not dream of falsifying results or inventing data. They take special pride in what they do and, in a curious paradox, remain detached from it whilst being totally involved with it. Once the science becomes a technology, things are not quite so clear-cut, and *corporate* responsibility is sometimes not quite so easy to define as *individual* responsibility. I have found it helpful to ask three questions when thinking about the ethics of a particular issue:

Public opinion, particularly with regard to consumer acceptance of a product, can have a very powerful effect on commercial success.

- Do I understand fully the issue and its implications?
- How will it affect the intended recipient(s) and the wider members of society?
- If I had to make a go/no-go decision, what would it be?

It may be helpful to keep these three questions in mind as you work your way through the rest of this book (another useful maxim here is to remember that just because we *can* do something does not necessarily mean that we *should* do it). In trying to answer the last question in the list, I imagine that you will at times see a fairly clear answer, and at others a very confused picture that may, in fact, not have an obvious answer at all. If so, you will be well prepared to contribute to the ethical debates that lie ahead.

Further Reading

Harvey, L. (2020). Research fraud: a long-term problem exacerbated by the clamour for research grants. *Quality in Higher Education*, 26 (3), 243–61. URL [www.tandfonline.com/doi/full/10.1080/13538322.2020.1820126]. DOI [https://doi.org/10.1080/13538322.2020.1820126].

Jensen, E. *et al.* (2021). Has the pandemic changed public attitudes about science? LSE Impact Blog. URL [https://blogs.lse.ac.uk/impactofsocialsciences/2021/03/12/has-the-pandemic-changed-public-attitudes-about-science/].

Mandal, J. *et al.* (2017). Bioethics: a brief review. *Trop. Parasitol.*, 7 (1), 5–7. URL [www.ncbi.nlm.nih.gov/pmc/articles/PMC5369276/]. DOI [doi: 10.4103/tp.TP_4_17].

National Human Genome Research Institute. Eugenics: its origin and development (1883–present). URL [www.genome.gov/about-genomics/educational-resources/timelines/eugenics].

World Health Organisation, Health Ethics and Governance Team. (2022). COVID-19 and mandatory vaccination: ethical considerations. WHO reference number: WHO/2019-nCoV/Policy_brief/Mandatory_vaccination/2022.1. URL [www.who.int/publications/i/item/WHO-2019-nCoV-Policy-brief-Mandatory-vaccination-2022.1].

Websearch

People

Aside from his association with eugenics, *Sir Francis Galton* made many significant contributions to a number of areas of science. Look at his career history to find out more. More recently, publishing related to COVID-19 has been causing some issues. Have a look at the story around *Dr Sapan Desai* and see what your views are on this.

Places

In January 2022, the *University of Maryland Medical Center* in Baltimore was in the news for a world first. What was this? Two types of ethical issue are relevant here – one to do with the procedure and its implications, and one to do with the history of the recipient. Find out more about this aspect.

Processes

The regulation and management of gene manipulation and related applications are a very important part of how the science is carried out. In your country, find out which body (or bodies) is responsible for regulating: (1) clinical trials for *gene therapy* applications, and (2) the generation and use of *GMOs*. Is the information clear and easily available?

Reflections

What topics in this chapter have you found most challenging? Look for resources that help to illustrate the key points.

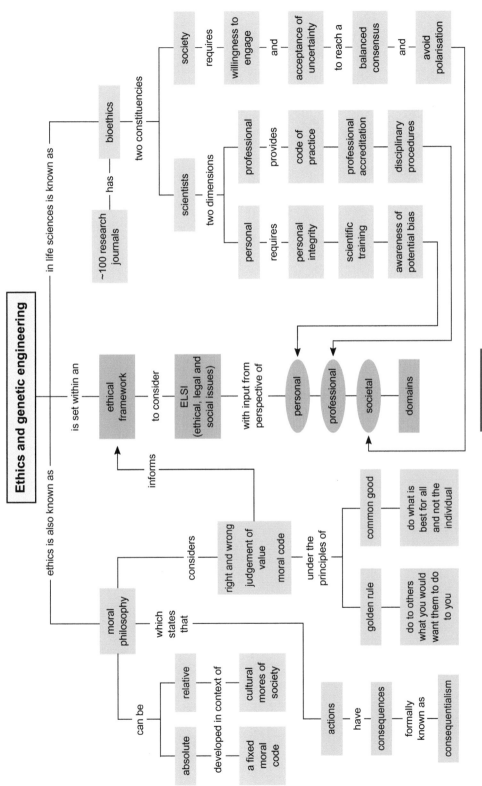

Ethics and genetic engineering

ethics is also known as — in life sciences is known as

moral philosophy

can be
- absolute — developed in context of — a fixed moral code
- relative — developed in context of — cultural mores of society

which states that — actions — have — consequences — formally known as — consequentialism

considers — right and wrong judgement of value moral code

under the principles of
- golden rule — do to others what you would want them to do to you
- common good — do what is best for all and not the individual

informs

ethical framework

to consider — ELSI (ethical, legal and social issues)

with input from perspective of — personal — professional — societal — domains

bioethics

~100 research journals — has — bioethics

two constituencies
- scientists
  - two dimensions
    - personal — requires — personal integrity — scientific training — awareness of potential bias
    - professional — provides — code of practice — professional accreditation — disciplinary procedures
- society
  - requires — willingness to engage — and — acceptance of uncertainty — to reach a — balanced consensus — and — avoid polarisation

is set within an

Concept Map 3

Part 2

The Basis of
Genetic Engineering

Chapter 4 Summary

Learning Objectives

When you have completed this chapter, you will be able to:

- Outline the organisation of living systems and the concept of emergent properties
- Review the chemistry of living systems
- Describe the genetic code and the flow of genetic information
- Summarise the structure and function of DNA and RNA
- Explain the need for regulation of gene expression
- Define the transcriptome and proteome

Key Words

Structure, function, emergent properties, carbon, covalent bonds, atoms, molecules, macromolecules, lipids, carbohydrates, proteins, nucleic acids, dehydration synthesis, hydrolysis, cell membrane, plasma membrane, cell wall, genetic material, prokaryotic, eukaryotic, nucleus, surface area-to-volume ratio, mitochondria, chloroplasts, adenine (A), guanine (G), cytosine (C), thymine (T), enzymes, heteropolymers, codons, minimum coding requirement, genetic code, double helix, redundancy, wobble, messenger RNA (mRNA), transcription, translation, Central Dogma, replication, reverse transcriptase, mutation, nucleotides, polynucleotide, deoxyribose, deoxyribonucleic acid, ribonucleic acid, phosphodiester, antiparallel, uracil, purines, pyrimidines, ribosomal RNA (rRNA), transfer RNA (tRNA), ribosomes, gene, chromosome, locus, homologous pairs, alleles, coding strand, template, non-template, non-coding, promoter, transcriptional unit, enhancers, silencers, operons, structural genes, operator, cyclic AMP (cAMP), cAMP receptor protein (CRP), Lac repressor, polycistronic, cistron, intervening sequences, introns, exons, dystrophin, primary transcript, RNA processing, RNA polymerase, consensus sequences, anticodon, polypeptide, N-terminus, C-terminus, adaptive regulation, differentiated, developmentally regulated, housekeeping genes, constitutive genes, catabolic, inducible, negative control, repressor protein, genome, bioinformatics, C-value, C-value paradox, C-value enigma, *Caenorhabditis elegans*, *Arabidopsis thaliana*, *Drosophila melanogaster*, *Mus musculus*, abundance classes, inverted repeats, palindromes, foldback DNA, highly repetitive sequences, moderately repetitive sequences, unique or single copy sequences, multigene families, transcriptome, proteome.

Chapter 4

Introducing Molecular Biology

This chapter presents a brief overview of the structure and function of DNA and its organisation within the genome (the total genetic complement of an organism). We will also have a look at how genes are expressed, and at the ways by which gene expression is regulated. The aim of the chapter is to provide an introduction to the molecular biology of cells, to ensure that the discussions of how DNA can be manipulated are based on a sound understanding of its structural and functional characteristics. It should also act as a useful refresher for those who have some background knowledge of DNA.

4.1 | How Living Systems Are Organised

Before we look at the molecular biology of the cell, it may be useful to think a little about what cells are, and how living systems are organised. Two premises are useful here. Firstly, there is a very close link between **structure** and **function** in biological systems. Secondly, living systems provide an excellent example of the concept of **emergent properties**. This is rather like the statement 'the whole is greater than the sum of the parts', in that living systems are organised in a hierarchical way, with each level of organisation becoming more complex. New functional features emerge as components are put together in more complicated arrangements. One often-quoted example is the reactive metal *sodium* and the poisonous gas *chlorine*, which combine to give *sodium chloride* (common table salt) that is of course not poisonous (although it can be if taken in excess). Thus it is often difficult or impossible to predict the properties of a more complex system by looking at its constituent parts, which is a general difficulty with the reductionist approach to experimental science.

Living systems are organised hierarchically, with close interdependence of structure and function.

The chemistry of living systems is based on the element **carbon**, which can form four covalent bonds with other **atoms**. By joining carbon atoms together, and incorporating other atoms, **molecules** can be built up, which in turn can be joined together to produce macromolecules. Biologists usually recognise four groups of macromolecules: **lipids**, carbohydrates, proteins and **nucleic acids**. The

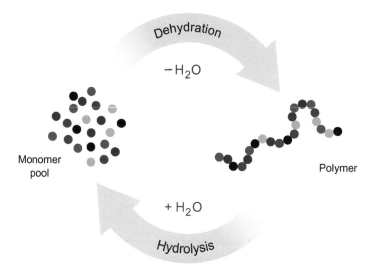

Monomer
pool

Polymer

Fig. 4.1 The monomer/polymer cycle. In this example, a representation of amino acids and proteins is shown. The amino acids (monomers) are joined together by removal of the elements of water (H_2O) during dehydration synthesis. When the protein is no longer required, it may be degraded by adding back the H_2O during hydrolysis. Although the cycle looks simple when presented like this, the synthesis of proteins requires many components, the functions of which are coordinated during the complex process of translation. The monomer/polymer cycle is a key part of many aspects of the metabolism of cells, not just protein synthesis.

Complex molecules (macromolecules) are made by joining smaller molecules together by dehydration synthesis.

synthesis of macromolecules involves a condensation reaction between functional groups on the molecules to be joined together. This **dehydration synthesis** forms a covalent bond by removing the elements of water. In the case of the large polymeric macromolecules of the cell (polysaccharides, proteins and nucleic acids), hundreds, thousands or even millions of individual monomeric units may be joined together in this way. The polymers can be broken apart into their constituent monomers by adding the elements of water back to re-form the original groups. This is known as **hydrolysis** (literally *hydro lysis*, water breaking). The monomer/polymer cycle and dehydration/hydrolysis are illustrated in Fig. 4.1.

The cell is the basic unit of organisation in biological systems. Although there are many different types of cell, there are some features that are present in all cells. There is a **cell membrane** (the **plasma membrane**) that is the interface between the cell contents and the external environment. Some cells, such as bacteria, yeasts and plant cells, may also have a **cell wall** that provides additional structural support. Some sort of **genetic material** (almost always DNA) is required to provide the information for cells to function, and the organisation of this genetic information provides one way of classifying cells. In **prokaryotic** cells (*e.g.* bacteria), the DNA is not compartmentalised, whereas in **eukaryotic** cells, the DNA is located within a membrane-bound **nucleus**. Eukaryotic cells also utilise membranes to increase the **surface area-to-volume ratio** (SA/V) of internal organelles such as **mitochondria** (plant and animal cells) and **chloroplasts** (plant cells). This increases the area available for localising enzyme reactions and thus increases the efficiency of metabolic processes. Prokaryotic cells are generally smaller in size than eukaryotic cells, but all cells have a maximum upper size limit. This is largely due to the limitations of diffusion as a mechanism for gas and nutrient exchange. Typical bacterial cells have a diameter of 1–10 μm, and plant and animal cells 10–100 μm. In multicellular eukaryotes, increase in size is achieved by using more cells rather than by making them bigger. As organismal complexity increases, cells diversify and become specialised for particular functions; this is a good illustration of the concept of emergent properties.

> The cell is the basic unit of organisation in biological systems; prokaryotic cells have no nucleus, whereas eukaryotic cells do.

4.2 | The Flow of Genetic Information

To set the structure of nucleic acids in context, it is useful to think a little about what is required, in terms of genetic information, to enable a cell to carry out its various activities. It is a remarkable fact that an organism's characteristics are encoded by a four-letter alphabet, defining a language of three-letter words. The letters of this alphabet are the nitrogenous bases adenine (A), guanine (G), cytosine (C) and thymine (T). So how do these bases enable cells to function?

> Life is directed by four nitrogenous bases: adenine (A), guanine (G), cytosine (C) and thymine (T).

The expression of genetic information is achieved ultimately *via* proteins, particularly the enzymes that catalyse the reactions of metabolism. Proteins are condensation heteropolymers synthesised from amino acids, of which 20 are used in natural proteins. Given that a protein may consist of several hundred amino acid residues, the number of different proteins that may be made is essentially unlimited, assuming that the correct sequence of amino acids can be specified from the genetic information. As the bases are critical informatic components, we can calculate that using the bases singly would not provide enough scope to encode 20 amino acids, as there are only four possible code 'combinations' (A, G, C and T). If the bases were arranged in pairs, that would give 4^2 or 16 possible combinations – still not enough. Triplet combinations provide 4^3 or 64 possible permutations, which is more than sufficient. Thus, great diversity of protein form and function can be achieved using an elegantly simple coding system, with sets of three nucleotides (codons) specifying the amino acids. A protein of 300 amino acids would therefore have a **minimum coding requirement** (MCR) of 900 nucleotides on a strand of DNA.

The genetic code or 'dictionary' is one part of molecular biology that, like the **double helix**, has become something of a biological icon. Although there are more possible codons than are required (64 as opposed to 20), three of these are 'STOP' codons. Several amino acids are specified by more than one codon, which accounts for the remainder, a feature that is known as redundancy of the code. An alternative term for this, where the first two bases in a codon are often critical, with the third less so, is known as wobble. These features can be seen in the standard presentation of the genetic code shown in Table 4.1.

Table 4.1 | The genetic code

First base (5')	Second base				Third base (3')
	U	C	A	G	
U	Phenylalanine	Serine	Tyrosine	Cysteine	U
	Phenylalanine	Serine	Tyrosine	Cysteine	C
	Leucine	Serine	**Stop**	**Stop**	A
	Leucine	Serine	**Stop**	Tryptophan	G
C	Leucine	Proline	Histidine	Arginine	U
	Leucine	Proline	Histidine	Arginine	C
	Leucine	Proline	Glutamine	Arginine	A
	Leucine	Proline	Glutamine	Arginine	G
A	Isoleucine	Threonine	Asparagine	Serine	U
	Isoleucine	Threonine	Asparagine	Serine	C
	Isoleucine	Threonine	Lysine	Arginine	A
	Methionine	Threonine	Lysine	Arginine	G
G	Valine	Alanine	Aspartic acid	Glycine	U
	Valine	Alanine	Aspartic acid	Glycine	C
	Valine	Alanine	Glutamic acid	Glycine	A
	Valine	Alanine	Glutamic acid	Glycine	G

Non-polar (hydrophobic)   Basic (+ve charge)

Polar (hydrophobic)   Acidic (−ve charge)

Note: This shows the standard table format for the genetic code. Codons are written as RNA bases (U substituted for T) and read 5' → 3'; thus, AUG specifies methionine, which is used as the start codon/amino acid for protein synthesis. Amino acids can be grouped by their chemical properties in a number of ways — one is the common polar/non-polar basic/acidic version shown here. Many amino acids are specified by more than one codon (redundancy of the code). Three termination, or stop, codons are UAA, UAG and UGA; these specify no amino acid and thus end translation. Other versions of the genetic code are available in circular format, including one where the importance of the second base is emphasised.

The flow of genetic information is unidirectional, from DNA to protein, with **messenger RNA (mRNA)** as an intermediate. The copying of DNA-encoded genetic information into RNA is known as **transcription** (T_C), with the further conversion into protein being termed **translation** (T_L). A linguistic analogy can be useful here — *transcribing* often means copying in the same language (nucleic acid language, DNA to mRNA),

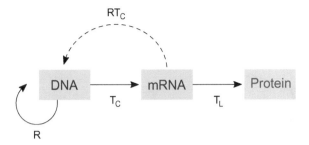

Fig. 4.2 The Central Dogma. This states that information flow is unidirectional, from DNA to mRNA to protein. The processes of transcription (T_C), translation (T_L) and DNA replication (R) follow this pattern of directional transfer. An exception is found in retroviruses (one group of RNA viruses), which have a single-stranded RNA genome and carry out a process known as reverse transcription (RT_C) to produce a DNA copy of the genome following infection of the host cell. Although this appears to reverse the flow of genetic information, the DNA copy is then used to integrate into the host cell's genome and is expressed in the 'normal' direction. Reverse transcription is therefore a workaround for a specific group of viruses, rather than a reversal of the Central Dogma, as transfer of information from proteins to nucleic acids does not occur.

whereas moving from nucleic acid to protein 'language' is analogous to *translating* a text into a different language. The concept of unidirectional flow of genetic information, and its conversion from nucleic acid to protein, is known as the **Central Dogma** of molecular biology and is an underlying theme in all studies on gene expression.

A further two aspects of information flow may be added to this basic model to complete the picture. Firstly, duplication of the genetic material prior to cell division represents a DNA~DNA transfer, and is known as DNA **replication**. A second addition, with important consequences for the genetic engineer, stems from the fact that some viruses have RNA, instead of DNA, as their genetic material. These viruses (chiefly members of the retrovirus group) have an enzyme called **reverse transcriptase** (an RNA-dependent DNA polymerase) that produces a double-stranded DNA molecule from the single-stranded RNA genome. Thus, the flow of genetic information is reversed with respect to the normal convention. The Central Dogma is summarised in Fig. 4.2.

The flow of genetic information is from DNA to RNA to protein, via the processes of transcription (T_C) and translation (T_L). This concept is known as the Central Dogma of molecular biology.

4.3 | The Structure of DNA and RNA

In most organisms, the primary genetic material is double-stranded DNA. What is required of this molecule? Firstly, it has to be *stable*, as genetic information may need to function in a living organism for up to 100 years or more. Secondly, the molecule must be capable of *replication*, to permit dissemination of genetic information as new cells are formed during growth and development. Thirdly, there should be the potential for limited alteration to the genetic material (**mutation**), to enable evolutionary pressures to exert their effects. The DNA molecule fulfils these criteria of stability, replicability and mutability and, when

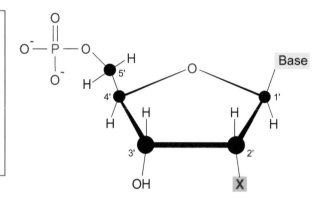

Fig. 4.3 The structure of a nucleotide. Carbon atoms are represented by black circles, numbered 1′ to 5′. In DNA, the sugar is deoxyribose (with a hydrogen atom at position X), and in RNA, the sugar is ribose (a hydroxyl group at position X). The base can be adenine (A), guanine (G), cytosine (C) or thymine (T) in DNA, and A, G, C or U (uracil) in RNA.

considered with RNA, provides an excellent example of the premises that we considered earlier – the very close relationship between structure and function, and the concept of emergent properties.

Nucleic acids are heteropolymers composed of monomers known as **nucleotides**, and a nucleic acid chain is therefore often called a **polynucleotide**. The monomers are themselves made up of three components: a sugar, a phosphate group and a nitrogenous base. The two types of nucleic acid (DNA and RNA) are named according to the sugar component of the nucleotide, with DNA having 2′-**deoxyribose** as the sugar (hence **Deoxyribo Nucleic Acid**) and RNA having ribose (hence **Ribo Nucleic Acid**). The sugar/phosphate components of a nucleotide are important in determining the structural characteristics of polynucleotides, with the nitrogenous bases determining their information storage and transmission characteristics. The structure of a nucleotide is summarised in Fig. 4.3.

Nucleotides can be joined together by a 5′ → 3′ **phosphodiester linkage**, which confers directionality on the polynucleotide. Thus, the 5′ end of the molecule will have a free phosphate group, and the 3′ end a free hydroxyl group; this has important consequences for the structure, function and manipulation of nucleic acids. In a double-stranded molecule such as DNA, the sugar–phosphate chains are found in an **antiparallel** arrangement, with the two strands running in different directions with respect to their 5′ → 3′ polarity.

The nitrogenous bases are the important components of nucleic acids in terms of their coding function. In DNA, the bases are as listed in Section 4.2, namely adenine (A), guanine (G), cytosine (C) and thymine (T). In RNA, the base thymine is replaced by **uracil (U)**, which is functionally equivalent. Chemically, adenine and guanine are **purines**, which have a double ring structure, whereas cytosine and thymine (and uracil) are **pyrimidines**, which have a single ring structure. In DNA, the bases are paired: A with T, and G with C. This pairing is determined both by the bonding arrangements of the atoms in the bases and by the spatial constraints of the DNA molecule, the only viable arrangement being a purine~pyrimidine base-pair. The bases are held together by hydrogen bonds: two in the case of an A·T base-pair, and three in the case of a G·C base-pair. The structure and base-pairing arrangement of the four DNA bases is shown in Fig. 4.4.

Nucleic acids are polymers composed of nucleotides; DNA is deoxyribonucleic acid, and RNA is ribonucleic acid.

In DNA, the bases are arranged as A·T and G·C base-pairs; this complementary base-pairing is the key to information storage, transfer and use.

(a)

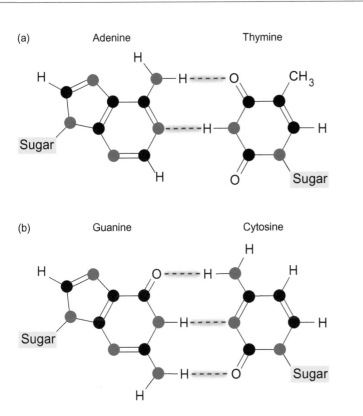

Adenine   Thymine

(b)

Guanine   Cytosine

● Carbon atom   ● Nitrogen atom   ----- Hydrogen bond

Fig. 4.4 Base-pairing arrangements in DNA. (a) An A·T base-pair is shown. The adenine and thymine bases are linked by two hydrogen bonds (dotted lines). (b) A G·C base-pair, with three hydrogen bonds linking the guanine and cytosine bases, is shown. The difference in the number of hydrogen bonds means that G·C base-pairs are more thermodynamically stable than A·T base-pairs. The melting temperature (T_m) of double-stranded DNA therefore varies with base composition; sequences with a higher G·C content have a higher T_m.

The DNA molecule *in vivo* usually exists as a right-handed double helix called the B-form. This is the structure proposed by Watson and Crick in 1953. Alternative forms of DNA include the A-form (right-handed helix) and the Z-form (left-handed helix). Although DNA structure is a complex topic, particularly when the higher-order arrangements of DNA are considered, a version of the standard representation of the B-form will suffice here, as shown in Fig. 4.5.

The structure of RNA is similar to that of DNA, the main chemical differences being the presence of ribose instead of 2′-deoxyribose and uracil instead of thymine. RNA is also most commonly single-stranded, although short stretches of double-stranded RNA may be found in self-complementary regions. With regard to information flow, there are three main types of RNA molecule found in cells: messenger RNA (mRNA), ribosomal RNA (rRNA) and transfer RNA (tRNA). Ribosomal RNA is the most abundant class of RNA molecule, making up some 85 per cent of total cellular RNA. It is associated with ribosomes, which are an essential part of the translational machinery. Transfer RNAs make up about 10 per cent of total RNA, and provide the essential specificity that enables the insertion of the correct amino acid into the protein that is being synthesised. Messenger RNA, as the name suggests, acts as the carrier of genetic information from DNA to the translational machinery, and is usually less than 5 per cent of total cellular RNA.

Three important types of RNA are ribosomal RNA (rRNA), messenger RNA (mRNA) and transfer RNA (tRNA).

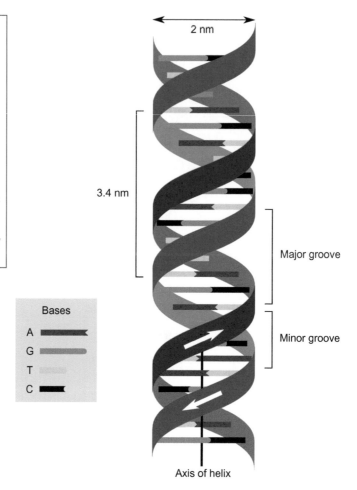

Fig. 4.5 The double helix. This is DNA in the commonly found B-form. The right-handed helix has a diameter of 2 nm and a pitch of 3.4 nm, with 10–10.5 base-pairs per turn. The sugar–phosphate 'backbones' are antiparallel (blue and red colour, arrowed) with respect to their $5' \rightarrow 3'$ orientations. The purine~pyrimidine base-pairs are formed across the axis of the helix. *Source:* Modified from Watanabe/DBCLS [www.dbcls.rois.ac.jp]. DOI: [https://doi.org/10.7875/togopic.2018.23]. Used under Licence CC-BY-4.0 [www.creativecommons.org/licenses/by/4.0/].

4.4 | Gene Organisation

The **gene** can be considered as the basic unit of genetic information. Although genes have been studied systematically since Mendel's experiments in the 1850s, the term was first used by Johannsen in 1909. Before the advent of molecular biology and the realisation that genes were made of DNA, the study of the gene was largely indirect; the effects of genes were observed in phenotypes and the 'behaviour' of genes was analysed. Despite the apparent limitations of this approach, a vast amount of information about how genes functioned was obtained, and the basic tenets of transmission genetics were formulated. The conceptual transition to the era of molecular biology began with the 'one gene one enzyme' hypothesis proposed by Beadle and Tatum in 1941. The work of Avery, MacLeod and McCarty (1944) and Hershey and Chase (1952) confirmed that DNA was the genetic material and completed this trio of classic experiments, paving the way for Watson and Crick to solidify the conceptual framework and mark the start of molecular biology as a discipline with their 1953 double-helical structure.

As the gene was studied in greater detail, the terminology associated with this area of genetics became more extensive, and the ideas about genes were modified to take account of developments. The term gene is usually taken to represent the genetic information transcribed into a single RNA molecule, which is in turn translated into a single protein. Exceptions are genes for RNA molecules (such as rRNA and tRNA), which are not translated. In addition, the nomenclature used for prokaryotic cells is slightly different due to the way that their genes are organised. Genes are located on chromosomes, and the region of the chromosome where a particular gene is found is called the locus of that gene. In diploid organisms, which have their chromosomes arranged as homologous pairs, different forms of the same gene are known as alleles.

The gene is the basic unit of genetic information. Genes are located on chromosomes at a particular genetic locus. Different forms of the same gene are known as alleles.

4.4.1 The Anatomy of a Gene

Although there is no such thing as a 'typical' gene, there are certain basic requirements for any gene to function. The most obvious is that the gene has to encode the information for the particular protein (or RNA molecule). The double-stranded DNA molecule has the potential to store genetic information in either strand, although in most organisms, only one strand is used to encode any particular gene. The nomenclature used to define the two DNA strands can be confusing, as these can be called coding/non-coding, sense/antisense, plus/minus, transcribed/non-transcribed or template/non-template. Recommendations from the International Union of Biochemistry (IUB; now the International Union of Biochemistry and Molecular Biology or IUBMB) and the International Union of Pure and Applied Chemistry (IUPAC) favour the terms coding/non-coding, with the coding strand of DNA taken to be the *mRNA-like* strand. This convention will be used in this book where coding function is specified. The terms template and non-template will be used to describe DNA strands when there is not necessarily any coding function involved, as in the copying of DNA strands during cloning procedures. Thus, genetic information is expressed by transcription of the non-coding strand of DNA, which produces an mRNA molecule that has the same sequence as the coding strand of DNA (although RNA has uracil substituted for thymine). The sequence of the coding strand is usually reported when dealing with DNA sequence data, as this permits easy reference to the sequence of the RNA. By convention, sequences are written $5' \rightarrow 3'$ from left to right.

In addition to the sequence of bases that specifies the codons in a protein-coding gene, there are other important regulatory sequences associated with genes (Fig. 4.6). A site for starting transcription is required, and this encompasses a region that binds RNA polymerase, known as the promoter (P), and a specific start point for transcription (T_C). A stop site for transcription (t_C) is also required. From T_C start to t_C stop is sometimes called the transcriptional unit, *i.e.* the DNA

Genes have several important regions. A promoter is necessary for RNA polymerase binding, with the transcription start and stop sites defining the transcriptional unit.

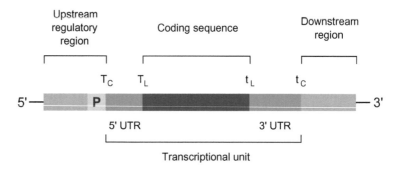

Upstream regulatory region | Coding sequence | Downstream region

T_C T_L t_L t_C

5' — P — 3'

5' UTR | 3' UTR

Transcriptional unit

Fig. 4.6 Gene organisation. The transcriptional unit produces the RNA molecule, and is defined by the transcription start site (T_C) and stop site (t_C). Within the transcriptional unit lies the coding sequence, from the translation start site (T_L) to the stop site (t_L). Note that eukaryotic genes often have intervening sequences (introns) within the coding sequence region. Regions transcribed but not translated are called untranslated regions or sequences (UTR or UTS). The upstream regulatory region may have controlling elements such as enhancers or operators, in addition to the promoter (P), which is the RNA polymerase-binding site.

region that is copied into RNA. Within this transcriptional unit, there may be regulatory sites for translation, namely a start site (T_L) and a stop signal (t_L). Other sequences involved in the control of gene expression, such as **enhancers** or **silencers**, may be present either *upstream* (towards the 5′ end of the coding strand) or *downstream* (towards the 3′ end) from the gene itself.

4.4.2 Gene Structure in Prokaryotes

In prokaryotic cells such as bacteria, genes are usually found grouped together in **operons**. The operon is a cluster of genes that are related (often coding for enzymes in a metabolic pathway) and under the control of a single promoter/regulatory region. Perhaps the best known example of this arrangement is the *lac* operon (Fig. 4.7), which codes for the enzymes responsible for lactose catabolism. Within the operon, there are three genes that code for proteins (termed **structural genes**) and an upstream control region encompassing the promoter and a regulatory site called the **operator**. In this control region, there is also a site that binds a complex of **cyclic AMP (cAMP)** and **cyclic AMP receptor protein (CRP)** (sometimes known as Catabolite gene Activator Protein or CAP), which is important in positive regulation (stimulation) of transcription. Lying outside the operon itself is the repressor gene (the *lac I* gene), which codes for a protein (the **Lac repressor**) that binds to the operator site and is responsible for negative control of the operon by blocking the binding of RNA polymerase.

The fact that structural genes in prokaryotes are often grouped together means that the transcribed mRNA may contain information for more than one protein. Such a molecule is known as a **polycistronic** mRNA, with the term **cistron** equating to the 'gene' as we have defined it (*i.e.* encoding one protein). Thus, much of the

Genes in prokaryotes tend to be grouped together in operons, with several genes under the control of a single regulatory region.

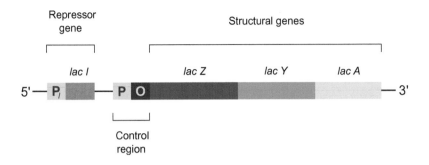

Fig. 4.7 The *lac* operon. The structural genes *lac Z*, *lac Y* and *lac A* encode β-galactosidase, galactoside permease and a transacetylase, respectively. The cluster is controlled by a promoter (P) and an operator region (O). The operator is the binding site for the repressor protein, encoded by the *lac I* gene. The repressor gene lies outside the operon itself and is controlled by its own promoter P_I.

genetic information in bacteria is expressed *via* polycistronic mRNAs, the synthesis of which is regulated in accordance with the needs of the cell at any given time. This system is flexible and efficient, and enables the cell to adapt quickly to changing environmental conditions.

4.4.3 Gene Structure in Eukaryotes

A major defining feature of eukaryotic cells is the presence of a membrane-bound nucleus, within which the DNA is stored in the form of chromosomes. Transcription therefore occurs within the nucleus and is separated from the site of translation, which is in the cytoplasm. The picture is complicated further by the presence of genetic information in mitochondria (plant and animal cells) and chloroplasts (plant cells only), which have their own separate genomes that specify many of the components required by these organelles. This compartmentalisation has important consequences for regulation, both genetic and metabolic, and thus gene structure and function in eukaryotes is more complex than in prokaryotes.

The most startling discovery concerning eukaryotic genes was made in 1977, when it became clear that eukaryotic genes contained 'extra' pieces of DNA that did not appear in the mRNA that the gene encoded. These sequences are known as intervening sequences, or introns, with the sequences that will make up the mRNA called exons (as in 'expressed' sequences). In many cases, the number and total length of the introns exceeds that of the exons, as in the chicken ovalbumin gene, which has a total of seven introns making up more than 75 per cent of the gene. As our knowledge has developed, it has become clear that eukaryotic genes are often extremely complex and may be very large indeed. Some examples of human gene complexity are shown in Table 4.2. This illustrates the tremendous range of sizes for human genes, the smallest of which may be only a few hundred base-pairs in length. At the other end of the scale, the dystrophin

Eukaryotic genes are more complex than prokaryotic genes, and often contain intervening sequences (introns). The introns form part of the primary transcript, which is converted to the mature mRNA by RNA processing.

Table 4.2 | Size and structure of some human genes

Protein	Gene locus	Size (kbp)	Exons	HGNC Symbol	ID
Insulin	11p15.5	1.4	3	INS	6081
β-globin	11p15.4	1.6	3	HBB	4827
Serum albumin	4q13.3	17	15	ALB	399
Blood clotting factor VIII	Xq28	187	26	F8	3546
CF transmembrane conductance regulator	7q31.2	189	27	CFTR	1884
Titin	2q31.2	281	363	TTN	12403
Dystrophin	Xp21.2-21.1	2 092	79	DMD	2928

Note: Gene locations are standard cytogenetic loci designations (chromosome, arm, region, band, sub-band). The *CFTR* gene is therefore located on the long (q) arm of chromosome 7, region 3, band 1 and sub-band 2. Gene sizes are shown as transcript sizes (the gene region may be longer than this, *e.g. CFTR* and *DMD* span around 230 and 2 400 kbp, respectively). Exons can include some non-coding exon regions. Symbols and ID numbers are as specified by the Human Genome Organisation (HUGO) Gene Nomenclature Committee (HGNC). Although the *DMD* gene is the largest human gene known, the *TTN* gene produces the largest protein (titin has 38 350 amino acids, with an M_r of 3 800 000). Further detail about human genes can be found at HGNC [www.genenames.org], NCBI [www.ncbi.nlm .nih.gov/gene] and Ensembl [www.ensembl.org]. Data collated from NCBI and EMBL/EBI (Ensembl).
Source: Reproduced with permission.

gene is spread over 2.4 Mbp of DNA on the X chromosome, with 79 exons generating a processed transcript of 14 kb, representing only 0.6 per cent of this length of DNA.

The presence of introns obviously has important implications for the expression of genetic information in eukaryotes, in that the introns must be removed before the mRNA can be translated. This is carried out in the nucleus, where the introns are spliced out of the **primary transcript**. Further intranuclear modification includes the addition of a 'cap' at the 5′ terminus and a 'tail' of adenine residues at the 3′ terminus. These modifications are part of what is known as **RNA processing**, and the end product is a fully functional mRNA that is ready for export to the cytoplasm for translation. The structures of the mammalian β-globin gene and its processed mRNA are outlined in Fig. 4.8 to illustrate eukaryotic gene structure and RNA processing.

4.5 | Gene Expression

As shown in Fig. 4.2, the flow of genetic information is from DNA to protein. Whilst a detailed knowledge of gene expression is not required in order to understand the principles of genetic engineering, it is important to be familiar with the main features of transcription and translation, and to have some knowledge of how gene expression is controlled. This will be particularly useful when we consider the applications of genetic engineering in Part 4.

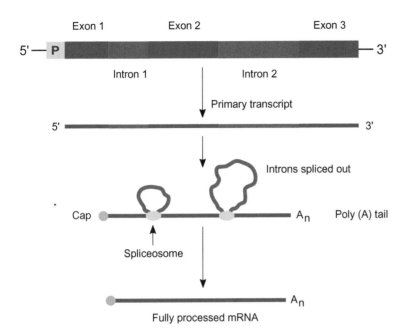

Fig. 4.8 Structure and expression of the mammalian β-globin gene. The gene contains two intervening sequences or introns. The expressed sequences (exons) are shaded blue and numbered 1–3. The two introns are shaded red. Note that the 5′ and 3′ untranslated regions (UTRs) are not shown. The primary transcript is processed by capping, polyadenylation and splicing (the spliceosome is a large ribonucleoprotein complex) to yield the fully functional mRNA.

4.5.1 From Genes to Proteins

At this point, it may be useful to introduce an analogy that I find helpful in thinking about the role of genes in determining cell structure and function. You may hear the term 'genetic blueprint' used to describe the genome. However, this is a little too simplistic, and I prefer to use the analogy of a *recipe* to describe how genes and proteins work. Let's consider making a cake – the recipe (gene) would be found in a particular book (chromosome), on a particular page (locus), and would contain information in the form of words (codons). One part of the recipe might read 'add 400 g of sugar and beat well', which is fairly clear and unambiguous. When put together with all the other ingredients and baked, the result is a cake in which you cannot see the sugar as an identifiable component. On the other hand, currants or blueberries would appear as identifiable parts of the cake. In a similar way, many of the characteristics of an organism are determined by multiple genes, with no particular single gene product being identifiable. Conversely, in single-gene traits, the product of a particular gene may be easily identifiable as a phenotypic characteristic (often as the causative agent in a disease).

Mutation can also be considered in the recipe context, to give some idea of the relative severity of effect that different mutations can have. If we go back to our sugar example, what would be the effect of the last 0 of 400 being replaced by a 1, giving 401 g as opposed to 400 g? This change would almost certainly remain undetected. However, if the 4 of 400 changed to a 9, or if an additional 0 was added to 400, then things would be very different (and much sweeter!). Thus, mutations in non-critical parts of genes may be of no consequence, whereas

> Genetic information is perhaps better thought of as a recipe rather than as a blueprint.

mutations in a critical part of a gene can have extremely serious consequences. In some cases, a single base insertion or substitution can have a major effect. Think of the impact of adding a 'k' in front of the 'g' in the 400 g part of the recipe …

The recipe analogy is a useful one, in that it defines the role of the recipe itself (specifying the components to be put together) and also illustrates that the information is only part of the story. If the cake is not mixed or baked properly, even with the correct proportions of ingredients, it will not turn out to be a success. Genes provide the information to specify the proteins, but the whole process must be controlled and regulated if the cell is to function effectively.

4.5.2 Transcription and Translation

These two processes are the critical steps involved in producing functional proteins in the cell. Transcription involves synthesis of an RNA from the DNA template provided by the non-coding strand of the transcriptional unit in question. The enzyme responsible is **RNA polymerase** (DNA-dependent RNA polymerase). In prokaryotes, there is a single RNA polymerase enzyme, but in eukaryotes, there are three types of RNA polymerase (I, II and III). These synthesise ribosomal, messenger and transfer/5 S ribosomal RNAs, respectively. All RNA polymerases are large multisubunit proteins with a relative molecular mass (M_r) of around 500 000 (M_r values are ratios, and thus dimensionless). You may see the *dalton* used as a unit for molecular mass; our polymerase would be 500 000 Da or 500 kDa).

Transcription has several component stages, these being: (1) DNA/RNA polymerase binding, (2) chain initiation, (3) chain elongation, and (4) chain termination and release of the RNA. Promoter structure is important in determining the binding of RNA polymerase, with the sequence of bases in the promoter region acting as a recognition site for RNA polymerase. Promoters have consensus sequences (we came across these in Section 2.1) that can help identification. Promoter locations are noted as regions lying upstream from the T_C start site, stated as negative numbers of nucleotides. In prokaryotes, consensus sequences at the −10 (TATAAT) and −35 (TTGACA) regions are found, although it is unusual for promoters to match these 'average' sequences exactly. As with many aspects, eukaryotic promoters are more complex. In addition to consensus sequences such as the TATA box (consensus sequence TATAAA, some 30 or 40 base-pairs upstream from the T_C start site), promoters in eukaryotic cells may have elements such as enhancers or silencers that lie hundreds or thousands of nucleotides away from the T_C start site and regulate transcription by looping back to the area of RNA polymerase binding.

Once the polymerase has bound at the promoter, chain elongation occurs with successive addition of nucleotides specified by complementary base-pairing. The non-coding (template) strand is read in a $3' \rightarrow 5'$ direction, with the mRNA therefore elongating in $5' \rightarrow 3'$ direction as transcription proceeds. When the mRNA molecule has been transcribed and released, it may be immediately available for translation (as in prokaryotes) or it may be processed in the nucleus and exported to the cytoplasm (eukaryotes) before translation occurs.

Translation requires an mRNA molecule, a supply of charged tRNAs (tRNA molecules with their associated amino acid residues) and ribosomes (composed of rRNA and ribosomal proteins). The ribosomes are the sites where protein synthesis occurs; in prokaryotes, ribosomes are composed of three rRNAs and some 52 different ribosomal proteins. The ribosome is a complex structure that essentially acts as a 'jig' which holds the mRNA in place, so that the codons may be matched up with the appropriate anticodon on the tRNA, thus ensuring that the correct amino acid is inserted into the growing polypeptide chain. The mRNA molecule is translated in a $5' \rightarrow 3'$ direction, corresponding to polypeptide elongation from N-terminus to C-terminus.

> The codon/anticodon recognition event marks the link between nucleic acid and protein.

Although transcription and translation are complex processes, the essential features (with respect to information flow) may be summarised as shown in Fig. 4.9. In conjunction with the brief descriptions presented above, this should provide enough background information about gene structure and expression to enable subsequent sections of the text to be linked to these processes where necessary.

4.5.3 Regulation of Gene Expression

Transcription and translation provide the mechanisms by which genes are expressed. However, it is vital that gene expression is controlled, so that the correct gene products are produced in the cell at the right time. Why is this so important? Let's consider two types of cell – a bacterial cell and a human cell. Bacterial cells need to be able to cope with wide variations in environmental conditions, and thus need to keep all their genetic material 'at the ready' in case particular gene products are needed. By keeping their genomes in this state of readiness, bacteria conserve energy (by not making proteins wastefully) and can respond quickly to any opportune changes in nutrient availability. This is an example of adaptive regulation of gene expression.

> Prokaryotic genes are often regulated in response to external signals such as nutrient availability.

In contrast to bacteria, human cells (usually) experience a very different set of environmental conditions. Cells may be highly specialised and differentiated, and their external environment is usually stable and controlled by homeostatic mechanisms to ensure that no wide fluctuations occur. Thus, cell specialisation brings more complex function, but requires more controlled conditions. Differentiation is a function of development, and genes in multicellular eukaryotes are often developmentally regulated. Gene regulation during the development and life cycle of a complex organism is, as you would expect, complex.

> Eukaryotic genes are often regulated in response to signals generated from within the organism.

In addition to genes that are controlled and regulated, there are many examples of gene products that are needed at all times during a cell's life. Such genes are sometimes called housekeeping genes or constitutive genes, in that they are essentially unregulated and

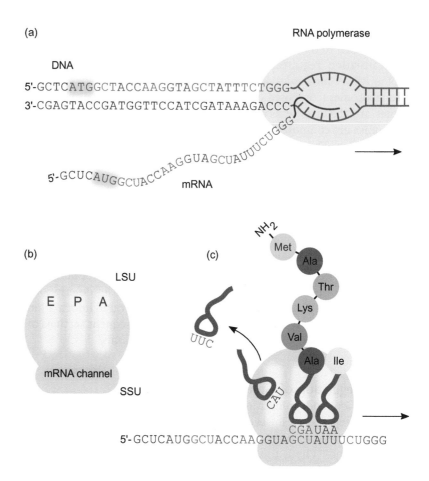

Fig. 4.9 Transcription and translation. (a) Transcription involves synthesis of mRNA by RNA polymerase. Part of the DNA/mRNA base sequence is shown. The mRNA has the same sequence as the coding strand in the DNA (the non-template strand), apart from U being substituted for T (codons are shown alternating red and blue). The start codon (ATG in DNA, AUG in mRNA) is highlighted in yellow. (b) The ribosome is the site of translation, and is made up of a large subunit (LSU) and a small subunit (SSU), each made up of ribosomal RNA molecules and many different proteins. There are three sites within the ribosome. The A (aminoacyl) and P (peptidyl) sites are involved in insertion of the correct tRNA~amino acid complex in the growing polypeptide chain. The E (exit) site facilitates the release of the tRNA after peptide bond formation has removed its amino acid. (c) The mRNA is being translated. The amino acid residue is inserted into the protein in response to the codon/anticodon recognition event in the ribosome. The first amino acid residue is encoded by AUG in the mRNA (tRNA anticodon UAC), which specifies methionine (see Table 4.1 for the genetic code). The remainder of the sequence is translated in a similar way. The ribosome translates the mRNA in a $5' \rightarrow 3'$ direction, with the polypeptide growing from its N-terminus. The residues in the polypeptide chain are joined together by peptide bonds.

encode proteins that are essential at all times (such as enzymes for primary **catabolic** pathways).

Although a detailed discussion of the control of gene expression is outside the scope of this book, the basic principles can be illustrated by considering how bacterial operons are regulated. A bacterial cell (living outside the laboratory) will experience a wide range of environmental conditions. In particular, there will be fluctuations in the availability of nutrients. If the cell is to survive, it must conserve energy resources, which means that wasteful synthesis of non-required proteins should be prevented. Thus, bacterial cells have mechanisms that enable operons to be controlled with a high degree of sensitivity. An operon that encodes proteins involved in a catabolic pathway (one that breaks down materials to release energy) is often regulated by being switched 'on' only when the substance becomes available in the extracellular medium. Thus, when the substance is absent, there are systems that keep catabolic operons switched 'off'. These are said to be **inducible** operons, and are usually controlled by a **negative control** mechanism involving a **repressor protein** that prevents access to the promoter by RNA polymerase. The classic example of a catabolic operon is the *lac* operon (the structure of which is shown in Fig. 4.7). When lactose is absent, the Lac repressor protein binds to the operator and the system is 'off'. The system is a little 'leaky', however, and thus the proteins encoded by the operon (β-galactosidase, permease and transacetylase) will be present in the cell at low levels. When lactose becomes available, it is transported into the cell by permease and binds to the repressor protein, causing a conformational (shape) change so that the repressor is unable to bind to the operator. Thus, the negative control is removed and the operon is accessible by RNA polymerase. A second level of control, based on the level of cAMP, ensures that full activity is only attained when lactose is present and energy levels are low. This dual-control mechanism is a very effective way of regulating gene expression, enabling a range of levels of expression that is a bit like a dimmer switch rather than an on/off switch for a light bulb. In the case of catabolic operons like the *lac* system, this ensures that the enzymes are only synthesised at maximum rate when they are really required.

> Gene regulation in bacteria enables a range of levels of gene expression to be attained, rather than a simple on/off switch.

4.6 | Genes and Genomes

When techniques for the examination of DNA became established, gene structure was naturally one of the first areas where efforts were concentrated. However, genes do not exist in isolation, but as part of the genome of an organism. Over the past few years, the emphasis in molecular biology has shifted, and today we are much more likely to consider the genome as a whole – almost as a type of cellular organelle – rather than just a collection of genes. The Human Genome Project, and the developments from this (considered in Chapter 13), is a good example of how development in the field of bioinformatics (Chapter 11) has been essential to enable the collation and interpretation of the staggering amount of sequence data that are generated by high-throughput sequencing of genomes.

> The genome is the total complement of DNA in the cell.

4.6.1 Genome Size and Complexity

The amount of DNA in the haploid genome is known as the **C-value**. It would seem reasonable to assume that genome size should increase with increasing complexity of organisms, reflecting the greater number of genes required to facilitate this complexity. The data shown in Table 4.3 show that, as expected, genome size does tend to increase with organismal complexity. Thus, bacteria, yeast, fruit fly and human genomes fit this pattern. However, mouse, tobacco and wheat have much larger genomes than humans – this seems rather strange, as intuitively we might assume that a wheat plant is not as complex as a human being. Also, as *Escherichia coli* has around 4 400 genes, it appears that the tobacco plant genome has the capacity to encode 4 000 000 genes, and this is certainly not the case, even allowing for the increased size and complexity of eukaryotic genes. This anomaly is sometimes called the **C-value paradox** or **C-value enigma**.

Table 4.3 | Genome size comparison

Genome	Size	RGS	Genes
HIV1 (human immunodeficiency virus; RNA)[1]	9.6 kb	0.002	9
Human mt DNA[2]	16.6 kbp	0.004	37
Bacteriophage lambda (λ)[1]	49.2 kbp	0.01	68
Escherichia coli (bacterium)[3]	4.6 Mbp	1	4 240
Saccharomyces cerevisiae (budding yeast)[3]	12.2 Mbp	2.7	6 600
Aspergillus nidulans (fungus)[3]	30.5 Mbp	6.6	10 534
Caenorhabditis elegans (nematode worm)[3]	100 Mbp	22	20 191
Arabidopsis thaliana (plant)[3]	135 Mbp	29	27 655
Drosophila melanogaster (fruit fly)[3]	144 Mbp	31	13 968
Mus musculus (mouse)[3]	2.7 Gbp	0.9	22 468
Homo sapiens (human)[3]	3.1 Gbp	1	20 442
Nicotiana tabacum (tobacco)[4]	4.6 Gbp	1.5	69 500
Triticum aestivum (wheat)[3]	17 Gbp	5.5	107 891
Neoceratodus forsteri (Australian lungfish)[5]	43 Gbp	13.9	31 120
Protopterus aethiopicus (marbled lungfish)[a,6]	130 Gbp	42	Uncertain
Paris japonica (Japanese canopy plant)[a,7]	149 Gbp	48	Uncertain

Note: Genome sizes are given in base-pairs (bases for the single-stranded RNA genome of HIV), prefixed k (kilo, 10^3 bp), M (mega, 10^6) or G (giga, 10^9) for the haploid genome (where applicable). Relative genome sizes (RGS) are set against the *E. coli* genome (for kb/kbp and Mbp genomes) and the human genome (for Gbp genomes) for comparison. The number of genes is often not known exactly, even where genomes have been sequenced multiple times; current best estimates for protein-coding genes are shown. The *Neoceratodus forsteri* genome is the largest sequenced to date. The genomes of *Protopterus aethiopicus* and *Paris japonica* (marked [a]) have been converted from C-values (haploid genome mass in picograms).

Sources: [1] NCBI [www.ncbi.nlm.nih.gov], [2] Zhang *et al.* (2017) DOI [https://10.1128/genomeA.01185-17], [3] Ensembl [www.ensembl.org], [4] Edwards *et al.* (2017) DOI [https://doi.org/10.1186/s12864-017-3791-6], [5] Meyer *et al.* (2021) DOI [https://doi.org/10.1038/s41586-021-03198-8], [6] Thomson (1972) DOI [https://doi.org/10.1002/jez.1401800307], [7] Pellicer *et al.* (2010) DOI [https://doi.org/10.1111/j.1095-8339.2010.01072.x]. Data from NCBI and EBML/EBI (Ensembl) reproduced with permission.

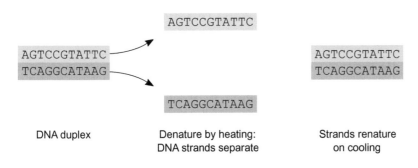

AGTCCGTATTC

AGTCCGTATTC
TCAGGCATAAG

AGTCCGTATTC
TCAGGCATAAG

TCAGGCATAAG

DNA duplex

Denature by heating:
DNA strands separate

Strands renature
on cooling

Fig. 4.10 The principle of nucleic acid hybridisation. This feature of DNA molecules is a critical part of many of the procedures involved in gene manipulation, and is also an essential feature of life itself. Thus, the simple G·C and A·T base-pairing (Fig. 4.4) has profound implications for living systems and for the applications of recombinant DNA technology.

In addition to size of the genome, genome complexity also tends to increase with more complex organisation. Before large-scale DNA sequencing became widely available, one way of studying genome complexity was to examine the renaturation of DNA samples. If a DNA duplex is denatured by heating the solution until the strands separate, the complementary strands will renature on cooling (Fig. 4.10). This feature can be used to provide information about the sequence complexity of the DNA in question, since sequences that are present as multiple copies in the genome will renature faster than sequences that are present as single copies only. By performing this type of analysis, eukaryotic DNA can be shown to be composed of four different abundance classes. Firstly, some DNA will form duplex structures almost instantly, because the denatured strands have regions such as inverted repeats, or palindromes, which fold back on each other to give a hairpin loop structure. This class is commonly known as foldback DNA. Second fastest to re-anneal are **highly repetitive sequences**, which occur many times in the genome. Following these are **moderately repetitive sequences**, and finally there are the **unique or single copy sequences**, which re-anneal very slowly under the conditions used for this type of analysis. We will consider how repetitive DNA sequence elements can be used in genome mapping and DNA profiling in Chapters 13 and 15.

Eukaryotic genomes may have a range of different types of repetitive sequences.

4.6.2 Genome Organisation

The C-value paradox and the sequence complexity of eukaryotic genomes raise questions about how genomes are organised. Viral and bacterial genomes tend to show very efficient use of DNA for encoding their genes, which is a consequence of (and an explanation for) their small genome size. However, in the human genome, only about 1–2 per cent of the total amount of DNA is actually coding sequence. Even when the introns and control sequences are added, the majority of the DNA has no obvious function. This is sometimes termed 'junk' DNA, although this is perhaps the wrong way to think

Most of the human genome is not involved in coding for proteins.

about this apparently redundant DNA. We are still trying to understand the function, if any, of much of this part of the genome.

Estimating the number of genes in a particular organism is not an exact science, and a number of different methods may be used. Even today, some 20 years after the completion of the human genome sequence, there is still debate about what constitutes a gene and how many of them there are in humans. There are many cases where gene coding sequences are recognised, but the protein products are unknown in terms of their biological function. We will look at this more closely in Chapter 13.

Many genes in eukaryotes are single copy genes, and tend to be dispersed across the multiple chromosomes found in eukaryotic cell nuclei. Other genes may be part of **multigene families**, and may be grouped at a particular chromosomal location or may be dispersed. When studying gene organisation in the context of the genome itself, features such as gene density, gene size, mRNA size, intergenic distance and intron/exon sizes are important indicators. Early analysis of human DNA indicated that the 'average' size of a coding region is around 1 500 base-pairs, and the average size of a gene is 10–15 kbp. Gene density is about one gene per 40–45 kbp, and the intergenic distance is around 25–30 kbp. However, as we have already seen, gene structure in eukaryotes can be very complex, and thus using 'average' estimates is a little misleading. What is clear is that genomes are now yielding much new information about gene structure and function as genome sequencing projects generate more data and interpretation methodologies become more sophisticated; we are now very firmly in the post-genomic era.

Genome sequencing has greatly improved our understanding of how genomes work.

4.6.3 The Transcriptome and Proteome

We finish this look at molecular biology by introducing two more '-omes' to complement the 'genome'. These terms have become widely used as researchers begin to delve into the bioinformatics of cells. The transcriptome refers to the population of transcripts at any given point in a cell's life. This expressed subset of the genomic information will be determined by many factors affecting the status of the cell. There will be general 'housekeeping' genes for basic maintenance of cell function, but there may also be tissue-specific genes being expressed, or perhaps developmentally regulated genes will be switched 'on' at that particular point. Analysis of the transcriptome therefore gives a good snapshot of what the cell is engaged in at that point in time.

Analysis of the transcriptome and proteome provides useful information about which genes a cell is expressing at any given time.

The proteome is a logical extension to the genome and transcriptome, in that it represents the population of proteins in the cell. The proteome will reflect the transcriptome to a much greater extent than the transcriptome reflects the genome, although there will be some transcripts that may not be translated efficiently, and there may be proteins that persist in the cell when their transcripts have been removed from circulation. Understanding the proteome is critical in developing a full picture of how cells work. The concept, as yet unrealised, of gaining a full and complete knowledge of all components of the cell is considered by some as the 'holy grail' of cell biology, comparing it with the search for the unifying theory in physics. The argument is that, if we understand how all the proteins of a cell work, then surely we have a complete understanding of cell structure and function? As with most things in biology, this is not a straightforward process, although great advances have been (and continue to be) made in this area as our ability to interrogate complex data sets has improved.

4.7 | Conclusion: Structure and Function

In this chapter, the central theme has been to illustrate how the structure of nucleic acids enables them to function as information-carrying molecules. Whilst the co-dependency of structure and function is not unique to biological systems, it is perhaps most elegantly illustrated in the workings of the cell. The concept of *emergent properties* is illustrated within the molecular architecture of cells (of which DNA is one part), with complexity arising from the interactions of simpler cell components. Underpinning this are two principles that are very simple but have profound consequences in the cell:

- The *monomer/polymer cycle* enables large polymeric molecules to be made by dehydration synthesis and broken down by hydrolysis
- The *base-pairing* arrangements in DNA mean that nucleic acid sequences can be complementary and therefore bind together

Almost all the activities of a cell, and arguably DNA technology itself, are dependent on these principles at some stage. The key to understanding the significance of the topics in this chapter is to see beyond the structures and consider how the systems demonstrate the interdependence of structure and function in producing the intended outcome. This may seem fairly straightforward but can sometimes be a real challenge when looking at multi-component systems that may be structurally complex.

In this chapter, we have seen how genetic information is encoded using a very simple system of four nitrogenous bases arranged in groups of three. This is able to generate all the complexity of living systems due to the amplification of potential combinations when the 20 amino acids are used to produce proteins that are essentially unlimited in terms of unique sequence possibilities. This aspect is not always easy to articulate, and some time spent on trying to get to grips with how the genetic code works in an informatics capacity is well worth the effort. The *Central Dogma* of molecular biology can help provide a reference for this, as its central tenet (that information flows from DNA to RNA to protein) helps to set a directional aspect to how the code works.

Having outlined the principles of encoding and transferring genetic information, we then looked at the structure of DNA and its role in gene structure and expression, noting the differences between *prokaryotic* and *eukaryotic* genes. The discovery that eukaryotic genes were interrupted by non-coding regions (*introns*), and that may genes have significantly more non-coding regions than coding sequences, illustrates the complexity of gene structure in eukaryotic organisms. For gene expression we looked at two aspects: the *mechanics* of the process (*transcription* and *translation*) and the *regulation* of gene expression in adaptive and developmental contexts, as well as the need for unregulated expression of certain housekeeping genes whose products are required at all times.

Our final section in this chapter looked at how genes were set within the *genome* and at some of the features of genomes across a range of organisms. In the era of genome sequencing, the study of genome organisation and complexity has been transformed, and the extension into the study of the *transcriptome* and *proteome* is providing insights into how the genome functions as a system rather than a collection of individual genes.

To conclude, this chapter has presented what might be considered basic (or even, these days, mundane) information about the structure and organisation of DNA. However, a sound grasp of the relationship between the structure and function of nucleic acids is essential if we are to make sense of the rest of this book, and we will revisit the themes covered in this chapter many times in the remaining chapters of Part 2, and in Parts 3 and 4.

Further Reading

Fraser, J. *et al.* (2015). An overview of genome organization and how we got there: from FISH to Hi-C. *Microbiol. Mol. Biol. Rev.*, 79 (3), 347–72. URL [https://journals.asm.org/doi/10.1128/MMBR.00006-15]. DOI [https://doi.org/10.1128/MMBR.00006-15].

Jones, S. (2000). *The Language of the Genes*. Flamingo/Harper Collins, London. ISBN 978-0-00-655243-7.

National Institutes of Health, National Library of Medicine. The discovery of the double helix, 1951–1953. The Francis Crick Papers. URL [https://profiles.nlm.nih.gov/spotlight/sc/feature/doublehelix].

National Institutes of Health, National Library of Medicine. Defining the genetic coding problem, 1954–1957. The Francis Crick Papers. URL [https://profiles.nlm.nih.gov/spotlight/sc/feature/defining].

National Institutes of Health, National Library of Medicine. Deciphering the genetic code, 1958–1966. The Francis Crick Papers. URL [https://profiles.nlm.nih.gov/spotlight/sc/feature/deciphering].

National Institutes of Health, National Library of Medicine. DNA as the 'stuff of genes': the discovery of the transforming principle, 1940–1944. The Oswald T. Avery Collection. URL [https://profiles.nlm.nih.gov/spotlight/cc/feature/dna].

Watson, J. D. (2012). *The Annotated and Illustrated Double Helix*. Edited by: A. Gann and J. Witkowski. Simon and Schuster, New York, NY. ISBN 978-1-4767-1549-0.

Websearch

People

Two pairs of people and two classic experiments for this chapter. Have a look at the experiments carried out by *Alfred Hershey/Martha Chase* and by *Matthew Meselson/Franklin Stahl*. They are masterpieces of design and execution, and illustrate the elegance of what George Wald later called 'a conversation with nature'.

Places

The *Institut Pasteur* in Paris has been the workplace of many scientists who have made significant contributions to molecular biology and related fields. Find out about the history of this famous institution, with particular emphasis on the work that led to the discovery of how the *lac* operon is regulated.

Processes

The discovery of intervening sequences, or 'split genes', in 1977 was a major milestone. Converting primary transcripts into functional mRNA molecules involves a number of stages, collectively known as *RNA processing* or *RNA splicing*. What is involved in these processes?

Reflections

What topics in this chapter have you found most challenging? Look for resources that help to illustrate the key points.

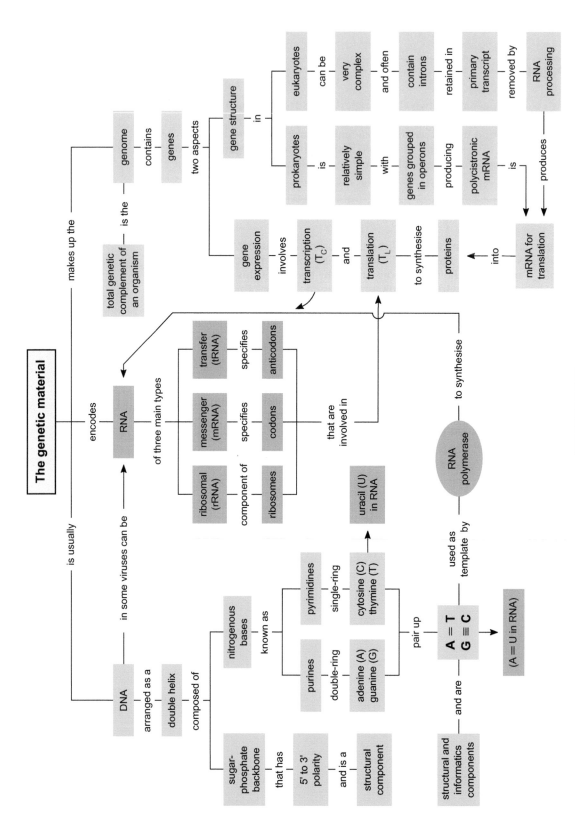

The genetic material

total genetic complement of an organism — is the — genome — contains — genes

genome — makes up the

genes — two aspects — gene structure

gene structure — in:
- eukaryotes — can be — very complex — and often — contain introns — retained in — primary transcript — removed by — RNA processing
- prokaryotes — is — relatively simple — with — genes grouped in operons — producing — polycistronic mRNA — is

gene expression — involves — transcription (T$_C$) — and — translation (T$_L$) — to synthesise — proteins

RNA processing — produces — mRNA for translation

polycistronic mRNA — is — mRNA for translation

mRNA for translation — into — proteins

DNA — is usually — in some viruses can be — RNA

DNA — encodes — RNA

RNA — of three main types:
- transfer (tRNA) — specifies — anticodons
- messenger (mRNA) — specifies — codons
- ribosomal (rRNA) — component of — ribosomes

that are involved in — transcription (T$_C$) and translation (T$_L$)

DNA — arranged as a — double helix — composed of:
- sugar-phosphate backbone — that has — 5' to 3' polarity — and is a — structural component
- nitrogenous bases — known as:
  - pyrimidines — single-ring — cytosine (C) thymine (T)
  - purines — double-ring — adenine (A) guanine (G)

cytosine (C) thymine (T) / adenine (A) guanine (G) — pair up — **A = T G ≡ C** — (A = U in RNA)

thymine (T) — uracil (U) in RNA

A = T G ≡ C — and are — structural and informatics components

RNA polymerase — used as template by — **A = T G ≡ C**

RNA polymerase — to synthesise — transcription (T$_C$) and translation (T$_L$)

Chapter 5 Summary

Learning Objectives

When you have completed this chapter, you will be able to:

- Outline the range of enzymes used in gene manipulation
- Describe the mode of action and uses of type II restriction enzymes
- Compare the mode of action and uses of a range of DNA-modifying enzymes, including endonucleases, exonucleases, polymerases and end-modifying enzymes
- Define the mode of action and use of DNA ligase

Key Words

Enzymes, ribozymes, restriction enzymes, restriction–modification system, methylation, recognition sequence, hemi-methylation, nucleases, endonucleases, *Escherichia coli*, *Bacillus amyloliquefaciens*, palindromes, inverted palindromes, mirror palindromes, isoschizomers, neoschizomers, blunt ends, flush ends, cohesive ends, sticky ends, restriction mapping, agarose gel electrophoresis, DNA modifying enzymes, exonucleases, BAL 31 nuclease, exonuclease III, deoxyribonuclease I, S1 nuclease, zinc-finger nucleases (ZFNs), transcription activator-like effector nucleases (TALENs), CRISPR-Cas 9 system (clustered regular interspersed short palindromic repeats), ribonucleases, processive, DNA polymerase I, Klenow fragment, reverse transcriptase, cDNA, complementary DNA, copy DNA, alkaline phosphatase, polynucleotide kinase (PNK), terminal transferase, bacterial alkaline phosphatase (BAP), calf intestinal alkaline phosphatase (CIP or CIAP), DNA ligase, thermolabile, thermostable, fluorophore, radioactive label.

Chapter 5

The Tools of the Trade

In Section 2.1, we saw that science needs both a *methodology* (Fig. 2.1) and an *infrastructure* (Fig. 2.2) to enable progress to be made. We now need to consider how these two aspects are realised when studying DNA. In this chapter, we will look at what we might call the 'biological methodology' needed to manipulate DNA, before expanding into more technologically based methods in Chapter 6.

The ability to cut and join DNA from different sources is the essence of creating recombinant DNA (rDNA) in the test tube. In addition, certain modifications may have to be carried out to the DNA during the various steps required to produce, clone and identify rDNA molecules. The tools that enable these manipulations to be performed are **enzymes**, which are almost always large and complex protein molecules (although there are some RNA-based enzymes known as **ribozymes**). They can be isolated from a wide range of organisms and are available commercially from various suppliers; these companies are a critical part of the sector that provides support infrastructure to enable the rDNA work to be carried out. In this chapter, we will have a look at some of the important classes of enzymes that make up the genetic engineer's toolkit.

5.1 | Restriction Enzymes – Cutting DNA

The restriction enzymes, which cut DNA at defined sites, represent one of the most important groups of enzymes for the manipulation of DNA. These enzymes are found in bacterial cells, where they function as part of a protective mechanism called the **restriction–modification** system. In this system, the restriction enzyme hydrolyses any exogenous DNA that appears in the cell. To prevent the enzyme from acting on the host cell DNA, the modification enzyme of the system (a methylase) modifies the host DNA by methylation of particular bases in the restriction enzyme's **recognition sequence**, which prevents the enzyme from cutting the DNA. Sometimes methylation of a single strand (hemi-methylation) is sufficient to prevent restriction. Restriction enzymes can be classified into a number of types (I, II, III, IV or V), and differ in the mode of action and/or the

Restriction enzymes act as a 'protection' system for bacteria, in that they hydrolyse exogenous DNA that is not methylated by the host modification enzyme.

Table 5.1	General properties of restriction enzymes
Type	Characteristics
I	The first restriction enzymes to be discovered. These enzymes are large, have combined restriction/modification functions and recognise asymmetric recognition sequences (RS). Cleavage sites (CS) are random sites some 1 000 bp from the RS. Of limited use in gene manipulation.
II	The most useful type for routine use. Loosely grouped into several subtypes, depending on specific characteristics. Type IIP (often just called type II) are the most common. These recognise an inverted palindrome RS and cut within it. Other types recognise asymmetrical RS and/or may cut a few base-pairs from the site, or may cut both ends of the RS. One of the most useful subtypes is type IIS. These enzymes have two domains, one for recognition and one for cleavage. They recognise an asymmetric RS and cut a small number of base-pairs from this. They have become a key part of a cloning method called Golden Gate cloning (see Section 8.4.5).
III	Type III are large combined restriction–modification systems and cut about 20 bp from the RS. Require two inverted RS and often do not cleave DNA fully.
IV	Cleave modified regions such as methylated DNA.
V	The Cas (CRISPR-associated protein) enzyme system is considered a form of restriction enzyme that utilises guide RNAs to cleave DNA. Most well known as the CRISPR-Cas9 editing system (see Chapter 12).
Artificial	Enzymes can be engineered by combining a DNA recognition domain with a functional nuclease domain. This approach opens up the design of specific recognition sequences that can be used for very precise manipulation of gene sequences. Examples include zinc-finger nucleases, TALENs and the CRISPR-Cas9 system (see Section 5.2.1 and Chapter 12).

structure of the recognition site. The key features of each are shown in Table 5.1.

Most of the enzymes commonly used in gene manipulation are type II enzymes, which have the simplest mode of action in that they cut the DNA at the recognition site. These enzymes are **nucleases**, and as they cut at an internal position in a DNA strand (as opposed to beginning degradation at one end), they are known as **endonucleases**. Thus, the correct designation of such enzymes is that they are *type II restriction endonucleases*, although they are often simply called restriction enzymes. In essence, they may be thought of as molecular scissors.

5.1.1 Type II Restriction Endonucleases

Restriction enzyme nomenclature is based on a number of conventions. The generic and specific names of the organism in which the enzyme is found are used to provide the first part of the designation, which comprises the first letter of the generic name and the first two letters of the specific name. Thus, an enzyme from a strain of *Escherichia coli* is termed *Eco*, one from *Bacillus amyloliquefaciens* is *Bam*, and so on. Further descriptors may be added, depending on the bacterial strain involved and on the presence or absence of extrachromosomal elements. Two widely used enzymes from the bacteria

Restriction enzymes are named according to the bacterium from which they are purified.

(a)　　　　　　　　　　　　　　(b)

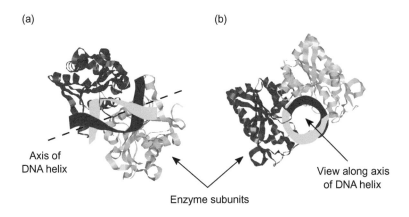

Axis of
DNA helix

View along axis
of DNA helix

Enzyme subunits

> **Fig. 5.1** Binding of the restriction enzyme *Bam* HI to the DNA helix. This shows how the
> enzyme wraps around the helix to facilitate hydrolysis of the phosphodiester linkages.
> (a) and (b) show different views with respect to the axis of the helix. This illustrates the
> very close relationship between structure and function in biology. Generated using *RasMol*
> molecular modelling software. *Source:* Roger Sayle, Public Domain.

mentioned above are *Eco* RI and *Bam* HI. The binding of *Bam* HI to its
recognition sequence is shown in Fig. 5.1.

The value of restriction endonucleases lies in their specificity. Each
particular enzyme recognises a particular sequence of bases in the
DNA, the most common recognition sequences being four, five or six
base-pairs in length. Thus, given that there are four bases in the DNA,
and assuming a random distribution of bases, the expected frequency
of any particular sequence can be calculated as 4^n, where n is the
length of the recognition sequence. This predicts that tetranucleotide
sites will occur every 256 base-pairs, pentanucleotide sites every 1 024
base-pairs and hexanucleotide sites every 4 096 base-pairs. There is, as
you might expect, considerable variation from these values, but generally the fragment lengths produced will lie at around the calculated
value. Thus, an enzyme recognising a tetranucleotide sequence (sometimes called a 'four-cutter') will produce shorter DNA fragments than
a six-cutter when the DNA is digested fully with the enzyme.

Recognition sites are generally **palindromes** (they read the same
backwards and forwards). Most often, these are **inverted palindromes**
where the sequence reads the same on each of the two strands of DNA
(*e.g.* 5'-GAATTC-3' and its complement), rather than **mirror palindromes**
where a sequence on a single strand is palindromic (*e.g.* 5'-TAGGAT-3').
Some of the most commonly used restriction enzymes are listed in
Table 5.2, with their recognition sequences and cutting sites.

> Different restriction enzymes
> generate different ranges of
> DNA fragment sizes; the size of
> fragment is linked to the
> frequency of occurrence of the
> recognition sequence.

5.1.2 Use of Restriction Endonucleases

Restriction enzymes are very simple to use – an appropriate amount of
enzyme is added to the target DNA in a buffer solution, and the
reaction is incubated at the optimum temperature (usually 37°C) for
a suitable length of time. Enzyme activity is expressed in units, with
one unit being the amount of enzyme that will cleave 1 μg of DNA in

Table 5.2 | Recognition sequences and cutting sites for some restriction endonucleases

Enzyme	Original source organism	Recognition sequence	Cutting sites	Ends
Bam HI	Bacillus amyloliquefaciens	5′–GGATCC–3′	G˅GATC C / C CTAG˄G	5′
Eco RI	Escherichia coli	5′–GAATTC–3′	G˅AATT C / C TTAA˄G	5′
Hae III	Haemophilus aegyptius	5′–GGCC–3′	GG˅CC / CC˄GG	Blunt
Hpa I	Haemophilus parainfluenzae	5′–GTTAAC–3′	GTT˅AAC / CAA˄TTG	Blunt
Pst I	Providencia stuartii	5′–CTGCAG–3′	C TGCA˅G / G˄ACGT C	3′
Sac I[a]	Streptomyces achromogenes	5′–GAGCTC–3′	G AGCT˅C / C˄TCGA G	3′
Sau 3A	Staphylococcus aureus	5′–GATC–3′	˅GATC / CTAG˄	5′
Sma I[b]	Serratia marcescens	5′–CCCGGG–3′	CCC˅GGG / GGG˄CCC	Blunt
Sst I[a]	Streptomyces stanford	5′–GAGCTC–3′	G AGCT˅C / C˄TCGA G	3′
Xma I[b]	Xanthomonas malvacearum	5′–CCCGGG–3′	C˅CCGG G / G GGCC˄C	5′

Note: The recognition sequences are given in single-strand form, written 5′ → 3′. Cutting sites are shown double-stranded to illustrate the type of ends produced by a particular enzyme; 5′ and 3′ refer to 5′ and 3′-protruding termini, respectively. The point at which the phosphodiester bonds are broken is shown by the arrow on each strand of the recognition sequence. Note that some enzymes (shown in colour) recognise the same sequence: [a] *Sac* I and *Sst* I are isoschizomers (they cut at the same positions in the recognition sequence) and [b] *Sma* I and *Xma* I are **neoschizomers** (they cut at different positions in the recognition sequence). Several hundreds of type II restriction enzymes are available commercially. Many of these are produced by expression of the cloned gene in *E. coli*, rather than by direct isolation from the source organism.

> One very useful feature of restriction enzymes is that they can generate cohesive or 'sticky' ends that can be used to join DNA from two different sources together to generate recombinant DNA molecules.

1 hour at 37°C. Although most experiments require complete digestion of the target DNA, there are some cases where various combinations of enzyme concentration and incubation time may be used to limit the amount of digestion. These *partial digests* can be performed to enable fragmentation of DNA in a pseudo-random way, which can be useful in generating genomic libraries (see Section 8.2.1).

The type of DNA fragment that a particular enzyme produces depends on the recognition sequence and on the location of the cutting site within this sequence. As we have already seen, fragment length is dependent on the frequency of occurrence of the recognition sequence. The actual cutting site of the enzyme will determine the

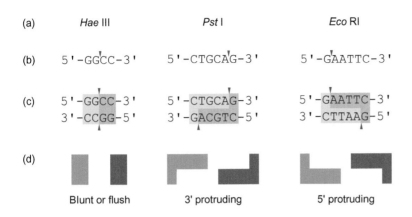

(a)   *Hae* III   *Pst* I   *Eco* RI

(b)   5'-GGCC-3'   5'-CTGCAG-3'   5'-GAATTC-3'

(c)   5'-GGCC-3'   5'-CTGCAG-3'   5'-GAATTC-3'
      3'-CCGG-5'   3'-GACGTC-5'   3'-CTTAAG-5'

(d)

    Blunt or flush   3' protruding   5' protruding

Fig. 5.2 Types of ends generated by different restriction enzymes. Three enzymes are shown. Names are listed in (a), derived from the original source bacteria *Haemophilus aegyptius*, *Providencia stuartii* and *Escherichia coli*. Recognition sequences and cutting sites are shown in (b) and (c). A schematic representation of the types of ends generated is shown in (d).

type of ends that the cut fragment has, which is important with regard to further manipulation of the DNA. Three types of fragment may be produced, these being: (1) **blunt** or **flush-ended** fragments, (2) fragments with protruding 3′ ends, and (3) fragments with protruding 5′ ends. An example of each type is shown in Fig. 5.2.

Enzymes such as *Pst* I and *Eco* RI generate DNA fragments with **cohesive** or '**sticky**' ends, as the protruding sequences can base-pair with complementary sequences generated by the same enzyme. Thus, by cutting two different DNA samples with the same enzyme and mixing the fragments together, rDNA can be produced, as shown in Fig. 5.3. This is one of the most useful applications of restriction enzymes and is a vital part of many manipulations in genetic engineering.

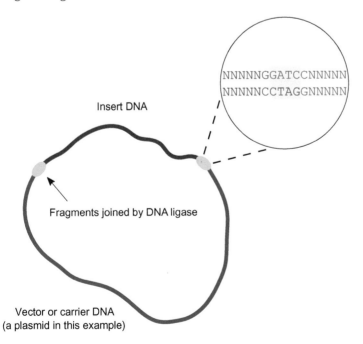

Insert DNA

NNNNNGGATCCNNNNN
NNNNNCCTAGGNNNNN

Fragments joined by DNA ligase

Vector or carrier DNA
(a plasmid in this example)

Fig. 5.3 Generation of recombinant DNA. DNA fragments from different sources can be joined together if they have cohesive ('sticky') ends, as produced by many restriction enzymes. On annealing the complementary regions together, the phosphodiester backbone is sealed using DNA ligase.

| Table 5.3 | Digestion of a 15 kbp DNA fragment with three restriction enzymes |

Single digests			Double digests			Triple digest
			Bam HI *Eco* RI	*Bam* HI *Pst* I	*Eco* RI *Pst* I	*Bam* HI *Eco* RI *Pst* I
Bam HI	*Eco* RI	*Pst* I				
14	12	8	11	8	7	6
1	3	7	3	6	5	5
			1	1	3	3
						1

Note: Data shown are lengths (in kbp) of fragments that are produced on digestion of a 15 kbp DNA fragment with the enzymes *Bam* HI, *Eco* RI and *Pst* I. Single, double and triple digests were carried out as indicated. Fragments produced by each digest are listed in order of length.

5.1.3 Restriction Mapping

Most pieces of DNA will have recognition sites for various restriction enzymes, and it is often beneficial to know the relative locations of some of these sites. The technique used to obtain this information is known as **restriction mapping**. This involves cutting a DNA fragment with a selection of restriction enzymes, singly and in various combinations. The fragments produced are separated by **agarose gel electrophoresis** (see Section 6.5), and their sizes determined. From the data obtained, the relative locations of the cutting sites can be worked out. A fairly simple example to illustrate the technique is outlined below.

Let's say that we wish to map the cutting sites for the restriction enzymes *Bam* HI, *Eco* RI and *Pst* I, and that the DNA fragment of interest is 15 kbp in length. Various digestions are carried out, and the fragments arising from these are analysed and their sizes determined. The results obtained are shown in Table 5.3. As each of the single enzyme reactions produces two DNA fragments, we can conclude that the DNA has a single cutting site for each enzyme. The double digests enable a map to be drawn up by working out the relative positions of the pairs of enzymes, and the triple digest confirms this. Construction of the map is outlined in Fig. 5.4. Some restriction maps can be very complex, and getting to the final map can seem almost impossible. However, by taking a simple stepwise approach, building up the complexity and cross-checking as you go, most mapping problems can eventually be solved.

A physical map of a piece of DNA can be assembled by determining where the restriction enzyme recognition sequences are relative to each other; this is known as restriction mapping.

5.2 | DNA Modifying Enzymes

In addition to restriction enzymes, there are many other enzymes used in genetic engineering. These may be loosely termed **DNA modifying enzymes**, with the term used here to include degradation, synthesis and alteration of DNA. Some of the most commonly used enzymes are described below. We will consider DNA ligase, which joins DNA molecules together, in Section 5.3.

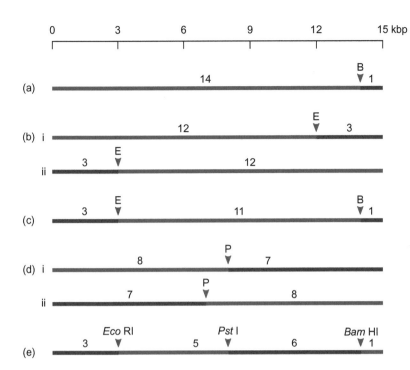

Fig. 5.4 Restriction mapping. (a) The 15 kbp fragment yields two fragments of 14 and 1 kbp when cut with *Bam* HI. (b) The *Eco* RI fragments of 12 and 3 kbp can be orientated in two ways with respect to the *Bam* HI site, as shown in (b)i and (b)ii. The *Bam* HI/*Eco* RI double digest gives fragments of 11, 3 and 1 kbp, and therefore, the relative positions of the *Bam* HI and *Eco* RI sites are as shown in (c). Similar reasoning with the orientation of the 8 and 7 kbp *Pst* I fragments (d) gives the final map (e).

5.2.1 Nucleases

Nuclease enzymes degrade nucleic acids by breaking the phosphodiester bond that holds the nucleotides together. As we have seen, restriction enzymes are good examples of endonucleases, which cut within a DNA strand. Nucleases that degrade DNA from the end of the molecule are part of a group known as exonucleases, defined as either 5′- or 3′-exonucleases, depending on the terminus that they attack.

Apart from restriction enzymes, there are around 30 or so different nucleases that are useful in genetic engineering. We will consider four examples to illustrate. These are BAL 31 nuclease and **exonuclease III** (exonucleases), and deoxyribonuclease I (DNase I) and S1 nuclease (endonucleases). These enzymes differ in their precise mode of action, and provide the genetic engineer with a variety of strategies for attacking DNA (the other variants may have similar functions but might vary in terms of source organism, specificity, conditions for use, *etc.*). The features of our four enzymes are summarised in Fig. 5.5.

The nucleases we have looked at so far (restriction enzymes and the four examples in Fig. 5.5) have a fairly simple mode of action, in that they generally recognise either a specific DNA sequence (restriction enzymes) or a functional group (such as a phosphate or hydroxyl group) at the end of a DNA strand. However, there are other nucleases that have been developed for genome editing procedures that involve sequence-specific recognition and complex modes of action. These are zinc-finger nucleases (ZFNs), transcription activator-like effector nucleases (TALENs) and the **CRISPR-Cas 9 system** (CRISPR = clustered

In addition to restriction endonucleases, there are several other types of nuclease enzymes that are important in the manipulation of DNA.

Fig. 5.5 Mode of action of various nucleases. (a) BAL 31 nuclease is a complex enzyme. Its primary activity is a fast-acting 3′ exonuclease, which is coupled with a slow-acting endonuclease. When BAL 31 is present at a high concentration, these activities effectively shorten DNA molecules from both termini. (b) Exonuclease III (Exo III) is a 3′ exonuclease that generates molecules with protruding 5′ termini. (c) DNase I cuts either single-stranded or double-stranded DNA at essentially random sites. (d) S1 nuclease is specific for single-stranded RNA or DNA. It can act on single-stranded molecules ((d)i) or on single-stranded regions (gaps) in a double-stranded helix ((d)ii). It can also be used to trim the single-stranded regions in hairpin loop structures, as shown in (d)iii.

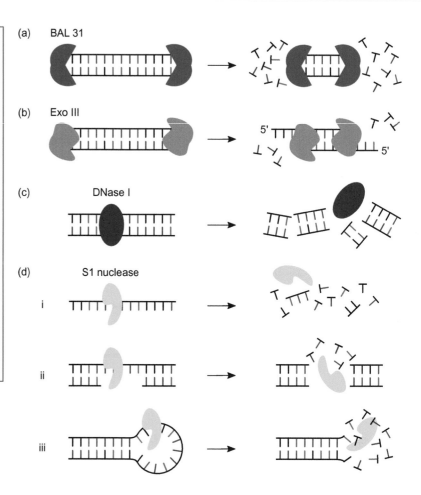

Not all nucleases are helpful! Ribonucleases can be a problem when working with purified preparations of RNA, and care must be taken to remove or inactivate RNase activity.

regular interspersed short palindromic repeats). We will look at how these complex enzyme systems are used in genome editing in Section 12.2.

In addition to DNA-specific nucleases, there are **ribonucleases (RNases)**, which act on RNA. These may be required for many of the stages in the preparation and analysis of recombinants, and are usually used to remove unwanted RNA in the preparation. However, as well as being useful, ribonucleases can pose some unwanted problems. They are remarkably difficult to inactivate and can be secreted in sweat. Thus, contamination with RNases can be a problem in preparing recombinant DNA, particularly where cDNA is prepared from an mRNA template. In this case, it is vital to avoid RNase contamination by wearing gloves and ensuring that all glass and plastic equipment is treated to reduce any ribonuclease activity.

5.2.2 Polymerases

Polymerase enzymes synthesise copies of nucleic acid molecules, and are used in many genetic engineering procedures. When describing a polymerase enzyme, the terms 'DNA-dependent' or 'RNA-dependent' may be used to indicate the type of nucleic acid template that the enzyme uses. Thus, a *DNA-dependent DNA polymerase* copies DNA into DNA; an *RNA-dependent DNA polymerase* copies RNA into DNA, and a *DNA-dependent RNA polymerase* transcribes DNA into RNA. These enzymes synthesise nucleic acids by joining together nucleotides whose bases are complementary to the template strand bases. The synthesis proceeds in a **processive** way in the $5' \rightarrow 3'$ direction, as each subsequent nucleotide addition requires a free 3'-OH group for the formation of the phosphodiester bond. This requirement also means that a short double-stranded region with an exposed 3'-OH (a primer) is necessary for synthesis to begin. We saw how this template-dependent processive mechanism works when we looked at how RNA polymerase generates mRNA (Fig. 4.9).

The enzyme DNA polymerase I has, in addition to its polymerase function, $5' \rightarrow 3'$ and $3' \rightarrow 5'$ exonuclease activities. The enzyme catalyses a strand replacement reaction, where the $5' \rightarrow 3'$ exonuclease function degrades the non-template strand as the polymerase synthesises the new copy. One use for this enzyme is in the nick translation procedure for labelling DNA (outlined in Section 6.3.3).

The $5' \rightarrow 3'$ exonuclease function of DNA polymerase I can be removed by cleaving the enzyme to produce what is known as the Klenow fragment. This retains the polymerase and $3' \rightarrow 5'$ exonuclease activities. The Klenow fragment is used where a single-stranded DNA molecule needs to be copied; because the $5' \rightarrow 3'$ exonuclease function is missing, the enzyme cannot degrade the non-template strand of double-stranded DNA during synthesis of the new DNA. The $3' \rightarrow 5'$ exonuclease activity is suppressed under the conditions normally used for the reaction. Uses for the Klenow fragment include labelling by primed synthesis and early developments in DNA sequencing by the dideoxy method (see Sections 6.3.4 and 6.6.3), in addition to the copying of single-stranded DNAs during the production of recombinants.

Reverse transcriptase (RTase) is an RNA-dependent DNA polymerase, and therefore produces a DNA strand from an RNA template. It has no associated exonuclease activity. The enzyme is used mainly for copying mRNA molecules in the preparation of cDNA (complementary or copy DNA) for cloning (see Section 8.2), although it will also act on DNA templates.

5.2.3   Enzymes That Modify the Ends of DNA Molecules

The enzymes alkaline phosphatase, polynucleotide kinase (PNK) and terminal transferase act on the termini of DNA molecules, and provide important functions that are used in a variety of ways. The phosphatase and kinase enzymes, as their names suggest, are involved in the removal or addition of phosphate groups. Bacterial alkaline phosphatase (BAP) (there is also a similar enzyme – calf intestinal alkaline phosphatase (CIP or CIAP)) removes phosphate groups from the 5' ends of DNA, leaving a 5'-OH group. The enzyme is used to

Polymerases are the copying enzymes of the cell; they are also essential parts of the genetic engineer's armoury. These enzymes are template-dependent and can be used to copy long stretches of DNA or RNA.

A modified form of DNA polymerase I, called the Klenow fragment, is a useful polymerase that is used widely in a number of applications.

Reverse transcriptase is a key enzyme in the generation of cDNA; the enzyme is an RNA-dependent DNA polymerase, which produces a DNA copy of an mRNA molecule.

In many applications, it is often necessary to modify the ends of DNA molecules using enzymes such as phosphatases, kinases and transferases.

prevent unwanted ligation of DNA molecules, which can be a problem in certain cloning procedures. It is also used prior to the addition of phosphate to the 5′ ends of DNA by polynucleotide kinase (see Section 6.3.2).

Terminal transferase (terminal deoxynucleotidyl transferase) repeatedly adds nucleotides to any available 3′ terminus. Although it works best on protruding 3′ ends, conditions can be adjusted so that blunt-ended or 3′-recessed molecules may be utilised. The enzyme is mainly used to add homopolymer tails to DNA molecules prior to the construction of recombinants (see Section 8.2).

5.3 | DNA Ligase – Joining DNA Molecules

DNA ligase is an important cellular enzyme, as its function is to repair broken phosphodiester bonds that may occur at random or as a consequence of DNA replication or recombination. In genetic engineering, it is used to seal nicks in the sugar–phosphate chains that arise when recombinant DNA is made by joining DNA molecules from different sources. It can therefore be thought of as molecular glue, which is used to stick pieces of DNA together. This function is crucial to the success of many experiments, and DNA ligase is therefore a key enzyme in genetic engineering.

DNA ligase is essentially 'molecular glue'; with restriction enzymes, this provides the tools for cutting and joining DNA molecules.

The enzyme used most often in experiments is T4 DNA ligase, which is purified from *E. coli* cells infected with bacteriophage T4. Although the enzyme is most efficient when sealing gaps in fragments that are held together by cohesive ends, it will also join blunt-ended DNA molecules together under appropriate conditions, albeit at a much lower efficiency. The enzyme works best at 37°C, but is often used at lower temperatures (4–15°C) to prevent thermal denaturation of the short base-paired regions that hold the cohesive ends of DNA molecules together. This is an example of a common problem in molecular biology, where optimal conditions for one enzyme (or process) may not be ideal for another enzyme/process, and thus a compromise is required. The mode of action of DNA ligase is shown in Fig. 5.6.

5.4 | Conclusion: The Genetic Engineer's Toolkit

The toolkit analogy used in this chapter is an appropriate way to think about what is needed to manipulate DNA. Inserting or removing a woodscrew or bolt/nut without the correct size and type of screwdriver or spanner is always difficult (and may be impossible). In a similar way, altering DNA molecules requires the correct enzyme to do the job. These enzymes have evolved naturally to carry out very specific functions *in vivo*, as we saw with the restriction–modification system in bacteria that enabled restriction enzymes to be identified and isolated. This biological specificity is what makes each enzyme

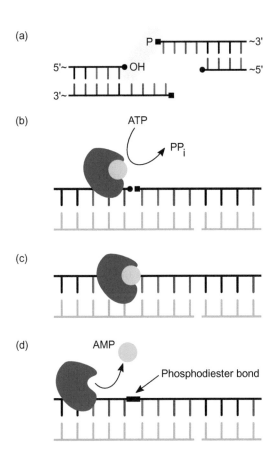

Fig. 5.6 Mode of action of DNA ligase. In (a), two DNA strands cut with the restriction enzyme *Eco* RI have sticky ends (5′-AATT-3′) that can self-anneal. The covalent sealing of the phosphodiester backbone requires joining the 5′-P (phosphate) to the 3′-OH (hydroxyl) group (highlighted). In (b), (c) and (d), the mode of action of DNA ligase on one DNA strand is shown (the other strand is omitted for clarity but would also be sealed by ligase in a similar way). The active site of the enzyme is adenylated by hydrolysis of ATP (b). The adenyl group then activates the 5′-P group (c). Catalysis is completed by formation of the phosphodiester bond by nucleophilic attack from the 3′-OH group, with the release of AMP (d).

uniquely suited to its task. We categorised the enzymes used in recombinant DNA work according to their functions:

- *Restriction endonucleases* cut DNA at defined sites (molecular 'scissors')
- Other *endo-* and *exonucleases* digest DNA and RNA in various ways
- *Polymerases* synthesise nucleic acids from a template
- *Phosphatases, kinases* and *transferases* modify the ends of nucleic acid molecules
- *DNA ligase* seals discontinuities in the phosphodiester backbone (molecular 'glue')

This basic set of tools enables manipulations to be carried out simply and efficiently by mixing the enzyme and the target DNA (or RNA) in an appropriately buffered solution and incubating at the correct temperature.

Stretching our toolkit analogy to breaking point, there are some cases where variants of tools of a similar type might be needed to cope with various tasks, or even times when a tool might be forced into use for a task for which it is not usually designed (it is possible, if not elegant or sensible, to drive in a woodscrew with a hammer!). Enzymes can show similar characteristics. Some may be **thermolabile** (heat-sensitive) and can be inactivated easily to stop the reaction by increasing the temperature in the test tube. Others may be **thermostable** and tolerate high temperatures (this is a key aspect of the polymerase chain reaction that we will look at in Chapter 9). Most enzymes do not discriminate between natural versions of molecules and those that are labelled with a **fluorophore** or **radioactive label** (thus facilitating labelling of nucleic acids and DNA sequencing), and the universal nature of the structure of DNA means that restriction enzymes from bacteria can be used to cut DNA from other sources. Some enzymes have more than one associated function (*e.g.* BAL 31 nuclease), and conditions can be adjusted to favour the preferred activity (exonuclease or endonuclease).

The *template-dependent* polymerases use the conceptually simple characteristics of base-pairing to produce copies of a nucleic acid sequence from a complementary template strand, which is one of the most important processes that the genetic engineer can use in manipulating DNA. In addition, annealing of complementary strands of nucleic acids is a feature of almost all manipulations at some stage, and ranges from the short base-pair overhangs from a restriction digest that anneal in a cloning procedure to generate recombinant DNA, to identification of a genome sequence using a nucleic acid hybridisation procedure.

In this chapter, we have seen how a range of enzymes can be used to modify DNA in a number of ways. Each of the few hundreds of key enzymes that are available commercially has a set of characteristics that enable it to carry out one (or sometimes more than one) function with a high level of specificity and efficiency. It is sometimes easy to take this astonishing level of functionality for granted when transferring a few microlitres of liquid from one small plastic tube into another as part of the day-to-day business of working with recombinant DNA.

Further Reading

Kornberg, A. (2000). Ten Commandments: lessons from the enzymology of DNA replication. *J. Bacteriol.*, 182 (13), 3613–18. URL [https://journals.asm.org/doi/10.1128/JB.182.13.3613-3618.2000]. DOI [https://doi.org/10.1128/jb.182.13.3613-3618.2000].

Pingoud, A. *et al.* (2014). Type II restriction endonucleases – a historical perspective and more. *Nucleic Acids Res.*, 42 (12), 7489–527. URL [https://academic.oup.com/nar/article/42/12/7489/1104749]. DOI [https://doi.org/10.1093/nar/gku447].

Roberts, R. J. (2005). How restriction enzymes became the workhorses of molecular biology. *Proc. Natl. Acad. Sci. U. S. A.*, 102 (17), 5905–8. URL [www.pnas.org/doi/10.1073/pnas.0500923102]. DOI [https://doi.org/10.1073/pnas.0500923102].

Tomkinson, A. E. and Sallmyr, A. (2013). Structure and function of the DNA ligases encoded by the mammalian *LIG3* gene. *Gene*, 531 (2), 150–7. URL [www.ncbi.nlm.nih.gov/pmc/articles/PMC3881560/]. DOI [https://doi.org/10.1016/j.gene.2013.08.061].

Yang, W. (2011). Nucleases: diversity of structure, function and mechanism. *Q. Rev. Biophys.*, 44 (1), 1–93. URL [www.ncbi.nlm.nih.gov/pmc/articles/PMC6320257/]. DOI [https://doi.org/10.1017/S0033583510000181].

Websearch

People

Who was *Arthur Kornberg* and what did he do? Although his work led to the award of a Nobel Prize in 1959, there is one other early achievement that is a notable part of his career – what is this? (Hint: look for information about his period of undergraduate study.)

Places

Two laboratories were involved in work that led to one of the most important discoveries of the later part of the twentieth century. One was at the *University of Wisconsin-Madison, WI*, and one at *Massachusetts Institute of Technology* in *Cambridge, Massachusetts*. Who were the laboratory leaders and what was the discovery which was published in the journal *Nature* in 1970?

Processes

For most routine procedures, restriction enzyme digestion is used to cut DNA samples to completion, where all the sites in the molecule are cut by the enzyme. Sometimes *partial digestion* can be a problem, but sometimes it is used deliberately. Can you find out more about these aspects of using restriction enzymes?

Reflections

What topics in this chapter have you found most challenging? Look for resources that help to illustrate the key points.

Cutting, joining and modifying DNA

RNAs (ribozymes) — but can be — catalytic proteins — that are usually — enzymes — requires

enzymes — isolated from — wide range of organisms — and are — available commercially

Three main groups

restriction endonucleases — that — cut DNA — at — base-specific recognition sites — *e.g.* — G↓AATTC / CTTAA↑G

base-specific recognition sites — and are used for:
- restriction mapping — that can help to — identify gene locations
- generating DNA fragments — that can be — joined to other fragments

DNA modifying enzymes:

nucleases — two classes:
- endonucleases — such as — restriction enzymes / DNase I / S1 nuclease
- exonucleases — such as — BAL 31 / Exo I and III / λ exo / Xm I

polymerases — such as — DNA polymerase / Klenow fragment / reverse transcriptase

enzymes that act on termini — such as — alkaline phosphatase / polynucleotide kinase / terminal transferase

together make up the 'DNA toolkit'

DNA ligase — that — joins DNA molecules — by — phosphodiester linkages — to generate — recombinant DNA molecules

to generate

Concept Map 5

Chapter 6 Summary

Learning Objectives

When you have completed this chapter, you will be able to:

- Outline the basic laboratory requirements for working with nucleic acids
- Identify the range of techniques for isolation, handling and processing of nucleic acids
- Describe the principles of nucleic acid hybridisation, gel electrophoresis and DNA sequencing
- Explain first- and next-generation DNA sequencing methods

Key Words

Erwin Chargaff, Rosalind Franklin, Ray Gosling, physical containment, biological containment, deproteinisation, proteinase K, ribonuclease (RNase), affinity chromatography, oligo(dT)-cellulose, gradient centrifugation, micrograms, nanograms, picograms, aqueous solutions, spectrophotometer, hyperchromic shift, microvolume spectrophotometers, photometry, fluorometric, ethidium bromide, precipitation, isopropanol, ethanol, probe, fluorophores, radiolabelling, deoxynucleoside triphosphate (dNTP), scintillation spectrometer (or counter), specific activity, phosphatase, polynucleotide kinase, terminal transferase, end labelling, DNA polymerase I, DNase I, oligonucleotides, Klenow fragment, nick translation, primer extension, oligolabelling, nucleic acid hybridisation, autoradiograph, phosphorimaging, photostimulated luminescence, excitation, emission, direct detection, indirect detection, haptens, bioconjugates, gel electrophoresis, polyanionic, agarose, polyacrylamide, SAGE, PAGE, restriction mapping, DNA sequencing, technology S-curve, Robert Holley, Walter Fiers, Allan Maxam, Walter Gilbert, chain termination, Sanger sequencing, nested fragments, primed synthesis, dideoxynucleoside triphosphates (ddNTPs), next-generation sequencing (NGS), massively parallel, high-throughput sequencing (HTS), sonication, polishing, adapters, base calling, sequencing by synthesis (SBS), sequencing by ligation (SBL), single-molecule real-time (SMRT), Illumina SBS, Ion Torrent, PacBio, nanopore, fragment library, whole genome sequencing (WGS), amplicon library, barcoding, multiplex, emulsion PCR (emPCR or ePCR), bridge amplification, reversible terminator, signal-to-noise ratio, zero-mode waveguide (ZMW), circular consensus sequencing (CCS), continuous long-read (CLR), MinION, motor protein, helicase, Arthur C. Clarke.

Chapter 6

Working with Nucleic Acids

In Chapter 5, we looked at what we called the *biological methodology* needed for gene manipulation – the enzymes that enable modification, copying, cutting and joining nucleic acid molecules. In this chapter, we will extend this to outline some of the key procedures that a typical laboratory might use on a day-to-day basis. We will cover a range of topics, from some mundane (but critical!) aspects of equipping a laboratory, through to the astonishing biological and technological complexity of next-generation DNA sequencing methods. It is sometimes difficult to make the link between theoretical and practical aspects of a subject, and an appreciation of the methods used in routine work with nucleic acids may be of help when the techniques needed for cloning and analysis of genes and genomes are described in Parts 3 and 4.

6.1 | Evolution of the Laboratory

One of the striking aspects of gene manipulation technology is that many of the procedures can be carried out within a fairly basic laboratory setup. Although applications such as large-scale DNA sequencing, production-scale biotechnology processes and work with transgenic plants and animals require large-scale facilities and significant investment (several millions of £/$), it is still possible to do high-level work within a 'normal' research laboratory. As might be expected, the evolution of the biological research laboratory since the early-1900s parallels the development of the subject itself, with techniques and methods being developed to address the technical requirements of the experiments that were needed to answer the increasingly complex questions that began to emerge. Most of these early laboratories were associated with university departments, although research institutes began to emerge in the late-1800s, mainly to investigate infectious diseases that were at that time poorly understood. Many are still extant and active today, prominent

Many of the techniques used in gene manipulation can be carried out with relatively modest laboratory facilities.

examples being the institutes Pasteur (1888), Koch (1891) and Lister (1891).

The concept of inter- and multidisciplinary research is sometimes thought of as a relatively recent addition to the life science landscape, and is often conflated with the large-scale developments of projects such as the Human Genome Project. Whilst it is true that large-scale projects tend to require inputs from a wider range of disciplines than smaller-scale ventures, interdisciplinary approaches have been around since the start of investigative research in the life sciences and health-related areas. A good example is the discovery of the double helix itself – whilst the deductive brilliance of Watson and Crick (who was originally a physicist) put the final pieces of the jigsaw together, the path to this point had been constructed by many others. Notable in the latter stages were the insights of the chemist **Erwin Chargaff** (who worked out the ratios of the bases in DNA), and the X-ray crystallography team at King's College London (**Rosalind Franklin**, **Ray Gosling** and **Maurice Wilkins**). These days, an interdisciplinary approach is often designed into a new laboratory venture, with examples such as the Broad and Francis Crick institutes (Table 2.2) established to encourage scientists to think beyond traditional discipline boundaries and open up new areas of research.

Interdisciplinary research has always had an important role to play in the history of gene manipulation, and is actively encouraged in research institutes.

The size of a laboratory, in terms of personnel, reflects the type of institution to a large extent. Thus, we can have small-scale university-based laboratories that might have one or two principal investigators, supported by a few postdoctoral research assistants, research students and technical staff. A group like this might be the only one focusing on its particular topic in the host university, and will often have external collaborations with other laboratories or institutes to ensure access to expertise in the subject and also perhaps to shared equipment where this may be too expensive for a small laboratory to purchase. As laboratory sizes get bigger, there is greater opportunity for shared expertise and resources, and this, in turn, attracts spin-off benefits such as incubator sites for start-up biotechnology companies that may be co-located with research facilities. Although this has changed the way in which much of current life science research is organised, the intellectual nucleus of a few scientists with a common interest is still the key driver for excellence in research; it may be just a little easier to generate new interdisciplinary ideas if the infrastructure is in place to support this. A research institute may have several hundreds of scientific staff (the Broad and Crick institutes have around 800 and 1 250, respectively), and thus finding an expert on a particular topic is likely to be, at least geographically, easier than in a smaller laboratory in a more remote location.

We have already seen that a laboratory needs certain core elements to enable it to function (Fig. 2.2). Providing all these on a

| Table 6.1 | Some funding agencies for life science and medical research |

Country/agency	Main funding areas	Budget (US $, billions)	Suggested search term (SST)
Australia			
Australian Research Council	S&T	0.60[a]	aus research council
National Health & Medical Research Council	BM/HC	0.62[a]	aus med res council
Japan			
Agency for Medical Research & Development	BM/HC	1.12[a]	amed japan
Science & Technology Agency	S&T	1.25[a]	jst japan
UK			
UK Research & Innovation (UKRI)	All	11.6[a]	ukri
Biotechnology & Biological Sciences Research Council (BBSRC)	BT/BS	0.31[b]	bbsrc
Medical Research Council (MRC)	BM/HC	0.39[b]	mrc
Natural Environment Research Council (NERC)	ES	0.18[b]	nerc
Wellcome	BM/HC	1.35[c]	wellcome
US			
National Institutes of Health	BM/HC	42[a]	nih
National Science Foundation	S&T	9[a]	nsf funding

Note: Areas are science and technology (S&T), environmental science (ES), biomedical/healthcare (BM/HC), biotechnology/biological sciences (BT/BS). UKRI covers all areas plus arts and humanities, economics and social sciences, and engineering and physical sciences. BBSRC, MRC and NERC are research councils that are part of UKRI. Data are for 2020−21, unless noted otherwise (although direct comparison may not be valid due to different financial reporting mechanisms). [a] Overall budget figures, [b] grant funding distributed and [c] grant funding for 2019−20. Further information can be found on the agency websites, accessed using the SSTs.

recurring and stable basis is expensive. Apart from any setup costs, which, for a new-build facility, can run into tens or hundreds of millions of £/$, annual running costs can be of a similar scale for a large institute or biotechnology company. Even a relatively small research group could easily exceed one million £/$ per year, with a high proportion of this dependent on grant income. Thus, a significant part of a senior research scientist's time may be taken up with securing grants for various projects, often with no guarantee of long-term funding being available. Some life science funding bodies are shown in Table 6.1.

> Annual running costs for even a small-scale laboratory are significant and usually need to be funded from a range of sources.

When we get into the actual laboratory, there will be some level of variation in terms of equipment (what is needed, how many of each and so on). There are, however, certain core requirements that we might summarise under three headings:

- General laboratory facilities
- Cell/organism culture and containment facilities
- Sample processing and analysis

General facilities include laboratory layout, capacity, workstations and furnishings and provision of essential services such as water (including distilled and/or deionised water sources), electrical power, piped gases (natural gas for burners, perhaps nitrogen and carbon dioxide, and potentially other specialised gases), compressed air, vacuum lines, drainage, cold storage (cold rooms, fridges, $-20°C$ and $-80°C$ freezers, liquid nitrogen storage), *etc.* IT equipment and networking capability are also essential and usually need detailed planning at the design stage. Most of the services would be provided as part of any laboratory establishment, and present no particular difficulty and expense beyond the normal setup and commissioning costs.

Cell culture and containment facilities are essential for growing the cell lines and organisms required for the research, with the precise requirements depending on the type of work being carried out. Most laboratories will require facilities for growing bacterial cells, with the need for equipment such as autoclaves, incubators (static and rotary), centrifuges and protective cabinets in which manipulations can be performed. Mammalian cell culture requires slightly more sophisticated facilities, and it is usually sensible to isolate these from bacterial culture areas to avoid cross-contamination. Plant tissue culture and algal culture usually require integration of lighting into the culture cabinets, with greenhouse and field growth capacity needed for larger plants.

> The biology of gene manipulation requires facilities for the growth, containment and processing of different types of cells and organisms.

In many cases, some form of **physical containment** is required, to prevent the escape of organisms during manipulation. The overall level of containment required depends on the type of host and vector being used, with the combination providing (usually) a level of **biological containment** in that the host is usually disabled and does not survive beyond the laboratory. The overall containment requirements will usually be specified by national bodies that regulate gene manipulation, and these may apply to bacterial and mammalian cell culture facilities. Thus, a simple cloning experiment with *Escherichia coli* may require only normal microbiological procedures, whereas an experiment to clone human genes using viral vectors in mammalian cell lines may require the use of more stringent safety systems.

For processing and analysis of cells and cell components such as DNA, there is a bewildering choice of different types of equipment. At the most basic level, the type of automatic pipette and microcentrifuge tube can be important (Fig. 6.1). A researcher struggling with pipettes that do not work properly, or with tubes that have caps that are very hard to open, will soon get frustrated! At the other end of the scale, equipment such as ultracentrifuges and automated DNA sequencers may represent a major investment for the laboratory and need to be chosen carefully. Much of the other equipment is (relatively) small and low-cost, with researchers perhaps having a particular brand preference.

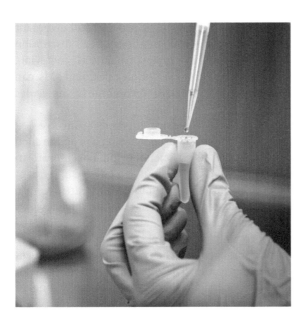

Fig. 6.1 Handling small volumes of liquids. One of the laboratory tools used most often is the mechanical micropipette, available in a range of sizes that can dispense volumes of less than 1 µL up to 5 or 10 mL (to visualise, a microlitre is a cubic millimetre). Plastic tips are used to avoid contamination, both of the samples and of the pipette body. The tube being used here is the ubiquitous 1.5 mL microcentrifuge or 'microfuge' tube. Smaller tubes (0.5 mL) are also popular, and specialist sizes are available for particular applications and equipment, such as 0.2 mL tubes for PCR thermal cyclers. Good-quality pipettes, tips and tubes are essential – anything that does not work well will quickly become frustrating.
Source: Photograph by Assembly / Stone / Getty Images.

6.2 | Isolation of DNA and RNA

Most gene manipulation experiments (apart from synthesis of sequences wholly *de novo*) require a source of nucleic acid, in the form of either DNA or RNA. It is therefore important that reliable methods are available for isolating these components from cells. There are three basic requirements: (1) opening the cells in the sample to expose the nucleic acids for further processing, (2) separation of the nucleic acids from other cell components, and (3) recovery of the nucleic acid in purified form. A variety of techniques may be used, ranging from simple procedures with few steps, up to more complex purifications involving several different stages. These days, most molecular biology supply companies sell kits that enable purification of nucleic acids from a range of sources.

Many of the routine methods required for rDNA work have been developed into reliable protocols and are available as 'off-the-shelf' kits from a range of suppliers.

The first step in any isolation protocol is disruption of the starting material, which may be viral, bacterial, plant or animal. The method used to open cells should be as gentle as possible, preferably utilising

Cells have to be disrupted to enable nucleic acids to be isolated – this should be done as gently as possible to avoid shearing large DNA molecules.

Once broken open, cell preparations can be deproteinised and the nucleic acids purified by a range of techniques. Some applications require highly purified nucleic acid preparations, and some may be able to use partially purified DNA or RNA.

enzymatic degradation of cell wall material (if present) and detergent lysis of cell membranes. If more vigorous methods of cell disruption are required (as is the case with some types of plant cell material), there is the danger of shearing large DNA molecules, and this can hamper the production of representative recombinant molecules during subsequent processing.

Following cell disruption, most methods involve a **deproteinisation** stage. This can be achieved by using a protease such as **proteinase K** and/or extraction using phenol or phenol/chloroform mixtures. On the formation of an emulsion and subsequent centrifugation to separate the phases, protein molecules partition into the phenol phase and accumulate at the interface. The nucleic acids remain mostly in the upper aqueous phase, and may be precipitated from solution using isopropanol or ethanol (see Section 6.3). More modern techniques that do not require the use of phenolic mixtures are available from several companies. These are safer and more pleasant to use than phenol-based extraction media, are generally quicker and may also be formulated to include sample clean-up for downstream applications that are contaminant-sensitive, such as DNA sequencing or the polymerase chain reaction (PCR).

If a DNA preparation is required, the enzyme ribonuclease (RNase) can be used to digest the RNA in the preparation. If mRNA is needed for cDNA synthesis, a further purification can be performed by **affinity chromatography** using **oligo(dT)-cellulose** to bind the poly(A) tails of eukaryotic mRNAs. This gives substantial enrichment for mRNA and enables most contaminating DNA, rRNA and tRNA to be removed.

The technique of **gradient centrifugation** is often used to prepare DNA, particularly plasmid DNA (pDNA). In this technique, a caesium chloride solution containing the DNA preparation is spun at high speed in an ultracentrifuge. Over a long period (up to 48 hours in some cases), a density gradient is formed and the pDNA forms a band at one position in the centrifuge tube. The band may be taken off, and the caesium chloride removed by dialysis to give a pure preparation of pDNA. As an alternative to gradient centrifugation, size exclusion chromatography (gel filtration) or similar techniques may be used.

6.3 | Handling and Quantification of Nucleic Acids

Solutions of nucleic acids are used to enable very small amounts to be handled, measured and dispensed easily.

It is often necessary to use very small amounts of nucleic acid (typically **micrograms**, **nanograms** or **picograms**) during a cloning, sequencing or PCR experiment. It is obviously impossible to handle these amounts directly, so most of the measurements and manipulations that are done involve the use of **aqueous solutions** of DNA and RNA. The concentration of a solution of nucleic acid can be determined by measuring the absorbance at 260 nm (nanometres), using a **spectrophotometer**. An A_{260} of 1.0 is equivalent to a concentration of 50 μg mL^{-1} for double-stranded DNA, 40 μg mL^{-1} for RNA and

$33\ \mu g\ mL^{-1}$ for single-stranded DNA (the higher absorbance of single-stranded DNA is due to what is called **hyperchromic shift**; the bases absorb less strongly when paired together in double-stranded DNA, so single-stranded DNA absorbs more ultraviolet (uv)). If the A_{280} is also determined, the A_{260}/A_{280} ratio indicates if there are contaminants present, such as residual phenol or protein. The A_{260}/A_{280} ratio should be around 1.8 for pure DNA, and 2.0 for pure RNA preparations. Standard spectrophotometers use a 1 cm path length cuvette for samples, but **microvolume spectrophotometers** (MVS) are available that can read sample volumes as small as 1 µL and detect concentrations from around $1\ ng\ \mu L^{-1}$. In addition to using **photometry** for direct absorbance measurements, **fluorometric** methods can be used to measure the fluorescent emissions of dyes that bind to nucleic acids. Whilst these methods have some advantages in terms of specificity, the direct absorbance method is easier to use (as no sample modification is required) and is the one most often used for routine quantification of nucleic acid concentrations.

In addition to spectrophotometric methods, the concentration of DNA may be estimated by monitoring the fluorescence of bound ethidium bromide in a simple spot assay. This dye binds between the DNA bases (intercalates) and fluoresces orange when illuminated with uv light. By comparing the fluorescence of the sample with that of a series of standards, an estimate of the concentration may be obtained. This method can detect as little as 1–5 ng of DNA, and is sometimes useful if uv-absorbing contaminants make spectrophotometric measurements difficult.

Having determined the concentration of a solution of nucleic acid, any amount (in theory) may be dispensed by taking the appropriate volume of solution. In this way, nanogram or picogram amounts may be dispensed with reasonable accuracy.

Precipitation of nucleic acids is an essential technique that is used in a variety of applications. The two most commonly used precipitants are **isopropanol** and **ethanol**, ethanol being the preferred choice for most applications. When added to a DNA solution in a ratio, by volume, of 2:1 in the presence of 0.2 M salt, ethanol causes the nucleic acids to come out of solution. Although it used to be thought that low temperatures ($-20°C$ or $-70°C$) were necessary, this is not an absolute requirement, and using an ice bath may be sufficient for routine applications. After precipitation, the nucleic acid can be recovered by centrifugation, which causes a pellet of nucleic acid material to form at the bottom of the tube. The pellet can be dried, and the nucleic acid resuspended in the buffer appropriate to the next stage of the experiment.

Nucleic acids can be concentrated by using alcohol to precipitate the DNA or RNA from solution; the precipitate is recovered by centrifugation and can then be processed as required.

6.4 | Labelling Nucleic Acids

Keeping track of the small amounts of nucleic acid involved in cloning experiments can be difficult. The amount of material usually diminishes after each step, as manipulations do not usually result in 100 per cent recovery of the sample. In other procedures, it may be necessary to use a nucleic acid probe that can identify cloned sequences in a library. Both these needs can be met by labelling the nucleic acid with a marker of some sort, so that the material can be identified at each stage of the procedure. So what can be used as the label?

6.4.1 Types of Label – Radioactive or Not?

Radioactive tracers have been used extensively in biochemistry and molecular biology for a long time, and procedures are now well established. The most common isotopes used in biological research are tritium (^{3}H), carbon-14 (^{14}C), sulphur-35 (^{35}S) and phosphorus-32 (^{32}P). Tritium is a low-energy emitter; ^{14}C and ^{35}S are classed as low-medium emitters (although ^{14}C is not often used to label nucleic acids), and ^{32}P is a high-energy emitter. Thus, ^{32}P is more radiologically hazardous than the other isotopes and requires particular care in use. There are also strict statutory requirements for the storage and use of isotopes, and the disposal of radioactive waste. Partly because of the inherent dangers of working with high-energy isotopes, the use of alternative technologies such as fluorescent dyes (**fluorophores**), or enzyme-linked labels, has become popular in recent years. Although these methods do offer advantages for particular applications (such as DNA sequencing and *in situ* hybridisation procedures), for routine tracing and probing experiments, a radioactive label may still be useful. In this case, the term radiolabelling is often used to describe the technique. Some radionuclides and fluorophores are listed in Table 6.2.

> Although radioactive isotopes can be used to label nucleic acids, they are more hazardous than non-radioactive labelling methods such as the use of fluorophores.

One way of tracing DNA and RNA samples is to label the nucleic acid with a radioactive molecule (usually a **deoxynucleoside triphosphate (dNTP)**, labelled with ^{3}H, ^{35}S or ^{32}P), so that portions of each reaction may be counted in a scintillation spectrometer (counter) to determine the amount of nucleic acid present. This is usually done by calculation, taking into account the amount of radioactivity present in the sample.

A second application of radiolabelling is in the production of highly radioactive nucleic acid molecules for use in hybridisation experiments. The difference between labelling for tracing purposes and labelling for probes is largely one of specific activity, i.e. the measure of how radioactive the molecule is. For tracing purposes, a low specific activity will suffice, but for probes, a high specific activity is necessary. In radioactive probe preparation, the label is usually the high-energy β-emitter ^{32}P, although both radioactive and fluorophore-based precursors can be used in enzyme-catalysed labelling methods. Examples of these are shown in Table 6.3; some are outlined in more detail in Sections 6.4.2 to 6.4.4.

> Probes labelled with radioactive or non-radioactive molecules can be used to identify target DNA or RNA sequences by complementary base-pairing.

Table 6.2 Radioactive and non-radioactive labelling compounds

Radionuclides

Isotope	Half-life	Emission (keV max/av)	Detection method(s)
Tritium (^{3}H)	12.3 years	β-decay (18.6/5.7)	Scintillation spectrometry
Carbon-14 (^{14}C)	5730 years	β-decay (156/49)	Phosphorimaging
Sulphur-35 (^{35}S)	87.4 days	β-decay (167/49)	Scintillation spectrometry
Phosphorus-32 (^{32}P)	14.3 days	β-decay (1710/700)	Phosphorimaging Autoradiography

Fluorescent labelling compounds

Fluorophore[a]	Emitted colour	$\lambda_{exc}/\lambda_{em}$ (nm)[b]	Detection equipment
Aminomethylcoumarin acetate (AMCA)	Blue	350/445	Fluorescence scanners
Fluorescein	Green	494/512	Microarray readers
Cy 3	Yellow	550/570	Microplate readers
Cy 5	Red	650/670	Spectrofluorometers

Note: [a] Fluorophores are incorporated into various formulations such as dNTPs and are available from biological suppliers. [b] λ_{exc} and λ_{em} are the excitation and emission wavelengths, respectively; the excitation laser stimulates the fluorophore, which then emits light of a longer wavelength as fluorescence. The peak excitation and emission wavelengths can vary slightly, depending on buffer composition and other factors.

6.4.2 End Labelling

In this technique, the label is added to either the 5′ or 3′ termini of the nucleic acid. For 5′ labelling, the terminal phosphate can be removed using a **phosphatase** (alkaline phosphatase or calf intestinal phosphatase). This favours the forward reaction catalysed by **polynucleotide kinase**, which transfers the terminal gamma (γ) phosphate group of ATP onto a 5′-OH terminus. If the ATP donor is radioactively labelled, this produces a labelled nucleic acid of relatively low specific activity, as only the termini of each molecule become radioactive. Non-radioactive labels can be used if a chemical group (such as a thiol group) has been added to the nucleoside triphosphate to enable the fluorophore to be incorporated.

A range of enzyme-catalysed methods can be used to label nucleic acids at various positions.

The enzyme **terminal transferase** is often used to label the 3′ ends of a nucleic acid. This sequentially adds nucleotides to the 3′-OH terminus. The reaction conditions can be adjusted to control the length of the addition, from a few nucleotides up to several hundreds. If the dideoxy form is used (ddNTP), a single nucleotide is added as the lack of the 3′-OH terminus does not permit chain elongation. Radioactive and non-radioactive labels can be added if the appropriate

Table 6.3 | Methods for enzymatic labelling of nucleic acids

Template	Method	Site of label incorporation	Labelled probe type	Enzyme used
RNA	5'-end labelling	5'-OH	RNA	Phosphatase and polynucleotide kinase
	3'-end labelling	3'-OH	RNA	Terminal transferase T4 RNA ligase
	Reverse transcription	Random sites	cDNA	Reverse transcriptase
DNA	5'-end labelling	5'-OH	DNA/oligonucleotide	Phosphatase and polynucleotide kinase
	3'-end labelling	3'-OH	DNA/oligonucleotide	Terminal transferase
	Nick translation	Random sites	DNA	DNase I and DNA polymerase I
	Primer extension	Random sites	DNA	Klenow fragment *Taq* polymerase
	PCR	Random sites	DNA	*Taq* polymerase
	Transcription *in vitro*	Random sites	RNA (cRNA)	T7 RNA polymerase

Note: Enzymatic labelling enables both radioactive and non-radioactive labels to be incorporated. The labelled compounds are usually very similar to the natural molecule (*e.g.* dNTP) in structure and size to ensure incorporation is not hampered by chemical or physical incompatibility. Labelling kits are available from many biological suppliers and are optimised for particular applications. Both radioactive and non-radioactive (*e.g.* fluorophore-based) labels can be incorporated. In some cases, specific requirements (*e.g.* T7 RNA polymerase requires a T7 promoter) increase the specificity and/or efficiency of the labelling process, but may affect ease of use.

chemical structure is available in the d/ddNTP form. A summary of **end labelling** is shown in Fig. 6.2.

6.4.3 Nick Translation

This method relies on the ability of the enzyme **DNA polymerase** I (see Section 5.2.2) to translate (move along the DNA) a 'nick' created in the phosphodiester backbone of the DNA double helix. Nicks may occur naturally, or may be caused by a low concentration of the nuclease **DNase I** in the reaction mixture. DNA polymerase I catalyses a strand replacement reaction that incorporates new dNTPs into the DNA chain. If one of the dNTPs supplied is radioactive, or is complexed with a fluorophore, the result is a highly labelled DNA molecule (Fig. 6.3).

6.4.4 Labelling by Primer Extension

This term refers to a technique that uses **oligonucleotides** (often hexadeoxyribonucleotide molecules – sequences of six nucleotides) to prime the synthesis of a DNA strand by DNA polymerase. The DNA to be labelled is denatured by heating, and the oligonucleotide

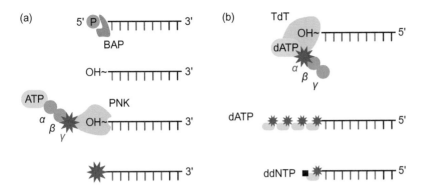

Fig. 6.2 End labelling DNA. Labels may be radioactive nucleosides or bioconjugated non-radioactive labels (e.g. fluorophores), and can be added to both the 5' and 3' ends of nucleic acids or oligonucleotides (for clarity, in this figure, only single strands are shown). In (a), DNA is dephosphorylated using bacterial alkaline phosphatase (BAP) to generate 5'-OH groups. The terminal phosphate of [γ-^{32}P] ATP (red star) is then transferred to the 5' terminus by polynucleotide kinase (PNK). The reaction can also occur as an exchange reaction with 5'-phosphate termini. In (b), the enzyme terminal deoxynucleotidyl transferase (TdT) is used to add nucleotides onto the 3'-OH terminus. In this example, [α-^{32}P] dATP enables sequential addition of the radiolabelled nucleotide. The length of the homopolymer can be controlled by varying the concentrations of the reaction components or by including a dideoxy NTP (ddNTP) to terminate chain elongation.

primers annealed to the single-stranded DNAs. The **Klenow fragment** of DNA polymerase (see Section 5.2.2) can then synthesise a copy of the template, primed from the 3'-OH group of the oligonucleotide. If a labelled dNTP is incorporated, DNA of very high specific activity (if radiolabelled) or high level of fluorescence (if a fluorophore is used) is produced (Fig. 6.4).

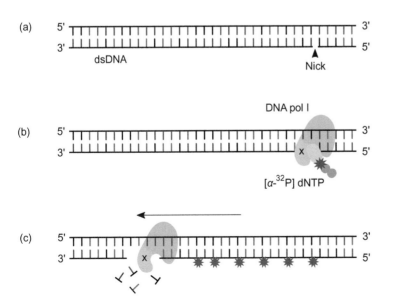

Fig. 6.3 Labelling DNA by nick translation. (a) A single-strand nick is introduced into the phosphodiester backbone of a DNA fragment using DNase I. (b) DNA polymerase I (DNA pol I) then synthesises a copy of the template strand, degrading the non-template strand with its 5' → 3' exonuclease activity (x). (c) If [α-^{32}P] dNTP is supplied, this will be incorporated into the newly synthesised strand. Usually only one radiolabelled dNTP is used and 'cold' (non-radioactive) dNTP is mixed with the radiolabelled version, so not all the nucleotides incorporated will be radioactive.

Fig. 6.4 Labelling DNA by primer extension (oligolabelling). (a) DNA is denatured to give single-stranded molecules (ssDNA). (b) An oligonucleotide primer is then added to give a short double-stranded region with a free 3'-OH group. (c) The Klenow fragment of DNA polymerase I can then synthesise a copy of the template strand from the primer, incorporating [α-^{32}P] dNTP to produce a labelled molecule with a very high specific activity. Hapten bioconjugates or fluorophore-labelled nucleosides can be used instead of radioactive dNTPs.

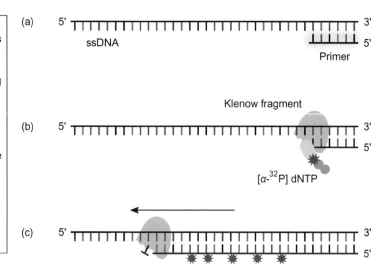

6.5 | Nucleic Acid Hybridisation

The simple base-pairing relationship between complementary sequences has very far-reaching consequences – for both the cell and its functioning, and also for the scientists who can use this feature to identify and manipulate specific sequences.

As we saw in Chapter 4 (in particular, in Figs. 4.4 and 4.10), the relatively simple concept of A·T (A·U in RNA) and G·C base-pairing is perhaps the most critical fundamental property of nucleic acids. All the cellular processes that are controlled by DNA and RNA through gene expression depend on the complementary nature of the base-pairs, and it is no surprise that recombinant DNA technology relies to a great extent on exploiting this. As with many protocols in molecular biology, the simple *concept* behind **nucleic acid hybridisation** becomes significantly more complex when used in the laboratory. Although there are many different techniques that use hybridisation at some stage, we can bring nucleic acid labelling and hybridisation together by looking at how we can detect specific sequences from a complex heterogeneous population such as a genomic library. To do this, we need three things:

- The population of target sequences
- A labelled probe complementary to the sequence of interest
- A detection method appropriate to the probe chemistry and physics

The technology used to visualise labelled probes has developed from simple film-based detection to sophisticated scanning of samples and digital capture of high-resolution images.

The target population is often fixed to a support such as a membrane or a glass slide, and the probe is hybridised to the target sequence in a solution that enables the formation of duplexes by base-pairing. The correct temperature and salt concentration are critical for this step, and usually need to be optimised by adjustment and repetition. Excess probe is washed away, and the probe detected using a range of methods. For radioactive probes, this can involve exposure to X-ray film to produce an **autoradiograph**, although this method has largely been replaced by **phosphorimaging**. The image is generated by radiation-stimulated trapping of electrons in a phosphor screen. When illuminated by a laser, the trapped electrons release their energy in the form of light in a process called **photostimulated luminescence**, and the image is captured digitally. Phosphorimaging is easier and more sensitive, and has a greater dynamic range than traditional autoradiography, and is a good illustration of how an established technique (that can still be useful) has been replaced by a more sophisticated alternative.

Detection of fluorophores requires a light source to activate the fluorophore (known as **excitation**), and a detector that can measure the light produced by fluorescence (**emission**). Filtering out unwanted wavelengths is important, as ideally only the fluorescence peak wavelength should be measured. Different excitation wavelengths, produced using different lasers or filters, can enable detection of multiple fluorophores. Multi-functional machines are available that can operate as phosphorimaging, fluorescence and optical scanners; these are extremely useful but, as might be expected, are expensive.

In addition to the chemistry and physics of labelling nucleic acid, the methods used to detect hybridised probes can be classified as either **direct** or **indirect**. Indirect methods use an intermediate in detection, usually with fluorophores. This can avoid some of the potential issues with physical or chemical interference, and can also amplify the signal. Small molecules known as haptens can be incorporated into nucleic acids instead of the fluorophore itself. The hapten can then be detected by binding a second molecule, to which the detection system (a fluorophore or enzyme) is bound. Using this approach, which is sometimes cited as an example of 'click chemistry', a set of molecules can be used as a flexible toolkit to generate a wide variety of bioconjugates that are useful in a range of applications. A summary of the techniques described in this section is shown in Fig. 6.5.

> Detection methods can be classed as direct or indirect; the indirect methods often use bioconjugated molecules to link the label and the probe.

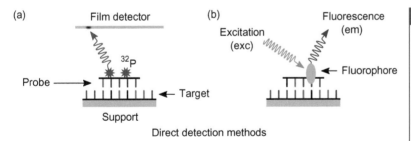

Direct detection methods

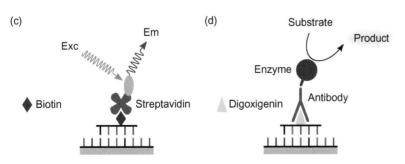

Indirect detection methods

> **Fig. 6.5** Summary of detection methods. Examples show target sequences fixed to a support membrane, with the specific target sequence identified using a complementary probe. Direct methods are shown in (a) and (b). In (a), the probe is labelled with ^{32}P and detected using X-ray film autoradiography. In (b), the probe is labelled with a fluorophore, detected using an imaging system. Indirect methods are shown in (c) and (d). In (c), biotin is used to label the probe, which is then detected using a streptavidin–fluorophore system. In (d), the probe is labelled with digoxigenin, and an antibody-linked enzyme used to generate a detectable product.

6.6 | Gel Electrophoresis

The technique of **gel electrophoresis** is one method that enables nucleic acid fragments to be visualised directly. The method relies on the fact that nucleic acids are **polyanionic** at neutral pH, *i.e.* they carry multiple negative charges due to the phosphate groups on the phosphodiester backbone of the nucleic acid strands. This means that the molecules will migrate towards the positive electrode when placed in an electric field. As the negative charges are distributed evenly along the DNA molecule, the charge-to-mass ratio is constant; thus, mobility depends on fragment length. The technique is carried out using a gel matrix which separates the nucleic acid molecules according to size. A typical nucleic acid electrophoresis setup is shown in Fig. 6.6.

The type of matrix used for electrophoresis has important consequences for the degree of separation achieved, which is dependent on the porosity of the matrix. Two gel types are commonly used, these being **agarose** and **polyacrylamide**. Agarose is extracted from seaweed, and can be purchased as a dry powder which is melted in buffer at an appropriate concentration, normally in the range of 0.3–2.0% (w/v). On cooling, the agarose sets to form the gel. Agarose gels are usually run in the apparatus shown in Fig. 6.6, with a technique called

Fig. 6.6 A typical system used for agarose gel electrophoresis. The gel is cast on a uv-transparent tray, placed in the gel tank and just covered with buffer (the technique is called *submerged agarose gel electrophoresis* or SAGE). Nucleic acid samples are placed in wells created in the gel using a plastic comb. DNA fragments in the gel will migrate towards the positive electrode and are separated on the basis of size. Agarose concentrations of 0.3–2.0% enable the separation of fragments of about 50 000–100 bp. A 1% gel is often used for routine separation of the most common fragment sizes generated in recombinant DNA manipulations.

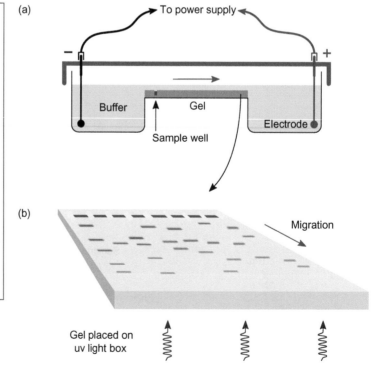

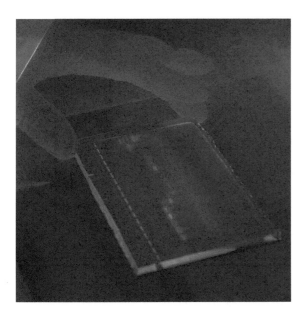

Fig. 6.7 Viewing an agarose gel. In this photograph, the researcher is examining an agarose gel on a uv-emitting light box called a transilluminator. The gel contains the dye ethidium bromide. This intercalates between the base-pairs of the DNA fragments and can be seen in the gel as bands of increased red-orange fluorescence. The sample wells in this gel are near the left-hand edge (the 'top' of the gel), and the DNA fragments have migrated towards the right-hand edge (the 'bottom' of the gel). *Source:* Photograph by pidjoe / E+ / Getty Images.

submerged agarose gel electrophoresis (or **SAGE**). Polyacrylamide gels are sometimes used to separate small nucleic acid molecules in applications such as DNA sequencing, as the pore size is smaller than that achieved with agarose. Polyacrylamide-based gel electrophoresis is called **PAGE**.

Electrophoresis is carried out by placing the nucleic acid samples in the gel and applying a potential difference across it. This is maintained until a marker dye (usually bromophenol blue, added to the sample prior to loading) reaches the end of the gel. The nucleic acids in the gel are usually visualised by staining with the intercalating dye ethidium bromide and examining under uv light. Nucleic acids show up as orange bands, which can be photographed to provide a record (Fig. 6.7). The data can be used to estimate the sizes of unknown fragments by construction of a calibration curve using standards of known size, as migration is inversely proportional to the $\log_{10}$ of the number of base-pairs. This is particularly useful in the technique of restriction mapping (Section 5.1.3).

In addition to its use in the analysis of nucleic acids, polyacrylamide gel electrophoresis is used extensively for the analysis of proteins. The methodology is different from that used for nucleic acids, but the basic principles are similar. One common technique is SDS-PAGE, in which the detergent SDS (sodium dodecyl sulphate) is used to denature multisubunit proteins and cover the protein molecules with negative charges. In this way, the inherent charge of the protein is masked, and the charge-to-mass ratio becomes constant. Thus, proteins can be separated according to their size in a similar way to DNA molecules.

As well as nucleic acids, proteins can be separated by electrophoresis using a polyacrylamide gel matrix rather than agarose.

6.7 | DNA Sequencing: The First Generation

The development of **DNA sequencing** is perhaps the most astonishing example of how technical innovation can drive research. At this point, it is worth looking at a well-known business analysis concept known as the technology S-curve. This will help frame developments in DNA sequencing, and will also be useful when we look at areas such as bioinformatics (Chapter 11), the applications of DNA sequencing (Chapter 13) and the biotechnology industry (Chapter 14). The premise is that any technological development goes through various stages – innovation, early adoption, growth, maturity and ultimately decline. Over time, there is a need to replace one technology with a new one, either by adapting an existing process or product, or developing a completely new technology (this is the equivalent of the paradigm shift/nudge we looked at in Section 2.3). An interesting exercise is to trace the developmental cycles of a technology in which you are interested (music reproduction is a good example, from wax cylinders to web-based streaming services). Although often presented as a series of regular cycles, technology S-curves are usually a bit more messy in reality. An outline of some key features is shown in Fig. 6.8.

The first nucleic acid molecules to be sequenced were RNA, not DNA. In 1965, the tRNA for alanine was sequenced by **Robert Holley**, and in 1972, the first gene was sequenced by **Walter Fiers** (the coat protein of the RNA virus MS2). The MS2 genome (3 569 nucleotides) was the first genome to be completely sequenced (by Fiers' group in 1976). Despite these successes, sequencing was slow and laborious, as the methods available did not easily address the need for rapid and sequential identification of nucleotide bases. At around the same time, work on techniques for sequencing DNA was ongoing and began to produce some notable results. Frederick Sanger's group had developed methods for using DNA polymerase-based methods. One variant of this (usually called the 'plus/minus' method, developed in 1975) enabled the 5 375-nucleotide genome of the bacteriophage ΦX174 to be sequenced in 1977 – the first DNA genome to be completed. In many ways, the year 1977 is the 'landmark' year for DNA sequencing; as well as the sequence of ΦX174, two methods appeared that changed the face of sequence determination significantly. One was a method devised by **Allan Maxam** and **Walter Gilbert**, based on chemical degradation of DNA at modified nucleotides, and the second was an extension of Sanger's work that led to the **chain termination** method. These two methods were the first rapid techniques for DNA sequencing (although, as we will see, rapid in 1977 is not the same as rapid in 2023!) and enabled the technology to be used widely across the scientific community. Over time, the Sanger protocol became the

The technology S-curve is a useful business analysis concept that can be used to track the development of many areas of science and technology.

DNA sequencing developed from early techniques using RNA, but the methods were not easily adapted to sequence long stretches of DNA.

Rapid DNA sequencing methods were first developed in the late-1970s and would lead the way towards sequencing of whole genomes.

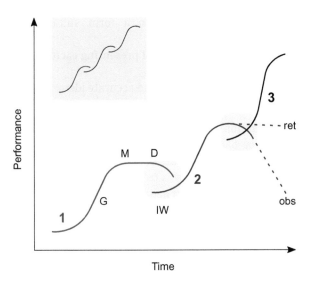

Fig. 6.8 The technology S-curve. This shows three cycles of the technology S-curve, where performance is mapped against time. Performance can be measured across different sectors and in a number of ways, including product output, market penetration, spread of use of a technology or the development of a new procedure or piece of equipment. The often portrayed idealised form is shown in the inset panel; in reality, this is rarely the case. The cycles shown in the main panel illustrate some typical features that could be applied to the development of DNA sequencing technology. Cycle 1 (blue line) shows a fairly typical emergence and growth (G) pattern, with the technology reaching maturity (M) at some point. In the absence of need or innovation, the technology might remain at the same 'performance' level for a period. It will, however, eventually need to be updated or replaced as demand grows, and new innovation/development of cycle 2 will begin as cycle 1 begins its decline phase (D). This period is sometimes called the innovation window (IW, highlighted yellow). When cycle 2 enters its decline phase, cycle 3 will replace it. Sometimes the development and growth period is faster for later technologies, as they are based on previous experience or may be a revolutionary technology that is adopted very quickly. The decline phase for cycle 2 in this example can lead to a very quick decline in use that makes the technology obsolete (obs). However, if the technology is still useful, it can be retained in the sector (ret) and continue to be used. This is particularly true where a high level of initial investment has been required to set up the system.

more widely used method, to the point where it gained the ultimate accolade of eponymous recognition and is universally known as **Sanger sequencing**. We will look at this in detail in Section 6.7.2.

6.7.1 Principles of First-Generation DNA Sequencing

By definition, the determination of a DNA sequence requires that the bases are identified in a sequential technique that enables the processive identification of each base in turn. There are three main requirements for this to be achieved:

Although the principles underpinning DNA sequencing are fairly simple, the procedures required to achieve the desired result are rather more complex.

- DNA fragments need to be prepared in a form suitable for sequencing
- The technique used must achieve the aim of presenting each base in turn in a form suitable for identification
- The detection method must permit rapid and accurate identification of the bases

Generation and preparation of DNA fragments is fairly simple on a purely technical level. For Sanger sequencing, the fragments have usually been cloned in a suitable vector; once this is achieved, the sequencing protocol essentially becomes a technical procedure rather than an experimental one. The aim is to generate a set of overlapping fragments that terminate at different bases and differ in length by one nucleotide. This is known as a set of **nested fragments**. These will be labelled (either radioactive or fluorophore-based labels can be used) and can then be separated by electrophoresis on a polyacrylamide gel. Slab gels, in which fragments are radioactively labelled, generate an autoradiograph. Automated sequencing procedures tend to use fluorescent labels and continuous electrophoresis in narrow capillary tubes to separate the fragments, which are identified as they pass a detector. So how do these techniques work?

6.7.2 Sanger (Dideoxy or Enzymatic) Sequencing

Sanger sequencing uses the method of **primed synthesis** by the Klenow fragment of DNA polymerase. The process is similar to one we have already encountered in Section 6.4.4 (and Fig. 6.4), where a primer is used on a single-stranded DNA template to generate a free $3'$-OH terminus to enable synthesis of a complementary copy of the template strand by the polymerase. The production of nested fragments is achieved by the incorporation of a modified dNTP in each reaction. These dNTPs lack a hydroxyl group at the $3'$ position of deoxyribose, which is necessary for chain elongation to proceed. Such modified dNTPs are known as **dideoxynucleoside triphosphates (ddNTPs)**. The four ddNTPs (A, G, T and C forms) are included in a series of four reactions, each of which also contains the four normal dNTPs. The concentration of the dideoxy form is such that it will be incorporated into the growing DNA chain infrequently. Each reaction therefore produces a series of fragments terminating at a specific nucleotide, and the four reactions together provide a set of nested fragments. If using radiolabelling, a radioactive dNTP is included in the reaction mixture. This is usually $[\alpha\text{-}^{35}S]$ dATP, which enables more sequence to be read from a single gel than ^{32}P-labelled dNTPs that were used previously. The dideoxy method is shown in Fig. 6.9.

Although replaced by other methods for large-scale sequencing, the Sanger method of sequencing by chain termination remains the 'gold standard' and is still used widely today.

6.7.3 Electrophoresis and Reading of Sequences

Separation of the DNA fragments created in Sanger sequencing reactions is achieved by gel electrophoresis. For the standard laboratory procedure based on the original technique (small-scale, non-

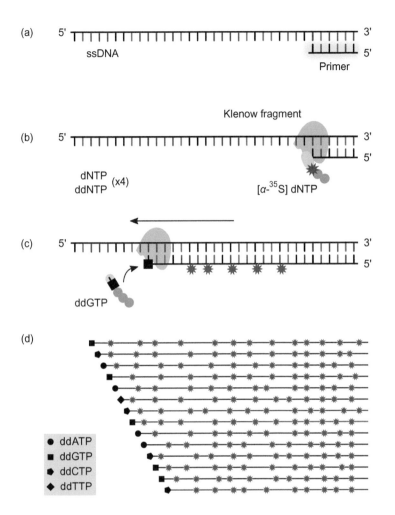

Fig. 6.9 DNA sequencing using the dideoxy chain termination (Sanger) method. (a) A primer is annealed to a single-stranded template and (b) the Klenow fragment of DNA polymerase I is used to synthesise a copy of the DNA. A radiolabelled dNTP (often [α-^{35}S] dNTP) is incorporated into the DNA. (c) Chain termination occurs when a dideoxynucleoside triphosphate (ddNTP; in this example, it is ddGTP) is incorporated. (d) A series of four reactions, each containing one ddNTP, in addition to the four dNTPs required for chain elongation, generate a set of radiolabelled nested fragments.

automated), a single gel system is used. The gels usually contain 6–20% polyacrylamide and 7 M urea, which acts as a denaturant to reduce the effects of DNA secondary structure. This is important because fragments that differ in length by only one base are being separated. The gels are very thin (0.5 mm or less) and are run at high power settings, causing them to heat up to 60–70°C. This also helps to maintain denaturing conditions. Sometimes two lots of samples are loaded onto the same gel at different times to maximise the amount of sequence information obtained.

When the gel has been run, it is removed from the apparatus and may be dried onto a paper sheet to facilitate handling. It is then exposed to X-ray film. The emissions from the radioactive label sensitise the silver halide grains, which are converted to metallic silver and look black when the film is developed and fixed. The result is known as an autoradiograph (Fig. 6.10a). Reading the autoradiograph is straightforward – the sequence is read from the smallest fragment upwards, as shown in Fig. 6.10b. Using this method, sequences of up

(a)

(b)

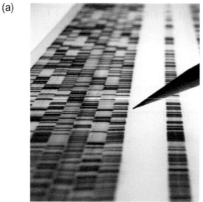

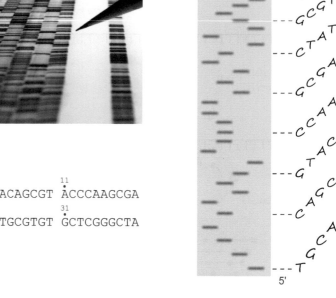

(c)

```
 1                11
TGCACAGCGT ACCCAAGCGA
21                31
CTATGCGTGT GCTCGGGCTA
```

Fig. 6.10 Reading a DNA sequence. (a) An autoradiograph of part of a sequencing gel, showing two sets of four lanes plus a control lane. (b) A representation of part of an autoradiograph. Each lane corresponds to a reaction containing one of the four ddNTPs used in the chain termination technique. The sequence is read from the bottom of the gel, each successive fragment being one nucleotide longer than the preceding one. Several hundred bases can be read from a good autoradiograph, although there may be regions that are not clear and thus need to be reprocessed. Usually any given piece of DNA is sequenced several times to ensure that the best data possible are obtained. The sequence can be presented in a number of ways – in (c), the sequence from (b) is listed in a single-stranded format as groups of ten nucleotides, using fixed-space capital typeface. *Source:* Photograph by Andrew Brookes / Image Source / Getty Images.

to several hundred bases may be read from single gels. The sequence data are then compiled (Fig. 6.10c) and studied using a computer, which can perform analyses such as translation into amino acid sequences and identification of restriction sites, regions of sequence homology and other structural motifs such as promoters and control regions.

6.7.4 Automation and Scale-Up of DNA Sequencing

One of the major advances in technology that enabled sequencing to move from single-gel laboratory-based systems up to large-scale 'production line' sequencing was the automation of many parts of

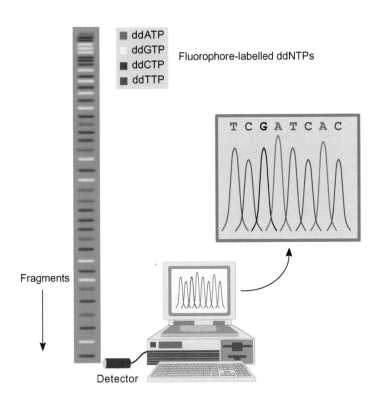

- ■ ddATP
- ☐ ddGTP
- ■ ddCTP
- ■ ddTTP

Fluorophore-labelled ddNTPs

T C G A T C A C

Fragments

Detector

Fig. 6.11 Automation of Sanger sequencing. The scaling-up required to facilitate genome sequencing was achieved by using fluorophore-labelled ddNTPs, each emitting a different colour of fluorescence on excitation. This enabled a sequencing reaction to be run on a single separating gel (slab gels, later in small-diameter capillaries), with the bases detected by continuous monitoring as fragments moved past a detector. This type of system could be scaled using standard 96-well microtitre plates and became the mainstay of the Human Genome Project. The convention is that the bases in the computerised display are coloured red (T), blue (C), green (A) and black (G) for clarity.

the process. Whereas a good laboratory scientist or technician could sequence maybe a few hundred bases per day, this was not going to solve the problem of determining *genome* sequences as opposed to *gene* sequences. Improving the technology by orders of magnitude was required. This was achieved by improvements in sample preparation and handling, with robotic processing enabling higher levels of sample throughput. In a similar way, automation of the sequencing reactions and linear continuous capillary electrophoresis techniques enabled scale-up of the sequence determination stage of the process. Thus, by the start of the 1990s, first-generation DNA sequencing machines (commonly known as 'sequencers') were available commercially, based on Sanger methodology with fluorophore labelling. An outline of this type of sequencing technology is shown in Fig. 6.11.

Sequencing technology had to be improved by orders of magnitude in order to tackle genome sequencing projects within realistic time frames.

6.8 | Next-Generation Sequencing Technologies

The development of automated first-generation sequencers enabled significant scale-up of DNA sequencing, and resulted in genome sequencing becoming established. However, this was largely driven by using more machines and bigger buildings, which was not a cost-effective way of continuing the expansion of sequencing to satisfy the demand that was stimulated by genome projects. Further development of sequencing capacity therefore required some novel thinking. This takes us into areas where the ingenuity of the scientists and engineers and the technical developments they generated are astonishing. A whole book could easily be written about DNA sequencing; in the final part of this chapter, we will try to get a sense of how the technology works by looking at the principles behind the main approaches to what became known as next-generation sequencing (NGS).

6.8.1 NGS – A Step-Change in DNA Sequencing

Instead of continuing to scale up first-generation techniques, new conceptual and technical approaches were needed to develop the next generation of sequencing technologies.

The development of NGS is a good example of the step-change/paradigm 'nudge' concept that we looked at in Section 2.3. The conceptual shift that drove the development of NGS is actually very simple – what if we could sequence lots of fragments at the same time, instead of just a relatively small number? (As it has turned out, 'lots of fragments' is a little bit of an understatement, as NGS platforms can now routinely deal with millions or billions of sequences at the same time.) This approach is common to all NGS methods, and is usually termed massively parallel or high-throughput sequencing (HTS). Many different technologies have been devised over the past 25 years, some of which have come and gone, are in development or are less commonly used than others. All have required impressive levels of intellectual innovation, and the development of fabrication technology to enable miniaturisation of the equipment needed. This reflects the technology S-curve cycle(s) of development (Fig. 6.8). The technologies that persist today have become essential parts of any sequencing laboratory, and range from large floor-standing machines with high throughput to bench-top machines (Fig. 6.12) and even portable sequencers (Fig. 6.18).

Most authors classify the various technologies involved in NGS as either second- or third-generation. Although there are still some differences of opinion as to exactly where boundaries lie, this is a useful way to make sense of the various methods, as navigating through the various technical details can be quite a task. In Section 6.8.2, we will look at the principles underpinning these methods, and then consider some of them in more detail in Section 6.8.3.

6.8.2 Principles of NGS

As fragment preparation is often simpler than for Sanger sequencing, and does not depend on the availability of cloned genomic or cDNA libraries, NGS can be used directly across a wider range of applications and samples.

The principles underpinning NGS are the same as those we looked at in Section 6.7.1 – fragment preparation, sequential presentation of the bases and a method for the detection and identification of each base. For most NGS methods, fragment preparation from the source DNA is simpler than for first-generation sequencing, where cloned fragments (or PCR-generated fragments) are required. The DNA is isolated and fragmented, usually by **sonication**, to ensure random

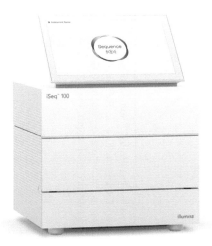

Fig. 6.12 Bench-top DNA sequencing technology. This is the iSeq100 sequencer from Illumina, which is the smallest version of their sequencing technology platform. This is ideal for small laboratories where small-scale genome sequencing might be carried out, or targeted sequencing from larger genomes. Larger machines with higher capacity are more suited to large-scale sequencing projects. *Source:* Photograph courtesy of Illumina [www.illumina.com]. Reproduced with permission.

fragmentation. Depending on the particular method, there may be other steps required, such as 'tidying up' the ends of the fragments (by filling in or trimming any overhangs, sometimes known as **polishing**), selecting fragments of a particular size range and adding **adapters** and other marker sequences to facilitate later processing steps.

In second-generation (2G) sequencing, fragments need to be amplified (using a modified form of the PCR) to provide clusters of identical sequences. This increases the strength of the output signal and also increases the fidelity of identification (sometimes termed base calling) by reducing the impact of detection errors. Sequencing is based on incorporation of labelled nucleotides, and is therefore similar to Sanger sequencing in that it is a **sequencing by synthesis (SBS)** method (a different technology called **sequencing by ligation (SBL)** was developed but is no longer used widely). Third-generation (3G) methods are able to use individual fragments to read the sequence (one method is called single-molecule real-time (SMRT) sequencing). These methods have the advantage that no amplification is needed, which helps reduce errors that can be introduced during this stage.

Other factors to assess when selecting a sequencing technology are the number of fragments that can be sequenced, the length of the 'reads', the time taken for a sequencing run, throughput per day and cost of equipment and consumables. At one point, around 15 different methods were available, although many of these have not survived the 'shake-down' that is a common feature of any new technology. A summary of some of the main surviving methods for sequencing DNA is shown in Table 6.4, using the three-generation approach commonly described in the literature.

Table 6.4 | Comparison of sequencing methodologies

Gen	Method	Company	Amp?	Reads	Read length (bases)	Accuracy (%)	Suggested search terms
1G	Sanger original	Various	N	Single[a]	600–900	99.9	sangerseq
	Sanger automated[b]	Applied Biosystems (ThermoFisher)	N	96–384	600–900	99.9	abisanger
2G	Ion Torrent	Life Technologies (ThermoFisher)	Y	2–130 million	200–600	99.6	iontorrent
	Illumina SBS	Illumina	Y	4 million–13 billion	150–250 single/paired	99.9	illuminasbs
3G	SMRT[c]	Pacific Biosciences	N	4 million	~50 k (CLR) ~20 k (CCS)	87 99.999	pacbio
	Nanopore[d]	Oxford Nanopore Technologies	N	512 or 2 675 channels per cell[e]	50–100k (Mb reads possible)	Up to 99	oxnano seq

Note: Amp = amplification of fragments. [a] A single fragment requires four lanes on the separating gel; modest scale-up is possible, but limited. [b] Some consider that automation of Sanger sequencing is the first 2G method. [c] SMRT PacBio sequencing has two modes. Continuous Consensus Sequencing (CCS or 'HiFi' reads) uses shorter repeated reads to generate very accurate data. Longer fragments can be sequenced, but accuracy is lower. [d] Some consider nanopore sequencing as the start of 4G technology (alternatively, 4G can refer to *in situ* sequencing of tissue samples). [e] Number of reads depends on user-defined read length. Useful videos of sequencing methodologies can be found using the SSTs.

6.8.3 NGS Methodologies

In this section, we will outline the basis of each of the four methods noted in Table 6.4 as 2G (Illumina **SBS** and **Ion Torrent**) and 3G (PacBio and **nanopore**) technologies. There are many excellent resources on the web that are useful in looking at NGS techniques (animations are particularly helpful), and the information provided by the companies listed in Table 6.4 is a good starting point.

The DNA (or RNA) for NGS is prepared from the sample using standard nucleic acid isolation protocols. These are often available in kit form, and may be automated to complement the technology of the sequencing application. This produces a **fragment library** that is used for sequencing. The characteristics of the library can be adjusted during isolation and fragmentation of the DNA, or during subsequent steps. Applications such as whole genome sequencing (WGS) obviously need a complete genomic library, but often a more targeted approach can be used to select or amplify a subset of the genome and generate an **amplicon library** (a useful method in the analysis of gene expression). Fragment length is one important feature that needs to be controlled or adjusted during preparation of the library, and NGS methods also need modifications to the ends of the fragments. These modifications include 'capture' sequences to facilitate processing, and primers to trigger the polymerase-based sequencing reactions. A unique identifying sequence can also be added at this point in a process known as barcoding. This is to enable efficient use of the sequencing capacity if the aim is to sequence samples from different sources (known as a **multiplex** approach). The data are generated in a single sequencing run and then separated at the data processing stage using the unique barcode sequence. An outline of fragment preparation for 2G sequencing is shown in Fig. 6.13.

For the 2G methods that use the SBS technique, the fragment library needs to be amplified to ensure that clusters of identical fragments are sequenced. There are two ways to achieve this. For the Ion Torrent method, the fragments are bound to small beads, with the DNA:bead ratio adjusted so that in most cases, a single fragment will bind to each bead. The fragment/bead combinations are then mixed with the components needed for PCR amplification and an oil that generates a water-in-oil emulsion by vigorous mixing. In essence, the beads and PCR components are formed into millions of tiny bioreactors that enable emulsion PCR (emPCR or ePCR) to amplify each fragment, producing thousands of identical copies of the fragment on each bead, and millions of different fragments in the mixture. For Illumina SBS, fragment amplification is achieved by a method known as bridge amplification. Capture oligonucleotides are immobilised on the surface of the flow cell (where the sequencing reactions will occur). Two capture oligos are used, complementary to the two adapters used to tail the DNA fragments during fragment production. The DNA is denatured and hybridised to the capture oligos, and a PCR amplification used to copy the strands. These are

As with any new commercial development, NGS saw a shake-down period as various methods became established and adopted widely. Others, despite equally innovative approaches, were less successful in what had quickly become a mature (and therefore highly competitive) market.

The methodology of NGS is astonishing and can only be appreciated fully by looking at some of the animations that are available to illustrate the various techniques.

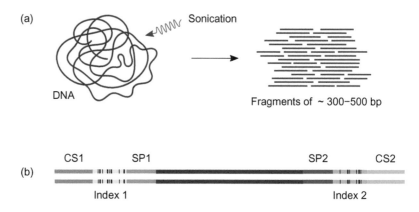

Fig. 6.13 Preparation of fragments for 2G sequencing. (a) High-molecular weight DNA is isolated and fragmented, ideally using a sequence-independent method such as sonication. A particular range of fragment sizes is required, depending on the sequencing method to be used, and can be generated by adjusting the level of fragmentation or selecting size ranges using a separation method. At this stage, the ends may be 'polished' by repairing any overhangs, and a single base extension can be added to help annealing of the adapters. (b) The fragments are modified by adding adapters to each end. These have primers for the sequencing polymerases (SP1 and SP2) and oligonucleotide capture sequences (CS1 and CS2). Short 'barcode' indexing sequences (Index 1 and Index 2) may be added if multiplex sequencing is to be carried out.

denatured, and the template washed away; the captured and copied strand 'bridges' to the second oligo sequence and another PCR round is carried out. This process is repeated many times and results in both forward and reverse strands being amplified; one of these is cleaved off and washed away, leaving a cluster of thousands of identical fragments ready for sequencing. This process is hard to envisage and sounds complex when described, but is much easier to grasp by looking at an animation – use the search term 'illumina clustering' to find one and all should be clear! Library amplification for 2G techniques is shown in Fig. 6.14.

When libraries have been prepared and amplified, the sequencing step is carried out. For both Ion Torrent and Illumina SBS methods, this involves sequential washes of reagents passed through the flow cell, so that the chemistry happens in a particular order. Addition of each base is captured, and the sequence inferred from the data stream. Although similar in concept, the detail of how this is done is different for each method.

The Ion Torrent method does not need labelled nucleotides to generate the sequence. Instead, the technique measures the release of hydrogen ions (H^+) that accompanies the incorporation of each nucleotide. The beads are washed over a flow cell that has tiny wells, each the right size to capture one bead. At the bottom of each well is a sensor that measures the change in pH that is generated by the sudden flux of H^+ ions (where the name 'Ion Torrent' comes from)

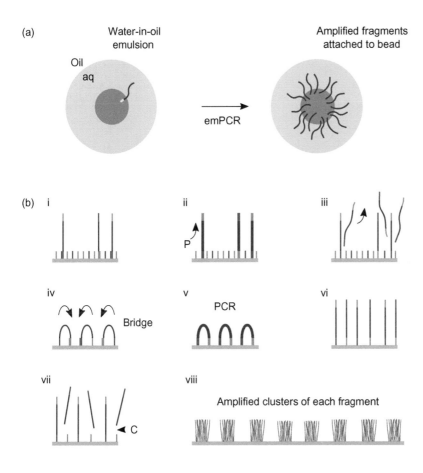

Fig. 6.14 Amplification of fragments for 2G sequencing. (a) For Ion Torrent sequencing, fragments are immobilised on small beads, with concentrations manipulated to favour binding of one fragment per bead. The bead mixture is emulsified in oil, with PCR components, to generate micelles. The fragments are amplified in the micelles to generate thousands of copies on each bead. (b) For Illumina sequencing, a 'lawn' of two types of capture oligos is generated in a flow cell by covalent binding to the inner surfaces of the cell. The fragments with the two adapter sequences are annealed to the oligos ((b)i, three fragments shown for clarity). Fragments are copied by a polymerase ((b)ii, marked P) and then denatured and the original fragments washed away ((b)iii). This leaves fragments tethered to the flow cell. The strands form bridges ((b)iv) by annealing the second adapter sequence to its complementary oligo. PCR amplification produces thousands of copies for each cluster ((b)v). Denaturation leaves two types of amplified fragments, the forward and reverse strands ((b)vi). The reverse strands are cleaved and washed away ((b)vii, marked C). The end result is a series of clusters containing amplified fragments of each sequence in the fragment library ((b)viii).

when the nucleotide is incorporated into the DNA by the polymerase. This is converted to a voltage readout, with the sensor acting as a very tiny pH meter. The flow cell is cleared by a wash step and the next nucleotide flows into the cell; by repeating the cycle of washes, the sequence can be built up by monitoring the nucleotides that generate

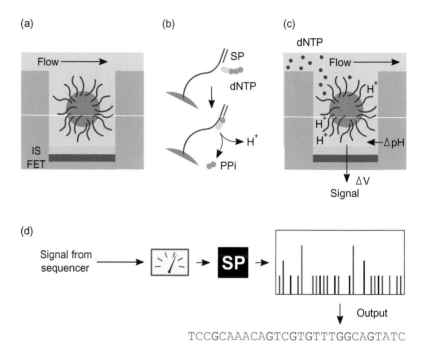

Fig. 6.15 Ion Torrent sequencing. (a) The sequencing chip has many very small microwells that can hold a single bead from the amplified fragment library. Under each well, a detector is formed from an ion-sensitive layer (IS) and an ion-sensitive field-effect transistor (ISFET, labelled as FET) to detect changes in voltage generated from change in pH of the solution in the well. (b) Ion release occurs when a dNTP is incorporated; the release of inorganic pyrophosphate (PPi) is accompanied by the release of a hydrogen ion (H^+). (c) For sequencing, each dNTP is washed across the flow cell in turn. If the base is complementary to the template base, it will be incorporated and the H^+ ions will be released (the 'ion torrent'). This sudden flood of H^+ ions changes the pH of the solution (Δ pH) and generates a change in voltage across the detector (Δ V). The flow cell is washed, and the next dNTP added to the wells. (d) The signal generated is captured and processed (SP) to generate an ionogram, which is then used to call the base identity and number of repeats. Repeated runs of the same base generate multiples of the basal level of signal for a single base incorporation; the example shows repeats as CC, AAA, TTT and GG from the left of the ionogram.

the ion flow. Repeated nucleotides in the sequence can be called as they will generate stronger signals (as more H^+ ions are released). However, longer stretches of the same base (homopolymer regions) cause difficulties as the relationship between ion release and the number of bases is not consistently linear (accurate reading of homopolymers is also a problem for other sequencing technologies and is not unique to this method). The basis of Ion Torrent sequencing is shown in Fig. 6.15.

The Illumina method uses a technology with which we are familiar from automated Sanger sequencing – the use of fluorophore-labelled precursors, excitation of the fluor and detection of the resulting emission. Methods can be based on four-colour chemistry (as with the Sanger method), although with clever design, two-colour, or even single-colour, detection can be used. The molecular biology of the system uses **reversible terminator** methodology and fluorophore cleavage to ensure that each round of incorporation generates information at one point in the sequence. Sequence is read by capturing the output from each round of excitation/fluorescence; the changing pattern of colours at each cluster location is assembled into the sequence. Single or paired-end reads (sequencing from each end of the fragment) can be carried out, increasing the range of protocols that can be used. Illumina sequencing and imaging is shown in Fig. 6.16.

The trade-off with massively parallel techniques such as Ion Torrent and Illumina is that whilst millions or billions of fragments can be read, the read lengths are generally limited to a few hundred bases. This can complicate the task of assembling whole genome sequences from these relatively short reads. Extending the fragment length and the number of wash/flow/detect cycles does not work because the system loses fidelity over time as the incorporations become less 'in step' and the **signal-to-noise ratio** decreases, making base calling more error-prone. In the development of NGS, it was therefore logical to extend the concept of sequencing lots of fragments at the same time by seeing whether fragment lengths could be extended. This led to the development of the 3G SMRT methods that can sequence fragments individually, yet retain a relatively high throughput capacity (thousands to millions rather than billions of fragments).

Fragment library preparation for 3G single-molecule sequencing is broadly similar to that for the massively parallel techniques, but fragments can be much longer. This is because the fragments are read as single molecules, and thus no amplification step is required to generate the clusters of fragments needed for 2G techniques. This means that there is no fall-off in read fidelity as a function of length (equivalent to cycles in the 2G techniques). Read length therefore essentially equates to fragment length, and the size distribution of fragments means that very long reads are possible for a percentage of molecules in the library (up to millions of bases in some cases for the nanopore method). As we saw for 2G sequencing, there are two main methods in common use that differ in the way that fragment bases are read.

The PacBio SMRT sequencing method is shown in Fig. 6.17. This uses a DNA polymerase that is fixed to the bottom of a very small well (100 nm in diameter) that is called a zero-mode waveguide (ZMW). The well's diameter is less than the wavelength of the light that is used to excite the fluorophores used in the chemistry of this technique, and the excitation is therefore restricted to the area at the

There are always limitations to any technology, and overcoming these is often part of making the next step in development. In NGS, the move to sequencing single molecules is usually taken to mark the development of 3G DNA sequencing.

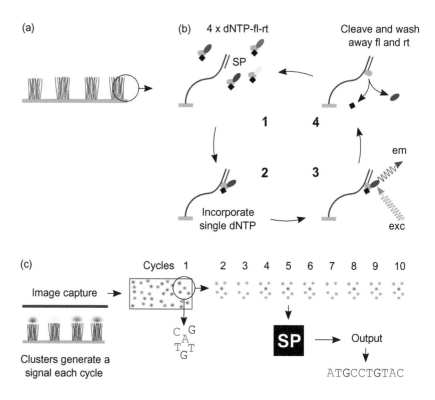

Fig. 6.16 Illumina four-channel SBS sequencing. (a) The library of amplified fragment clusters ready for sequencing. (b) A cycle of four operations is repeated to produce the sequence. A sequencing primer (SP) provides the 3′-OH group for chain growth. All four dNTPs are available, labelled with fluorophores and a reversible chain terminator (dNTP-fl-rt, black square). A single dNTP is incorporated in part 2 of the cycle, with further extension prevented by the chain terminator. Excess dNTPs are washed away, and the base called by excitation of, and emission from, the fluorophore in part 3 of the cycle. In part 4, the terminator and fluorophore are cleaved and washed away to make the 3′-OH on the incorporated nucleotide available for the next cycle. (c) A set of four photographs (one for each fluorophore) is taken at point 3 of each cycle, producing a synchronous array of different colours corresponding to the four bases. In the example shown, the circled area has six bases called as shown. The central green signal is followed through the next nine cycles (other colours greyed out for clarity). The output signals are processed (SP), and the sequence assembled.

bottom of the well, reducing background noise and improving the accuracy of the base calling. The sequencing chemistry is similar to the Illumina 2G method in that fluorophores are used and detected at incorporation, then removed to reduce noise. The main difference is that the template is read in real-time mode, rather than through a series of cyclical washes as in the 2G method, and thus sequencing is faster.

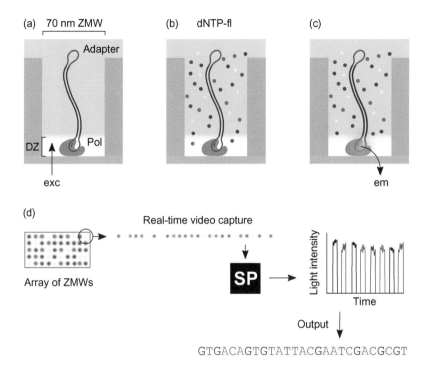

Fig. 6.17 PacBio SMRT sequencing. (a) Very small pores called zero-mode waveguides (ZMWs) are created in a supporting matrix. These are large enough to take a single DNA polymerase/DNA complex, but are smaller than the wavelength of the light used to excite the fluorophores (exc). The light therefore penetrates only a short distance into the waveguide (the detection zone, DZ). DNA polymerase is bound to a fragment that has been circularised by adding adapters, and the library added to the flow cell, so that ideally there is one DNA complex per ZMW. (b) All four fluorophore-labelled dNTPs (dNTP-fl) are added. (c) As the polymerase incorporates each dNTP, the fluorescence is emitted and captured (em). Because the ZMW detection zone is so small, the signal is mostly generated from the incorporated dNTP and is relatively free from background noise. (d) The array of ZMWs generates a continuous asynchronous pattern of fluorescence that is captured as a real-time video. Output from a single ZMW is shown; the signal is processed (SP) and used to produce the base calls that generate the sequence.

Fragments for PacBio sequencing are converted to a circular form using adapters that generate hairpin loop structures at each end of the fragment (Fig. 6.17a). Two sequencing modes can then be used. The circular consensus sequencing (CCS) mode (also known as HiFi sequencing) processes each strand of the insert fragment multiple times and therefore builds a consensus sequence, resulting in a more accurate outcome, but with fragment length limited to about 20 kb (still much longer than the short-fragment 2G methods). Alternatively, the continuous long-read (CLR) method generates a

Fig. 6.18 Portable NGS. The MinION is the smallest DNA sequencing platform (it is just over 10 cm long). The device is powered by a laptop computer and uses the nanopore sequencing method for single-molecule sequencing. It can be used in the laboratory and in clinical or field-based situations. It has also been used on the International Space Station. *Source:* Photograph courtesy of Oxford Nanopore Technologies [www .nanoporetech.com]. Reproduced with permission.

single readthrough over potentially much longer fragment lengths of around 50–100 kb, with lower accuracy.

The final method we will consider is nanopore sequencing. This technology is the most recent method to be fully developed, and is proving popular due to a number of features. One of these is size and portability – the smallest version from Oxford Nanopore Technologies (the **MinION**) is about the size of a TV remote control and plugs into a laptop computer (Fig. 6.18). The basis of nanopore sequencing is the nanopore itself. Although nanopores can be constructed using non-organic methods, those used in nanopore sequencing are usually protein molecules. These are embedded in a high-resistance polymer membrane and bathed in an electrophysiological solution. When a voltage is applied across the membrane, an ionic current flows through the nanopore and establishes the baseline or steady-state reading. When a DNA (or RNA) strand passes through the nanopore (which is of a comparable diameter to the molecule), each base disrupts the ion flow in a different way, caused by the different shapes of the bases. This disruption of the current can be read by an electronic sensor grid and interpreted to call the base sequence.

Fragment preparation is fairly straightforward, although if long read lengths are required, care is needed to ensure that long fragments are not broken by mechanical shearing. An adapter sequence that aids recognition and entry into the nanopore is added to the end of each fragment, along with a **motor protein** (this is often a polymerase or **helicase**) that controls the rate of passage of the DNA through the pore, so that the data generated can be read in real-time mode. Nanopore sequencing is outlined in Fig. 6.19.

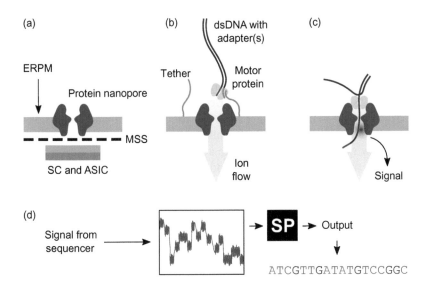

Fig. 6.19 Nanopore sequencing. (a) A protein nanopore is embedded in an electrically resistant polymer membrane (ERPM), supported by a microscaffold support (MSS) with a sensor chip (SC) and an application-specific integrated circuit (ASIC) detection system. An electrolyte solution is used as the hydration medium. (b) DNA is prepared by binding an adaptor sequence and a tether oligo (to help guide the fragment to the area of the nanopore), and the motor protein (shown in yellow). A potential difference across the membrane generates an ion current through the nanopore (green arrow). (c) The DNA fragment enters the nanopore and passes through the pore. Each base disrupts the ion flow as it passes through the pore (shown in red), altering the current and generating a signal. (d) The signal is captured and processed by a signal processor (SP) to call the bases and produce the sequence.

6.9 | Conclusion: Essential Techniques and Methods

A sound understanding of the key methods needed to work with nucleic acids is important. In this chapter, we have considered a range of essential requirements and techniques. Although not in itself a technique, we looked at how important a good laboratory setup is, not just the physical arrangement of location, space, infrastructure and so on, but also the need to have the right mix of scientists and support staff. Laboratories show a great deal of variation in terms of scale, from small university-based laboratories with few staff up to large biotechnology companies or multidisciplinary research institutes. All have a role to play in providing a stimulating environment for creative research and development.

Methods for isolating, handling and quantifying nucleic acids are essential. Very small amounts (in terms of both mass and volume) can be managed as aqueous solutions of DNA and RNA, and can be measured and dispensed using accurate small-volume automatic pipettes. Automated methods for many routine procedures have become more common to support technology-driven methodology, and a modern molecular biology laboratory is a much more 'high-tech' environment than was the case 20 years ago. Despite this more advanced technology, *enzymes* are still the key tools used to perform the various manipulations required when dealing with DNA and RNA molecules. We saw how enzymes are used to add labels to the ends of molecules (*kinases* and *transferases*), or to add label along a DNA strand (techniques such as nick translation and primer extension using polymerases).

Improved imaging technologies have enabled *fluorophores* to replace radiolabelling for many applications. The range of fluorophores available provides great flexibility in designing methods of incorporation, binding and detection, and the use of a 'click chemistry' approach using *haptens* and *bioconjugates* has helped to make the complex chemistry and biochemistry easier to use.

As well as complex chemistry, we also considered some simpler concepts. We looked at how *gel electrophoresis* is used to visualise fragments of DNA, exploiting the simplicity of negatively charged molecules moving under the influence of an electric field. Various types of gel matrix can be used in a wide range of routine applications for routine analysis of DNA fragments. However, if there is one thing that underpins almost all aspects of working with DNA, at least at some stage, it is *nucleic acid hybridisation*. Whilst very simple in concept, A·T and G·C base-pairing (or A·U in RNA) enables a whole range of procedures, from probing genomes to look for particular gene sequences, to single-base additions to DNA fragments that help to add adapters to the ends of the fragments prior to sequencing.

In the final part of the chapter, we looked at how methods for determining the sequence of a DNA molecule have evolved, which took us back into a more complex conceptual territory! This area is a good example of how technology and science are interdependent, with engineering and fabrication supporting the implementation of more complex techniques. The *technology S-curve* is one useful way to map how the methods developed over time. The first widely used rapid technique was developed in the late-1970s and became known as *Sanger sequencing*. This method, with modification and automation, enabled the human genome to be sequenced and brought 'big science' to biology for the first time. However, scaling this further required new thinking (rather than simply building bigger and bigger sequencing facilities with enormous resource implications, both in terms of staff and recurrent consumable costs). This led to the development of *NGS*. The 2G techniques favoured a *massively parallel* approach, with millions of fragments being sequenced at the same time; the expansion of scale therefore happened at the level of the DNA molecule itself through some astonishingly innovative thinking. The 3G methods have taken this further into the realm of *single-molecule sequencing*.

The various techniques used in DNA sequencing are based on a complex mix of chemistry, molecular biology and technology, and are perhaps the best biological examples of the well-known phrase from the scientist and author **Arthur C. Clarke** that 'Any sufficiently advanced technology is indistinguishable from magic.'

Further Reading

Heather J. M. and Chain, B. (2016). The sequence of sequencers: the history of sequencing DNA. *Genomics*, 107 (1), 1–8. URL [www.ncbi.nlm.nih.gov/pmc/articles/PMC4727787/]. DOI [https://doi.org/10.1016/j.ygeno.2015.11.003].

Kulski, J. K. (2015). Next-generation sequencing – an overview of the history, tools, and 'omic' applications. In: J. K. Kulski, editor. *Next Generation Sequencing – Advances, Applications and Challenges*. IntechOpen, London. URL [www.intechopen.com/chapters/49602]. DOI [https://doi.org/10.5772/61964].

Lee P. Y. *et al.* (2012). Agarose gel electrophoresis for the separation of DNA fragments. *J. Vis. Exp.*, 62, 3923. URL [www.ncbi.nlm.nih.gov/pmc/articles/PMC4846332/]. DOI [https://doi.org/10.3791/3923].

Michel, B. Y. *et al.* (2020). Probing of nucleic acid structures, dynamics, and interactions with environment-sensitive fluorescent labels. *Front. Chem.*, 8, 112. URL [www.frontiersin.org/articles/10.3389/fchem.2020.00112/full]. DOI [https://doi.org/10.3389/fchem.2020.00112].

Slatko, B. E. *et al.* (2018). Overview of next-generation sequencing technologies. *Curr. Protoc. Mol. Biol.*, 122 (1), e59. URL [www.ncbi.nlm.nih.gov/pmc/articles/PMC6020069/]. DOI [https://doi.org/10.1002/cpmb.59].

Websearch

People

What contribution did *Erwin Chargaff* make to the discovery of the structure of DNA? Can you find out why he was frustrated with the approach of Watson and Crick? What were Chargaff's views on the potential impact of genetic manipulation? As another task, look up the life story of *Frederick Sanger* to see how this modest and unassuming man changed biology forever – twice!

Places

In Section 6.1, we considered the importance of scientists communicating and working together, and the development of inter- and multidisciplinary research. Find out what the *Broad* and *Crick* institutes do – what areas they cover, their funding and management, the sort of working environment they provide and their impact.

Processes

The gel electrophoresis techniques we looked at in this chapter (SAGE, PAGE and capillary) are not very good at separating large DNA molecules (over ~15 kbp or so). Have a look at the technique called *pulsed-field gel electrophoresis (PFGE)* to see how this can help address this problem.

Reflections

What topics in this chapter have you found most challenging? Look for resources that help to illustrate the key points.

Working with nucleic acids

Working with nucleic acids

requires

- **advanced techniques**
  - such as

DNA sequencing

is

ultimate in reductionist structural information

two main divisions

first generation

now mostly

Sanger sequencing

by

automated capillary sequencing

can be scaled up for

genome sequencing

but is

limited and expensive

next generation (NGS)

classed as

2nd gen (2G)

uses

massively parallel methods

such as

Illumina Ion Torrent

producing

short sequence reads

3rd gen (3G)

uses

single DNA molecules

such as

PacBio SMRT Nanopore

producing

long sequence reads

the various protocols provide a flexible range of methods for different applications

---

requires

a research laboratory

may be

university institute hospital company private

requires

location staff infrastructure funding

involved in developing

- **science**
- **technology**

that leads to

new methods, approaches and applications

such as

- **basic techniques**

such as

isolating

by

disrupting cells

then

removing other cell components

leaving

nucleic acids

that are

precipitated

dissolved in buffer to give

an aqueous solution

that can be

handled easily using pipettes

analysing

by

measuring concentration

and

fragment sizing

using

agarose gel electrophoresis

labelling

using

enzymes

e.g.

polymerases transferases kinases

to label

3' ends 5' ends DNA strands

labels can be

radioactive

such as

^3H, ^{32}P, ^{35}S

can produce

high energy and specific activity

detected by

scintillation and fluorography

but present

safety issues

fluorescent

called

fluorophores

detected by

stimulated light emission

are

less hazardous

thus fewer

(→ safety issues)

Concept Map 6

Part 3

The Methodology of Gene Manipulation

Chapter 7 Summary

Learning Objectives

When you have completed this chapter, you will be able to:

- List the types of host cell used in gene manipulation
- Describe the features of plasmid and bacteriophage vectors
- Identify vectors for use in eukaryotic hosts
- Outline the range of methods available to get recombinant DNA into host cells, including *in vitro*, microinjection, and biolistic delivery methods

Key Words

Gene cloning, vector, host, primary host, *Escherichia coli*, prokaryotic, nucleoid, shuttle vectors, transgenic, *Saccharomyces cerevisiae*, *Aspergillus nidulans*, *Neurospora crassa*, *Chlamydomonas reinhardtii*, origin of replication, selectable marker, dispensable, conjugative, non-conjugative, conjugation, copy number, stringent, relaxed, Francisco Bolivar, deletion derivative, ampicillin, tetracycline, insertional inactivation, polylinker, multiple cloning site (MCS), α-peptide, X-gal, Addgene, genomic library, bacteriophages, Max Delbrück, phage, phages, capsid, virulent, temperate, lytic, lysogenic, *cos* site, adsorption, multiplicity of infection, prophage, packaged, bacterial lawn, plaque, plaque-forming units, replicative form, insertion, replacement, substitution, stuffer fragment, cosmids, phagemids, yeast episomal plasmids (YEps), yeast integrative (YIps), yeast replicative (YRps), yeast centromere (YCps), auxotrophic, yeast artificial chromosomes (YACs), bacterial artificial chromosomes (BACs), P1-based artificial chromosomes (PACs), human artificial chromosomes (HACs), transformation, transfection, growth transformation, Frederick Griffith, competent, transformants, packaging *in vitro*, concatemer, packaging extract, protoplasts, electroporation, microinjection, biolistic, microprojectiles, macroprojectile.

Chapter 7

Host Cells and Vectors

Construction of recombinant DNA is carried out in the test tube, and generates molecules that must be processed further to enable selection of the required sequence. To achieve this, we need to move away from a purely *in vitro* process and begin to use the properties and characteristics of living systems. Three things have to be done to isolate a gene from a collection of recombinant DNA sequences:

- The individual recombinant molecules have to be physically separated from each other
- The recombinant sequences have to be amplified to provide enough material for further analysis
- The fragment of interest has to be selected by some sort of sequence-dependent method

In this chapter, we will look at the first two of these requirements, which in essence represent the systems and techniques involved in gene cloning. Although some procedures such as PCR amplification and DNA sequencing do not require cloned fragments, standard cloning and sub-cloning procedures are still a key part of many projects. The biology of gene cloning is concerned with the selection and use of a suitable carrier molecule or vector, and a living system or host in which the vector can be propagated. In this chapter, the various types of host cell will be described first, followed by vector systems and methods for getting DNA into cells.

> Gene cloning utilises the characteristics of living systems to propagate recombinant DNA molecules; in essence, this can be considered as a form of molecular agriculture.

> Gene cloning is achieved by using a vector (carrier) to propagate the desired sequence in a host cell. Choosing the right vector/host combination is one of the critical stages of a cloning procedure.

7.1 | Types of Host Cell

The type of host cell used for a particular application will depend mainly on the purpose of the cloning procedure. If the aim is to isolate a gene for structural analysis, the requirements may call for a simple system that is easy to use. If the aim is to express the genetic information in a higher eukaryote such as a plant, a more specific system will be required. These two aims are not necessarily mutually exclusive; often a simple **primary host** is used to isolate a sequence that is then introduced into a more complex system for expression. The main types of host cell are shown in Table 7.1.

Table 7.1 Some host cells used for genetic engineering

Major group	Prokaryotic/eukaryotic	Type	Examples
Bacteria	Prokaryotic	Gram −	*Escherichia coli*
		Gram +	*Bacillus subtilis*
			Streptomyces spp.
Fungi	Eukaryotic	Microbial	*Saccharomyces cerevisiae*
		Filamentous	*Aspergillus nidulans*
Plants	Eukaryotic	Protoplasts	Various types
		Intact cells	Various types
		Whole organism	Various types
Animals	Eukaryotic	Insect cells	*Drosophila melanogaster*
		Mammalian cells	Various types
		Oocytes	Various types
		Whole organism	Various types

Note: Bacteria and fungi are generally cultured in liquid media and/or agar plates, using relatively simple growth media. Plant and animal cells may be subjected to manipulation either in tissue culture or as cells in the whole (or developing) organism. Growth requirements for these cells are often more exacting than for microbial host cells.

7.1.1 Prokaryotic Hosts

An ideal host cell should be easy to handle and propagate, should be available as a wide variety of genetically defined strains and should accept a range of vectors. The bacterium *Escherichia coli* fulfils these requirements and is used in many cloning protocols. *E. coli* has been studied in great detail, and many different strains were isolated by microbial geneticists as they investigated the genetic mechanisms of this **prokaryotic** organism. These studies provided the essential background from which genetic engineering would emerge.

E. coli is a Gram-negative bacterium with a single chromosome packed into a compact structure known as the **nucleoid**. Genome size is 4.6×10^6 base-pairs, with around 4 200 genes (Table 4.3). Transcription and translation are coupled, with no post-transcriptional modification. *E. coli* can therefore be considered as one of the simplest host cells, and its many strains are used for much of the routine gene manipulation carried out in laboratories.

In addition to *E. coli*, other bacteria may be used as hosts for gene cloning experiments, including species of *Bacillus*, *Pseudomonas* and *Streptomyces*. There are, however, some drawbacks with most of these, and it is often more sensible to perform an initial cloning procedure in *E. coli*, isolate the required sequence and then introduce the purified DNA into the target host. Many of the drawbacks can be overcome by using this approach, particularly when vectors that can function in the target host and in *E. coli* (**shuttle vectors**) are used.

7.1.2 Eukaryotic Hosts

One disadvantage of using a prokaryote such as *E. coli* as a host for cloning is that it lacks the membrane-bound nucleus (and other

The bacterium *E. coli* is the most commonly used prokaryotic host cell, with a wide variety of different strains available for particular applications.

organelles) found in eukaryotic cells. This means that eukaryotic genes may not function in *E. coli* as they would in their normal environment, particularly if RNA processing is required. If the production of a eukaryotic protein is the desired outcome of a cloning experiment, it may not be easy to ensure that a prokaryotic host produces a fully functional protein.

Eukaryotic cells range from microbes, such as yeast and algae, to cells from complex multicellular organisms, such as ourselves. Microbial cells have many of the characteristics of bacteria with regard to ease of growth and availability of mutants. Higher eukaryotes present a different set of problems to the genetic engineer, many of which require specialised solutions. Often the aim of a gene manipulation experiment in a higher plant or animal is to alter the genetic makeup of the organism by creating a transgenic, rather than to isolate a gene for further analysis or to produce large amounts of a particular protein. Transgenesis is discussed further in Chapter 16.

The yeast *Saccharomyces cerevisiae* is one of the most commonly used eukaryotic microbes in genetic engineering. It has been used for centuries in the production of bread and beer, and has been studied extensively. The organism is amenable to classical genetic analysis, and a range of mutant cell types is available. In terms of genome complexity, *S. cerevisiae* has about 3.5 times more DNA than *E. coli*. Other fungi that may be used for gene cloning experiments include *Aspergillus nidulans* and *Neurospora crassa*.

Plant and animal cells may also be used as hosts. Unicellular algae, such as *Chlamydomonas reinhardtii*, have all the advantages of microorganisms plus the structural and functional organisation of plant cells. Plant and animal cells from multicellular organisms are usually grown as cell cultures, which are much easier to manipulate than cells in a whole organism.

> Prokaryotic host cells have certain limitations when the cloning and expression of genes from eukaryotes are the aims of the procedure.

> Microbes (such as yeast) and mammalian cell lines are two examples of eukaryotic host cells that are used widely in gene manipulation.

7.2 | Plasmid Vectors for Use in *E. coli*

All vectors have certain essential features. Ideally they should be fairly small DNA molecules, to facilitate isolation and handling. There must be an **origin of replication**, so that their DNA can be copied and thus maintained in the cell population as the host organism grows and divides. It is desirable to have some sort of selectable marker that will enable the vector to be detected, and the vector must also have at least one unique restriction endonuclease recognition site, to enable DNA to be inserted during the production of recombinants. Plasmids have these features, and are used extensively as vectors in cloning experiments.

7.2.1 What Are Plasmids?

Many types of plasmid are found in nature and in bacteria and some yeasts. They are circular DNA molecules, relatively small when compared to the host cell chromosome, that are maintained mostly in an extrachromosomal state. Although plasmids are generally **dispensable** (*i.e.* not essential for cell growth and division), they often confer

Table 7.2 | Properties of some naturally occurring plasmids

Plasmid	Size (kb)	Conjugative?	Copy number	Selectable markers
ColE1	7.0	No	10–15	E1^{imm}
RSF1030	9.3	No	20–40	Ap^r
CloDF13	10.0	No	10	DF13^{imm}
RK6	42	Yes	10–40	Ap^r, Sm^r
F	103	Yes	1–2	—
R1	108	Yes	1–2	Ap^r, Cm^r, Sm^r, Sn^r, Km^r
RK2	56.4	Yes	3–5	Ap^r, Km^r, Tc^r

Note: Antibiotic abbreviations: Ap = ampicillin; Cm = chloramphenicol; Km = kanamycin; Sm = streptomycin; Sn = sulphonamide; Tc = tetracycline. E1^{imm} and DF13^{imm} represent immunity to the homologous, but not to the heterologous, colicin. Thus, plasmid ColE1 is resistant to the effects of its own colicin (E1), but not to colicin DF13. Copy number is the number of plasmids per chromosome equivalent.

Plasmids are extrachromosomal genetic elements that are not essential for bacteria to survive, but often confer advantageous traits (such as antibiotic resistance) on the host cell.

traits (such as antibiotic resistance) on the host organism, which can be a selective advantage under certain conditions. The antibiotic resistance genes encoded by plasmid DNA (pDNA) are often used in the construction of vectors for genetic engineering, as they provide a convenient means of selecting cells containing the plasmid. When plated on growth medium that contains the appropriate antibiotic, only the plasmid-containing cells will survive. This is a very simple and powerful selection method.

Plasmids can be classified into two groups, termed **conjugative** and **non-conjugative**. Conjugative plasmids can mediate their own transfer between bacteria by the process of **conjugation**. A further classification is based on the number of copies of the plasmid found in the host cell, a feature known as the **copy number**. Low copy number plasmids tend to exhibit **stringent** control of DNA replication, with replication of the pDNA closely tied to host cell chromosomal DNA replication. High copy number plasmids are termed **relaxed** plasmids, with DNA replication not dependent on host cell chromosomal DNA replication. In general terms, conjugative plasmids are large, show stringent control of DNA replication and are present in low copy numbers, whilst non-conjugative plasmids are small, show relaxed DNA replication and are present in high copy numbers. Some examples of plasmids are shown in Table 7.2.

7.2.2 Basic Cloning Plasmids

pBR322 is a very 'famous' plasmid and has all the essential requirements for a cloning vector – relatively small size, useful restriction enzyme sites, an origin of replication and antibiotic resistance genes.

For genetic engineering, naturally occurring plasmids have been extensively modified for use as vectors. In naming plasmids, p is used to designate plasmid, and this is usually followed by the initials of the worker(s) who isolated or constructed the plasmid. Numbers may be used to classify the particular isolate. An important plasmid in the history of gene manipulation is pBR322, which was developed by **Francisco Bolivar** and his colleagues. Construction of pBR322

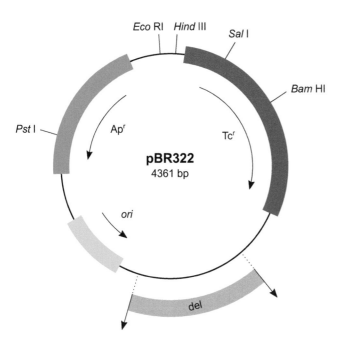

Fig. 7.1 Map of plasmid pBR322. Important regions indicated are the genes for ampicillin and tetracycline resistance (Ap^r and Tc^r) and the origin of replication (*ori*). Some unique restriction sites are given. The grey shaded region shows the two fragments that were removed from pBR322 to generate pAT153.

involved a series of manipulations to get the right pieces of DNA together, with the final result containing DNA from three sources. The plasmid has all the features of a good vector, such as low molecular weight, antibiotic resistance genes, an origin of replication, and several single-cut restriction endonuclease recognition sites. A map of pBR322 is shown in Fig. 7.1.

One variant of pBR322 is worth describing. The plasmid is pAT153, which is still widely used and has some advantages over its progenitor. It is a **deletion derivative** of pBR322 (Fig. 7.1). The plasmid was isolated by removal of two *Hae* II fragments (704 bp) of DNA from pBR322. The effect was to increase the copy number and to remove sequences necessary for mobilisation (a requirement for conjugal transfer). Thus, pAT153 is in some respects a 'better' vector than pBR322, as it is present as more copies per cell and has a greater degree of biological containment because it is not mobilisable.

New plasmid vectors can be constructed by rearranging various parts of the plasmid. This can involve addition or deletion of DNA to change the characteristics of the vector. In this way, a wide range of vectors can be constructed with relative ease.

In vectors such as pBR322, the presence of two antibiotic resistance genes (Ap^r and Tc^r) enables selection for cells harbouring the plasmid, as they will be resistant to both ampicillin and tetracycline. The unique restriction sites within the antibiotic resistance genes permit selection of cloned DNA by what is known as insertional inactivation, where the inserted DNA interrupts the coding sequence of the resistance gene and makes the cell sensitive to the antibiotic.

7.2.3 Slightly More Exotic Plasmid Vectors

Although basic plasmid vectors are still used for some applications, other vectors have been constructed with additional features such as a wider range of restriction sites for cloning DNA fragments. They may contain specific promoters for the expression of inserted genes, or

Fig. 7.2 Map of plasmid pUC18. (a) The physical map, with the positions of the origin of replication (*ori*) and the ampicillin resistance gene (Ap^r) indicated. A multiple cloning site (MCS) or polylinker lies within the *lac Z'* gene that encodes the α-peptide fragment of β-galactosidase. Its promoter (P) and operator (O) lie upstream of the MCS. (b) The polylinker region. This has multiple restriction sites immediately downstream from the promoter (P). Plasmid pUC19 is identical to pUC18, apart from the orientation of the polylinker region, which is reversed.

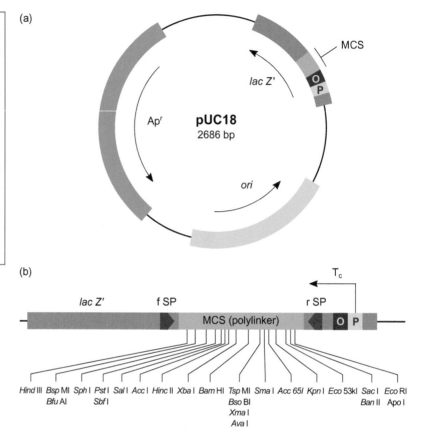

(a)

(b)

Multiple cloning sites (polylinkers) increase the flexibility of vectors by providing a range of restriction sites for cloning.

they may offer other advantages such as direct selection for recombinants. One series that has proved popular is the pUC family. These plasmids have a region that contains several unique restriction endonuclease sites in a short stretch of the DNA. This region is known as a **polylinker** or **multiple cloning site (MCS)**, and is useful because of the choice of site available for insertion of DNA fragments during recombinant production. A map of one of the early pUC vectors, with the restriction sites in its polylinker region, is shown in Fig. 7.2. In addition to the multiple cloning sites in the polylinker region, the pUC plasmids have a region of the β-galactosidase gene that codes for what is known as the **α-peptide**. This sequence contains the polylinker region, and insertion of a DNA fragment into one of the cloning sites results in a non-functional α-peptide. This forms the basis for a powerful direct recombinant screening method using the chromogenic substrate **X-gal** (outlined in Sections 10.1.1 and 10.1.2).

Today there are many different plasmids available for specific purposes, often from commercial sources. These vectors are sometimes provided as part of a 'cloning kit' that contains all the essential components to conduct a cloning experiment. This has made the technology much more accessible to a greater number of scientists, although it has not yet become totally foolproof! Some commercially available plasmids are listed in Table 7.3. An excellent resource for scientists is the not-for-profit company **Addgene**, which curates and

Vector	Features	Applications	Supplier
Table 7.3 Examples of plasmid cloning vectors			
pBR322	Ap^r Tc^r Single cloning sites	General cloning and sub-cloning in *E. coli*	Various
pAT153	Ap^r Tc^r Single cloning sites	General cloning and sub-cloning in *E. coli*	Various
pGEM-3Z	Ap^r MCS SP6/T7 promoters *Lac Z* α-peptide	General cloning and *in vitro* transcription in *E. coli* and single-stranded DNA production	Promega
pCI	Ap^r MCS T7 promoter CMV enhancer/promoter	Expression of genes in mammalian cells	Promega
pET-3	Ap^r MCS T7 promoter	Expression of genes in bacterial cells	Agilent
pCMV-Script	Neo^r Large MCS CMV enhancer/promoter	High-level expression of genes in mammalian cells Cloning of PCR products	Agilent

Note: There are hundreds of variants available from many different suppliers. A good source of information is the supplier's catalogue or website. Ap^r = ampicillin resistance; Tc^r = tetracycline resistance; Neo^r = neomycin resistance (selection using kanamycin in bacteria, G418 in mammalian cells); MCS = multiple cloning site; SP6/T7 = promoters for *in vitro* transcription; *lac Z* = β-galactosidase gene; CMV = human cytomegalovirus.

distributes over 100 000 different plasmids from around 5 000 laboratories across the world.

7.3 | Bacteriophage Vectors for Use in *E. coli*

One issue with plasmid vectors is that cloning large DNA fragments (> ~15 kbp) is not very efficient. Applications such as the generation of a **genomic library**, in which all the sequences present in the genome of an organism are represented, need vectors that can accept larger pieces of DNA. This led to the development of vectors based on **bacteriophages**, particularly lambda (λ) and M13. Although in many ways more specialised than plasmids, phage vectors fulfil essentially the same function by acting as carrier molecules for fragments of DNA.

Plasmid vectors have an upper size limit for efficient cloning, which can sometimes restrict their use where a large number of clones is required. In this case, it makes sense to clone longer DNA fragments, and a different vector system is needed.

7.3.1 What Are Bacteriophages?
In the 1940s, **Max Delbrück** and his 'phage group' laid the foundations of modern molecular biology by studying bacteriophages. These are literally 'eaters of bacteria', and are viruses that are dependent on bacteria for their propagation. The term bacteriophage is often

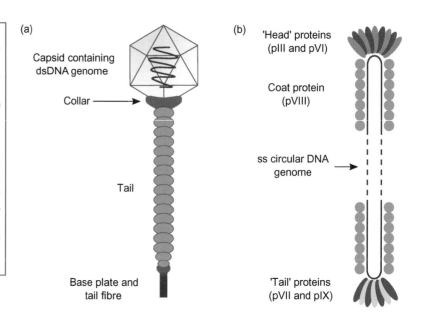

Fig. 7.3 Structure of bacteriophages λ and M13. Phage λ has a capsid or head that encloses the double-stranded DNA genome. The tail region is required for adsorption to the host cell (in some strains, there are multiple tail fibres). M13 has a simpler structure, with the single-stranded DNA genome enclosed in a protein coat. The pIII protein at the 'head' end is important in both adsorption and extrusion of the phage. Phage λ is approximately 200 nm in length, and M13 is not drawn to scale; it is a long, thin structure about 880 nm long.

Bacteriophages are essentially bacterial viruses, and usually consist of a DNA genome enclosed in a protein head (capsid). As with other viruses, they depend on the host cell for their propagation and do not exist as free-living organisms.

shortened to **phage**, and can be used to describe either one or many particles of the same type. Thus, we might say that a test tube contained one λ phage or 2×10^6 λ phage particles. The plural term **phages** is used when different types of phage are being considered; we therefore talk of T4, M13 and λ as being *phages*.

Structurally, phages fall into three main groups: (1) tailless, (2) head with tail, and (3) filamentous. The genetic material may be single or double-stranded DNA or RNA, with double-stranded DNA (dsDNA) being found most often. In tailless and tailed phages, the genome is encapsulated in an icosahedral protein shell called a **capsid** (sometimes known as a phage coat or head). In typical dsDNA phages, the genome makes up about 50 per cent of the mass of the phage particle. Thus, phages represent relatively simple systems when compared to bacteria, and for this reason, they have been used extensively as models for the study of gene expression. The structure of phages λ and M13 is shown in Fig. 7.3.

Phages may be classified as either **virulent** or **temperate**, depending on their life cycles. When a phage enters a bacterial cell, it can produce more phage and kill the cell (this is called the **lytic** growth cycle), or it can integrate into the chromosome and remain in a quiescent state without killing the cell (this is the **lysogenic** cycle). Virulent phages are those that exhibit a lytic life cycle only. Temperate phages exhibit lysogenic life cycles, but most can also undergo the lytic response when conditions are suitable. The best-known example of a temperate phage is λ, which has been the subject of intense research effort and is now more or less fully characterised in terms of its structure and mode of action.

The genome of phage λ is 48.5 kb in length, and encodes some 60 genes (Fig. 7.4). The λ genome was the first major sequencing project to be completed, and represents one of the milestones of

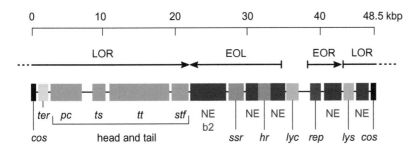

Fig. 7.4 Map of the phage λ genome. The *cos* sites enable circularisation of the genome. Three of the major transcripts are shown above the map; the late operon right (LOR), early operon left (EOL) and early operon right (EOR). Some of the genes/gene regions shown are as follows: *rep* (DNA replication), *lyc* and *lys* (lysogenic conversion and lysis), *ter* (encodes the terminase), *ssr* (site-specific recombination), *hr* (homologous recombination) and *pc*, *ts*, *tt* and *stf* (head and tail components). Non-essential regions are marked NE, with b2 being the longest contiguous non-essential region. The NE regions can be manipulated to increase vector capacity.

molecular genetics. At the ends of the linear genome, there are short (12 bp) single-stranded regions that are complementary. These act as cohesive or 'sticky' ends, which enable circularisation of the genome following infection. The region of the genome that is generated by the association of the cohesive ends is known as the *cos* site.

Phage infection begins with **adsorption**, which involves the phage particle binding to receptors on the bacterial surface (Fig. 7.5). When the phage has adsorbed, the DNA is injected into the cell and the life cycle can begin. The genome circularises and the phage initiates either the lytic or lysogenic cycle, depending on a number of factors that include the nutritional and metabolic state of the host cell and the **multiplicity of infection** (MOI – the ratio of phage to bacteria during adsorption). If the lysogenic cycle is initiated, the phage genome integrates into the host chromosome and is maintained as a prophage. It is then replicated with the chromosomal DNA and passed on to daughter cells in a stable form. If the lytic cycle is initiated, a complex sequence of transcriptional events essentially enables the phage to take over the host cell and produce multiple copies of the genome and the structural proteins. These components are then assembled or **packaged** into mature phage, which are released following lysis of the host cell.

To determine the number of bacteriophage present in a suspension, serial dilutions of the phage stock are mixed with an excess of indicator bacteria (MOI is very low) and plated onto agar using a soft agar overlay. On incubation, the bacteria will grow to form what is termed a **bacterial lawn**. Phage that grow in this lawn will cause lysis of the cells that the phage infect, and as this growth spreads, a cleared area or plaque will develop (Fig. 7.6). Plaques can then be counted to determine the number of **plaque-forming units** (PFU) in the stock suspension, and may be picked from the plate for further growth and analysis. Phage may be propagated in liquid

Bacteriophage λ has played a major role in the development of bacterial genetics and molecular biology. In addition to fundamental aspects of gene regulation, λ has been used as the basis for a wide variety of cloning vectors.

Fig. 7.5 Life cycle of bacteriophage λ. Infection occurs when a phage particle is adsorbed and the DNA is injected into the host cell. In the lytic response, the phage utilises the host cell replication mechanism and produces copies of the phage genome and structural proteins. Mature phage particles are then assembled and released by lysis of the host cell. In the lysogenic response, the phage DNA integrates into the host genome as a prophage, which can be maintained through successive cell divisions. The lytic response can be induced in a lysogenic bacterium in response to a stimulus such as ultraviolet light.

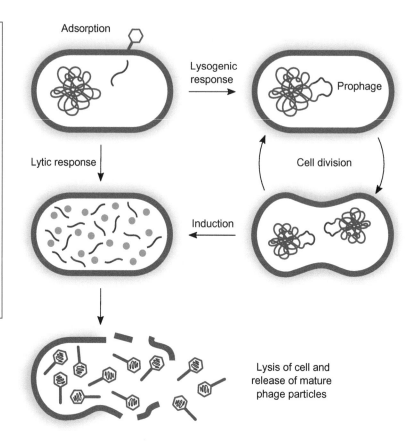

Fig. 7.6 Bacteriophage plaques. To propagate phage, particles are mixed with a strain of *E. coli* and plated using a soft agar overlay. After overnight incubation, the bacterial cells grow to form a lawn, in which regions of phage infection appear as cleared areas or plaques. The plaques are areas where lysis of the bacterial cells has occurred. *Source:* Photograph by Madboy. Used under Licence CC-BY-SA-3.0 [https://creativecommons.org/licenses/by-sa/3.0].

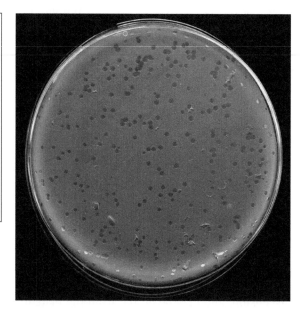

culture by infecting a growing culture of the host cell and incubating until cell lysis is complete, with the yield of phage particles depending on the MOI and the stage in the bacterial growth cycle at which infection occurs.

The filamentous phage M13 differs from λ both structurally (Fig. 7.3) and in its life cycle. The M13 genome is a single-stranded circular DNA molecule, 6 407 bp in length. The phage will infect only *E. coli* that have F-pili (thread-like protein 'appendages' found on conjugation-proficient cells). When the DNA enters the cell, it is converted to a double-stranded molecule known as the **replicative form (RF)**, which replicates until there are about 100 copies in the cell. At this point, DNA replication becomes asymmetric, and single-stranded copies of the genome are produced and extruded from the cell as M13 particles. The bacterium is not lysed and remains viable during this process, although growth and division are slower than in non-infected cells.

7.3.2 Vectors Based on Bacteriophage λ

Not all of the λ genome is essential for the phage to function, and thus there is scope for the introduction of exogenous DNA. These non-essential (NE) regions of the λ genome (shown on the map in Fig. 7.4) are largely dispensable, with region b2 being the longest, so no complex rearrangement of the genome *in vitro* is required if only this region is being removed. Wild-type λ phage have multiple recognition sites for the restriction enzymes commonly used in cloning procedures, so some manipulation was needed to generate the desired combination of restriction enzyme recognition sites. In later versions of λ-based vectors, polylinker sites were included to provide more flexibility in selecting the insert site.

One of the drawbacks of λ vectors is that the capsid places a physical constraint on the amount of DNA that can be incorporated during phage assembly. During packaging, viable phage particles can be produced from DNA that is between approximately 37 kb and 52 kb in length. Thus, a wild-type phage genome could accommodate only around 3.5 kb of cloned DNA before becoming too large for viable phage production. This limitation has been minimised by careful construction of vectors to accept pieces of DNA that are close to the theoretical maximum for the particular construct. These are of two types: (1) **insertion vectors**, and (2) **replacement or substitution vectors**. The difference between these types is outlined in Fig. 7.7. As with plasmids, there is now a bewildering variety of λ vectors available for use in cloning experiments, each with slightly different characteristics. To illustrate, we will look at an early example of each type.

The vector λgt10 (Fig. 7.8(a)) has an *Eco* RI site into which DNA can be inserted. This cuts the 43.3 kb genome and generates left and right 'arms' of 32.7 and 10.6 kb, respectively, which can accept insert DNA fragments up to approximately 7.6 kb in length. The *Eco* RI site lies within the *cI* gene (λ repressor), and this forms the basis of a selection/

The λ genome has to be modified and rearranged to produce the right combination of features for a cloning vector.

Insertion vectors, as the name suggests, are vectors into which DNA fragments are inserted without removal of part of the vector DNA.

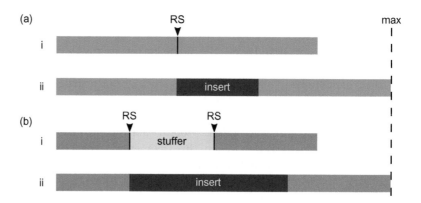

Fig. 7.7 Insertion and replacement phage vectors. (a) An insertion vector. In (a)i, the vector and a single-cut restriction site (RS) is shown. To generate a recombinant, the target DNA is inserted into this site ((a)ii). The size of fragment that may be cloned is therefore determined by the difference between the vector size and the maximum packagable fragment size (max). (b) A replacement vector. These vectors have two restriction sites (RS) that flank a region known as the stuffer fragment (shown in yellow). This section of the phage genome is replaced during cloning into this site ((b)ii). This approach enables larger fragments to be cloned than is possible with insertion vectors.

screening method based on plaque formation and morphology (see Section 10.1.2).

As insertion vectors offered limited scope for cloning large pieces of DNA, replacement vectors were developed in which a central **stuffer fragment** is removed and replaced with the insert DNA. The EMBL4

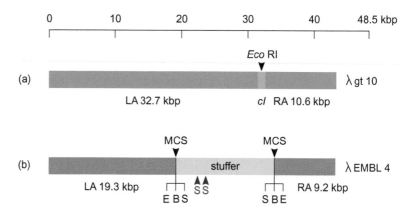

Fig. 7.8 Insertion and replacement vectors based on bacteriophage λ. Lengths of the left and right arms (LA and RA) are shown. (a) The insertion vector λgt10 has a single *Eco* RI cloning site within the *cl* gene. (b) The λ replacement vector EMBL4. The stuffer fragment is flanked by inverted polylinkers (MCS) containing the sites for *Eco* RI (E), *Bam* HI (B) and *Sal* I (S). A further two *Sal* I sites lie within the stuffer fragment; these can be useful as they help avoid re-ligation of the cleaved stuffer fragments during cloning protocols.

vector (Fig. 7.8(b)) is 41.9 kb and has a central 13.2 kb stuffer fragment flanked by inverted polylinker sequences containing sites for the restriction enzymes *Eco* RI, *Bam* HI and *Sal* I. Two *Sal* I sites are also present in the stuffer fragment. DNA may be inserted into any of the cloning sites, with the choice depending on the method of preparation of the fragments. Often a partial *Sau* 3A or *Mbo* I digest is used in the preparation of a genomic library (see Section 8.3.2), which enables insertion into the *Bam* HI site. Inserts may be excised from the recombinant by digestion with *Eco* RI. During preparation of the vector for cloning, the *Bam* HI digestion (which generates sticky ends for accepting the insert DNA) is often followed by a *Sal* I digestion. This cleaves the stuffer fragment at the two internal *Sal* I sites and also releases short *Bam* HI/*Sal* I fragments from the polylinker region. This is helpful because it prevents the stuffer fragment from re-annealing with the left and right arms and generating a viable phage that is non-recombinant.

DNA fragments between approximately 9 and 23 kb may be cloned in EMBL4, with the lower limit representing the minimum size required to form viable phage particles and the upper limit the maximum packagable size of around 52 kb. These size constraints can act as a useful initial selection method for recombinants, although an additional genetic selection mechanism can be employed with EMBL4 (the Spi$^-$ phenotype; see Section 10.1.4).

> Replacement vectors have restriction sites flanking a region of non-essential DNA that can be removed and replaced with a DNA fragment for cloning. This increases the size of insert that can be accepted by phage-based vectors.

7.3.3   Vectors Based on Bacteriophage M13

Two aspects of M13 infection make it useful as a vector. Firstly, the RF is essentially similar to a plasmid, and can be isolated and manipulated easily. Secondly, the single-stranded DNA produced during infection is useful in a range of techniques; this aspect alone made M13 immediately attractive as a potential vector.

Unlike phage λ, M13 does not have any non-essential genes. The 6 407 bp genome is also used very efficiently in that most of it is taken up by gene sequences, so that the only part available for manipulation is a 507 bp intergenic region. This has been used to construct the M13mp series of vectors, by inserting a polylinker/*lac Z* α-peptide sequence into this region (Fig. 7.9). This enables the X-gal screening system to be used for the detection of recombinants, as is the case with the pUC plasmids. When M13 is grown on a bacterial lawn, 'plaques' appear due to the reduction in growth of the host cells (which are not lysed), and these may be picked for further analysis.

A second disadvantage of M13 vectors is that they do not function efficiently when long DNA fragments are inserted into the vector. Although in theory there should be no limit to the size of clonable fragments, as the capsid structure is determined by the genome size (unlike phage λ), there is a marked reduction in cloning efficiency with fragments longer than about 1.5 kb. In practice, this was not a major problem, as the main use of the early M13 vectors was in sub-cloning small DNA fragments for sequencing.

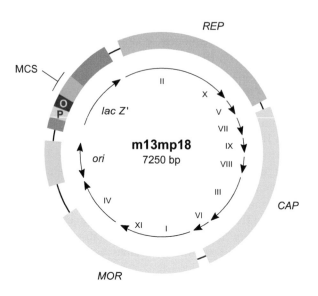

Fig. 7.9 Map of the filamentous phage vector M13mp18. The double-stranded replicative form is shown. DNA replication is bidirectional from the origin of replication (*ori*). The polylinker and β-galactosidase α-peptide regions (MCS and *lac Z'*) are derived from pUC18 (Fig. 7.2). Transcription of phage genes proceeds clockwise, with the transcripts indicated in Roman numerals. Genes in the *REP* region encode proteins that are important for DNA replication. The *CAP* and *MOR* regions contain genes that specify functions associated with capsid formation and phage morphogenesis, respectively. The vector M13mp19 is identical, except for the orientation of the polylinker region.

7.4 | Other Vectors

So far, we have concentrated on what we might call 'basic' plasmid and bacteriophage vectors for use in *E. coli* hosts. Although these vectors still represent a major part of the technology of gene manipulation, there has been continued development of more sophisticated bacterial vectors, as well as vectors for other organisms. One driving force in this has been the need to clone and analyse ever larger pieces of DNA, as the emphasis in molecular biology has shifted towards the analysis of genomes, rather than simply genes in isolation. In addition, the commercial development of integrated approaches to cloning procedures has required new vectors. Such kit-based products are often marketed as 'cloning technologies'. Cloning kits have been a successful addition to the gene manipulator's armoury, often reducing the time taken to achieve a particular outcome. In this section, we will look at the features of some additional bacterial vectors, and some vectors for use in other organisms.

As cloning methodology developed, the limitations of plasmid and phage-based vectors placed some constraints on what could be achieved, and there was an increasing need for vectors with increased cloning capacities.

7.4.1 Hybrid Plasmid/Phage Vectors

One feature of phage vectors is that the technique of packaging *in vitro* (see Section 7.5.2) is sequence-independent, apart from the requirement of having the *cos* sites separated by DNA of packagable size (37–52 kb).

This has been exploited in the construction of vectors that are made up of plasmid sequences joined to the *cos* sites of phage λ. These vectors are known as cosmids. They are small (4–6 kb) and can therefore accommodate cloned DNA fragments up to some 47 kb in length. As they lack phage genes, they behave as plasmids when introduced into *E. coli* by the packaging/infection mechanism of λ. Cosmid vectors therefore offer an apparently ideal system – a highly efficient and specific method of introducing the recombinant DNA into the host cell, and a cloning capacity some twofold greater than the best λ replacement vectors. However, they are not without disadvantages, and often the gains of using cosmids instead of phage vectors are offset by losses in terms of ease of use and further processing of cloned sequences.

> Cosmids and phagemids incorporate some of the characteristics of both plasmid and phage-based vectors.

Hybrid plasmid/phage vectors in which the phage functions are expressed and utilised in some way are known as phagemids. One series of vectors is the λZAP family. Features of these include the potential to excise cloned DNA fragments *in vivo* as part of a plasmid. This automatic excision is useful in that it removes the need to sub-clone inserts from λ into plasmid vectors for further manipulations. Other variants of hybrid plasmid/phage vectors have been developed to overcome the size limitation of the M13 cloning system, and can be used for applications such as DNA sequencing and the production of probes for hybridisation. These vectors are essentially plasmids, which contain the f1 phage origin of replication. When cells containing the plasmid are superinfected with helper phage, they produce single-stranded copies of the plasmid DNA and secrete these into the medium as M13-like particles. Some examples of vectors based on bacteriophages are listed in Table 7.4.

7.4.2 Vectors for Use in Eukaryotic Cells

The unicellular yeast *S. cerevisiae* is one of the model organisms that has played a major part in molecular biology, and a range of vectors for use in yeast cells have been developed. **Yeast episomal plasmids** (YEps) are based on the naturally occurring yeast 2 μm plasmid, and may replicate autonomously or integrate into a chromosomal location. **Yeast integrative plasmids** (YIps) are designed to integrate into the chromosome in a similar way to YEp plasmids, and **yeast replicative plasmids** (YRps) remain as independent plasmids and do not integrate. Plasmids that contain sequences from around the centromeric region of chromosomes are known as **yeast centromere plasmids** (YCps), and these behave essentially as mini-chromosomes. Vector-borne selection mechanisms are often nutritional, using **auxotrophic** host cells (mutants deficient in a biosynthetic pathway) that are complemented by the marker gene on the vector. Yeast vectors are often shuttle vectors that can be used in *E. coli.*

> A range of plasmid-based vectors for the yeast *Saccharomyces cerevisiae* was developed from the naturally occurring yeast 2 μm plasmid.

When manipulating multicellular eukaryotes such as plants and animals, the aims are usually either to express cloned genes in plant and animal cells in tissue culture or to generate a transgenic organism. The latter aim in particular can pose technical difficulties, as the recombinant DNA has to be introduced very early in development or in some sort of vector that will promote the spread of the recombinant sequence throughout the organism.

Table 7.4 Some commercially available bacteriophage-based vectors

Vector	Features	Applications	Supplier
λGT11	λ insertion vector Insert capacity 7.2 kbp *lac Z* gene	cDNA library construction Expression of inserts	Various
λEMBL3/4	λ replacement vectors Insert capacity 9–23 kbp	Genomic library construction	Various
λZAP II	λ-based insertion vector Insert capacity 12 kbp *In vivo* excision of inserts Expression of inserts	cDNA library construction Also genomic/PCR cloning	Agilent
λFIX II	λ-based replacement vector, capacity 9–23 kbp Spi$^+$/P2 selection system T3 and T7 promoters	Genomic library construction	Agilent
pBluescript II	Phagemid vector Produces single-stranded DNA	*In vitro* transcription DNA sequencing	Agilent

SuperCos 1	Cosmid vector with Ap^r and
	  Neo^r markers, plus T3 and T7
	  promoters
	Capacity 30–42 kbp
		Generation of cosmid-based
		  genomic DNA libraries
		T3/T7 promoters allow end-
		  specific transcripts to be
		  generated for chromosome
		  walking techniques
			Agilent

Note: As with plasmid vectors, there are many variants available from a range of different suppliers. A good source of information is the supplier's catalogue or website. Ap^r = ampicillin resistance; Neo^r = neomycin resistance (selection using kanamycin in bacteria, G418 in mammalian cells); T3/7 = promoters for *in vitro* transcription; *lac Z,* = β-galactosidase gene; SV40 = promoter for expression in eukaryotic cells.

> Vectors for use in plant and animal cells have properties that enable them to function in these cell types; they are often more specialised than the basic primary cloning vectors such as λ.

Vectors used for plant and animal cells may be introduced into cells directly by techniques such as those described in Section 7.5.3, or they may have a biological entry mechanism if based on viruses or other infectious agents such as *Agrobacterium*. Examples of the types of system that have been used to develop vectors for plant and animal cells are shown in Table 7.5. The use of specific vectors is described further in Part 4 of this book when considering some of the applications of gene manipulation technology in eukaryotes.

7.4.3 Artificial Chromosomes

The development of vectors for cloning very large pieces of DNA was essential to enable large genome sequencing projects to proceed at a reasonable rate, although genomes such as *S. cerevisiae* have been sequenced mainly by using cosmid vectors to construct the genomic libraries required. However, even insert sizes of 40–50 kb are too small to cope with projects such as the Human Genome Project. The development of **yeast artificial chromosomes (YACs)** (Fig. 7.10) has

Table 7.5 Some vectors for plant and animal cells

Cell type	Vector type	Genome	Examples
Plant cells	Plasmid	DNA	Ti plasmids of *Agrobacterium tumefaciens* are well-established vector systems
	Viral	DNA	Cauliflower mosaic virus Geminiviruses
		RNA	Tobacco rattle virus
Animal cells	Plasmid	DNA	Many vectors available commercially, often using sequences/promoters/origins of replication from Simian virus 40 (SV40) and/or cytomegalovirus (CMV)
	Viral	DNA	Baculoviruses for insect cells Papilloma viruses Adenovirus SV40 Vaccinia virus
	Viral	RNA	Retroviruses
	Transposon	DNA	P elements in *Drosophila melanogaster*

Note: Certain aspects of 'vectorology' are more advanced than others. Often a particular type of system will be developed as a vector, becoming extensively modified in the process as different versions are generated. In areas where vector technology is not well developed, techniques such as PCR can sometimes be used to overcome any limitations of the system.

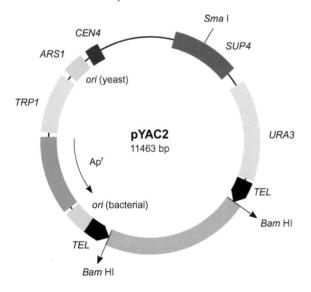

Fig. 7.10 Map of the yeast artificial chromosome vector pYAC2. This has two origins of replication, one bacterial (*ori*, from pBR322) and one yeast (autonomously replicating sequence, *ARS1*). The ampicillin resistance gene (Ap^r) enables selection in bacterial hosts, with the *TRP1* and *URA3* genes providing selectable markers for yeast host cells. The cloning site in the *SUP4* gene enables a form of insertional inactivation based on a red-coloured mutant cell; when the *SUP4* gene is active, it suppresses the 'red' mutation and colourless colonies are produced. When the *SUP4* gene is inactivated by cloning a fragment into the *Sma* I site, the suppressor no longer functions and colonies with inserts are red. Chromosomal elements are the centromere (*CEN4*) and telomere regions (*TEL*). The telomeres are separated by a fragment flanked by *Bam* HI sites that is removed to generate the left and right arms of the chromosomal construct.

enabled DNA fragments in the megabase range to be cloned, although there can be problems with insert instability. YACs are the most sophisticated yeast vectors and, to date, represent the largest capacity vectors available. They have centromeric and telomeric regions, and the recombinant DNA is therefore maintained essentially as a yeast chromosome.

A further development of artificial chromosome technology came with the construction of **bacterial artificial chromosomes** (BACs). These are based on the F plasmid, which is much larger than the standard plasmid cloning vectors, and therefore offers the potential of cloning larger fragments. BACs can accept inserts of around 300 kb, and many of the instability problems of YACs can be avoided by using the bacterial version. Much of the sequencing of the human genome was accomplished by using a library of BAC recombinants. Vectors based on the phage P1 have also been developed, both as phage vectors and also as **P1-based artificial chromosomes** (PACs). **Human artificial chromosomes** (HACs) have also been constructed, with applications in transgenesis and the production of therapeutics.

> Artificial chromosomes are elegantly simple vectors that mimic the natural construction of chromosomal DNA, with telomeres, a centromere and an origin of replication, in addition to features designed for ease of use, such as selectable markers.

7.5 | Getting DNA into Cells

Following manipulation of vector and insert DNAs in the test tube, the recombinant DNA has to be put into the host cell for propagation. The efficiency of this step is often a crucial factor in determining the success of a given cloning experiment, particularly when a large number of recombinants is required. Efficiency may not be an issue where a sub-cloning procedure is used, as the target sequence is likely to have been cloned (or perhaps generated using the PCR; see Chapter 9) and available in relatively large amounts. The methods available for getting recombinant DNA into cells depend on the type of host/vector system, and range from very simple procedures to more complicated and esoteric ones.

7.5.1 Transformation and Transfection

The techniques of transformation and transfection represent the simplest methods available for getting recombinant DNA into cells. In the context of cloning in *E. coli*, transformation refers to the uptake of plasmid DNA, and transfection to the uptake of phage DNA. For animal cells, transformation is used in a different context (when talking about **growth transformation** when a cell becomes a cancerous cell), so transfection has become the preferred term for DNA uptake.

> Transformation of *Escherichia coli* cells with recombinant plasmid DNA is one of the classic techniques of gene manipulation.

Transformation in bacteria was first demonstrated in 1928 by **Frederick Griffith** in his famous 'transforming principle' experiment that paved the way for the discoveries that eventually showed that genes were made of DNA. However, not all bacteria can be transformed easily, and it was not until the early 1970s that transformation was demonstrated in *E. coli*, the mainstay of gene manipulation technology. To effect transformation of *E. coli*, the cells need to be made **competent** by soaking in an ice-cold solution of calcium chloride. The recombinant DNA is mixed with the cells and incubated on ice for 20–30 minutes, and the mixture given a brief heat shock (2 minutes at 42°C is often used) to enable the DNA to enter the cells. The transformed cells are usually incubated in a nutrient broth at 37°C for 60–90 minutes to enable the plasmids to become established and permit phenotypic expression of their traits. The cells can then be plated out onto selective media for propagation of cells harbouring the plasmid.

Transformation is an inefficient process in that only a very small percentage of competent cells become transformed, representing uptake of a fraction of the plasmid DNA. Despite this, up to 10^9 transformed cells (**transformants**) per microgram of input DNA can be achieved, although transformation frequencies of around 10^6 or 10^7 transformants per microgram are more usual in practice. Many biological supply companies offer a variety of competent cell strains that have been pretreated to produce high transformation frequencies. Transfection is a similar process to transformation, the difference being that phage DNA is used instead of plasmid DNA. It is again a somewhat inefficient process, and it has largely been superseded by **packaging** *in vitro* for applications that require the introduction of phage DNA into *E. coli* cells.

> Transformation efficiency is often a limiting factor in using the technique, and this may be critical if the aim of the procedure is to prepare a representative clone bank.

7.5.2 Packaging Phage DNA *In Vitro*

During the lytic cycle of phage λ, the phage DNA is replicated to form what is known as a **concatemer**. This is a very long DNA molecule composed of many copies of the λ genome, linked together by the *cos* sites (Fig. 7.11(a)). When the phage particles are assembled, the DNA is packaged into the capsid, which involves cutting the DNA at the *cos* sites using a phage-encoded endonuclease. Mature phage particles are thus produced, ready to be released on lysis of the cell and capable of infecting other cells. This process normally occurs *in vivo*, with the particular functions being encoded by the phage genes. However, it is possible to carry out the process in the test tube, which enables recombinant DNA that is generated as a concatemer to be packaged into phage particles. Two strains of bacteria are used to produce a lysate known as a **packaging extract**. Each strain is a mutant in one function of phage morphogenesis, so that the packaging extracts will not work in isolation. When the two are mixed with the concatemeric recombinant DNA, phage particles are produced. These can be used to

> Packaging recombinant bacteriophage DNA *in vitro* mimics the normal process that occurs during phage maturation and assembly, and has proven to be a very useful method for the construction of genomic libraries.

Fig. 7.11 Packaging bacteriophage DNA *in vitro*. (a) A concatemeric DNA molecule composed of wild-type phage DNA (λ+). The individual genomes are joined at the *cos* sites. (b) Recombinant genomes (λ rec) are shown being packaged *in vitro*. A mixed lysate from two bacterial strains supplies the head and tail precursors and the proteins required for the formation of mature λ particles. On adding this mixture to the concatemer, the DNA is cleaved at the *cos* sites (arrowed) and packaged into individual phage particles, each containing a recombinant genome.

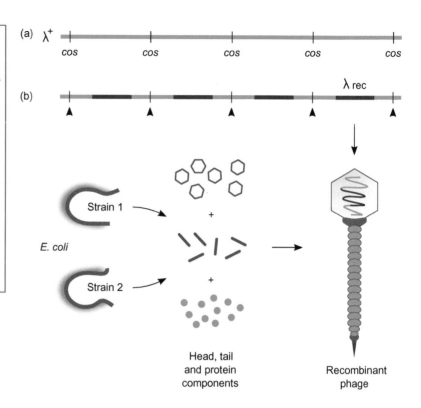

infect *E. coli* cells, which are plated out to obtain plaques. The process is summarised in Fig. 7.11(b).

7.5.3 Alternative DNA Delivery Methods

The methods available for introducing DNA into bacterial cells are not easily transferred to other cell types. The phage-specific packaging system is not available for other systems, and transformation by normal methods may prove impossible or too inefficient to be a realistic option. However, there are alternative methods for introducing DNA into cells. Often these are more technically demanding and less efficient than the bacterial methods, but reliable results have been achieved in many situations where there appeared to be no hope of getting recombinant DNA molecules into the desired cell.

Most of the problems associated with getting DNA into non-bacterial cells have involved plant cells, which have a rigid cell wall that is a barrier to DNA uptake. This can be alleviated by the production of **protoplasts**, in which the cell wall is removed enzymatically. The protoplasts can then be transformed using a technique such as **electroporation**, where an electrical pulse is used to create transient holes in the cell membrane, through which the DNA can pass. The protoplasts can then be regenerated. In addition to this application,

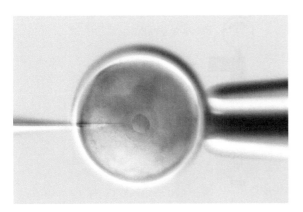

Fig. 7.12 Microinjection of DNA into an animal cell. The cell is held on a glass capillary (on the right of the photograph) by gentle suction. The microinjection needle is made by drawing a heated glass capillary out to a fine point. Using a micromanipulator (a mechanical device for fine control of the capillary), the needle has been inserted into the cell (on the left of the photograph), where its tip can be seen approaching the cell nucleus. *Source:* Photograph by Chad Baker/Thomas Northcut / Photodisc / Getty Images.

protoplasts also have an important role to play in the generation of hybrid plant cells by fusing protoplasts together.

An alternative to transformation procedures is to introduce DNA into the cell by some sort of physical method. One way of doing this is to use a very fine needle and inject the DNA directly into the nucleus. This technique is called microinjection (Fig. 7.12), and has been used successfully with both plant and animal cells. The cell is held on a glass tube by mild suction, and the needle is used to pierce the membrane. The technique requires a mechanical micromanipulator and a microscope, and plenty of practice! One obvious disadvantage is that this technique is labour-intensive and not suitable for primary cloning procedures where large numbers of recombinants are required. However, in certain specialised cases, it is an excellent method for targeting DNA delivery once a suitable recombinant has been identified and developed to the point where microinjection is feasible.

An ingenious and somewhat bizarre development has proved extremely useful in transformation of plant cells. The technique, which is called biolistic DNA delivery, involves literally shooting DNA into cells (Fig. 7.13). The DNA is used to coat microscopic tungsten particles known as **microprojectiles**, which are then accelerated on a **macroprojectile** using compressed gas. At one end of the 'gun', there is a small aperture that stops the macroprojectile but allows the microprojectiles to pass through. When directed at cells, these microprojectiles carry the DNA into the cell and, in some cases, stable transformation will occur.

> Introducing recombinant DNA into eukaryotic cells can involve biological methods or one of a range of techniques such as electroporation, microinjection or biolistics.

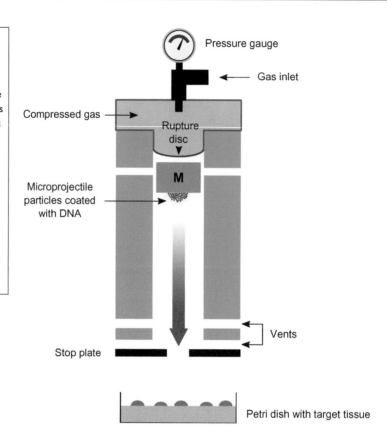

Fig. 7.13 Biolistic apparatus. The DNA is coated onto microprojectiles, usually tungsten or gold particles. These are accelerated by the macroprojectile on firing the 'gene gun'. Helium gas is used to build up pressure that is released when a rupture disc reaches its breaking point. The sudden pressure release accelerates the macroprojectile (shown as the red arrow). At the stop plate, the macroprojectile is retained in the chamber and the microprojectiles carry on to the target tissue. Desktop and hand-held versions of the apparatus are available.

7.6 | Conclusion: From *In Vitro* to *In Vivo*

In this chapter, we began to focus on using living systems to work with recombinant DNA, moving on from the *in vitro* techniques that generate recombinant DNA into the area of *host cells* and *vectors*. A range of types of cell can be used as hosts for cloning, ranging from prokaryotic and eukaryotic microbes (*E. coli* and *S. cerevisiae*) to plant and animal cells from multicellular organisms. Many cloning systems are now available from commercial suppliers, in which the components have been optimised for use in a particular application.

We saw that the first vectors were based on naturally occurring *plasmids* and *bacteriophages* of *E. coli*, modified and manipulated to generate vectors with essential features such as an *origin of replication*, unique *cloning sites* (or perhaps *polylinkers*) and *selectable markers* to help identify cells containing the vector or recombinant. We looked at some examples of basic cloning vectors, and some more highly developed ones with additional functionality such as *cosmids* and *phagemids*.

The cloning capacity of plasmid and phage-based vectors is limited by a number of constraints, and this led to the development of vectors that could clone much larger fragments. These *artificial chromosomes* were developed for yeast (*YACs*) and bacteria (*BACs*), and provided the means to move into much larger projects such as mapping and sequencing genomes.

Plant and animal cells pose different challenges when compared to microbial cells. They tend to be used to produce *recombinant-derived proteins*, or in the generation of *transgenic* organisms, rather than in primary cloning procedures. Despite the change in emphasis, the key requirements for vector and host are similar to microbial systems.

The final stage in cloning using a vector/host combination is to get the recombinant DNA into the host cell. For bacterial systems, *transformation* (plasmids) and *transfection* (phages) are direct methods, with DNA taken up by the cell in a passive process. In contrast, *packaging* recombinant phage particles *in vitro* is a good example of adapting a biological process to develop a more efficient process for getting DNA into cells. Mechanical methods can also be used for plant and animal cells, such as *electroporation*, *microinjection* and *biolistic* techniques.

Further Reading

Casjens, S. R. and Hendrix, R. W. (2015). Bacteriophage lambda: early pioneer and still relevant. *Virology*, 479–80, 310–30. URL [www.ncbi.nlm.nih.gov/pmc/articles/PMC4424060/]. DOI [https://doi.org/10.1016/j.virol.2015.02.010].

Godiska, R. *et al.* (2009). Linear plasmid vector for cloning of repetitive or unstable sequences in Escherichia coli. *Nucleic Acids Res.*, 38 (6), e88. URL [https://academic.oup.com/nar/article/38/6/e88/3112557]. DOI [https://doi.org/10.1093/nar/gkp1181].

Lacroix, B. and Citovsky, V. (2020). Biolistic approach for transient gene expression studies in plants. *Methods Mol. Biol.*, 2124, 125–39. URL [www.ncbi.nlm.nih.gov/pmc/articles/PMC7217558/]. DOI [https://doi.org/10.1007/978-1-0716-0356-7_6].

Nora, L. C. *et al.* (2019). The art of vector engineering: towards the construction of next-generation genetic tools. *Microb. Biotechnol.*, 12 (1), 125–47. URL [www.ncbi.nlm.nih.gov/pmc/articles/PMC6302727/]. DOI [https://doi.org/10.1111/1751-7915.13318].

Osoegawa, K. *et al.* (2001). A bacterial artificial chromosome library for sequencing the complete human genome. *Genome Res.*, 11 (3), 483–96. URL [www.ncbi.nlm.nih.gov/pmc/articles/PMC311044/]. DOI [https://doi.org/10.1101/gr.169601].

Websearch

People

Have a look at the biographies of two scientists whose work led to discoveries that underpinned the development of cloning using bacteria as host cells. Firstly, what contribution did *Frederick Griffith* make in the late-1920s? Secondly, the 'phage group' established by *Max Delbrück* made many fundamental discoveries from its inception in the early-1940s. Have a look at his life story.

Places

Herbert Boyer's laboratory at the *University of California San Francisco (UCSF)* was one of the key centres for the early development of gene cloning. Which vector was developed there in the late-1970s? Construct a lineage map of this vector and its progenitors and descendants.

Processes

We considered a few techniques for getting DNA into cells but did not look at all of them in depth. Fill in some of the details about the techniques of *electroporation*, *microelectroporation* and *nanotransfection*.

Reflections

What topics in this chapter have you found most challenging? Look for resources that help to illustrate the key points.

Host cells and vectors

methods for / **need** / **living systems**

host cells

may be

prokaryotic — such as — *E. coli*

eukaryotic

- **microbial** — such as — *S. cerevisiae*
- **multicellular** — *animals* / *plants*

E. coli and *S. cerevisiae* that are **easy to handle** and have (relatively) **simple genomes** thus used for **primary cloning gene expression** that facilitate → **gene expression transgenesis genome editing**

animals / plants that are **harder to grow** and have **more complex genomes** thus used for **gene expression transgenesis genome editing**

vectors

may be

plasmids — e.g. — **pBR322 pAT153 pUC18/19 pGEM** — can be — **shuttle vectors**

phages — e.g. — **lambda M13** — can be — **insertion or replacement**

artificial chromosomes — e.g.
- **BACs** — for cloning in — **bacteria**
- **YACs** — for cloning in — **yeast**

with the **largest cloning capacity**

or **hybrids** — such as — **cosmids phagemids**

all vectors have unique or multiple cloning sites, origin of replication, selectable markers, and may have promoters for gene expression

lowest ——— vector cloning capacity ——— *highest*

getting DNA into cells

three ways

mechanical — e.g. — **electroporation microinjection biolistics** — that require — **complex apparatus**

DNA uptake — by — **transformation transfection** — that are — **passive processes**

biological — e.g. — **packaging recombinant phage DNA *in vitro***

that generate → **bacterial, yeast, plant or animal cells containing rDNA**

can be used to → **isolate specific genes / map and sequence genomes / express recombinant proteins / generate transgenic organisms**

Concept Map 7

Chapter 8 Summary

Learning Objectives

When you have completed this chapter, you will be able to:

- Outline the range of strategies that may be employed to clone cDNA and genomic DNA
- Describe the rationale and methodology involved in generating genomic and cDNA libraries
- Discuss how the PCR, bioinformatics and gene synthesis have contributed to the development of cloning technology
- Explain how DNA fragments can be joined to vectors
- Review the range of cloning methods available, including restriction-free methods for sub-cloning and fragment assembly

Key Words

Polymerase chain reaction, genome sequence data, clone banks, genomic libraries, cDNA libraries, transcriptome, exome, bioinformatics, *in silico*, reference genome, reverse transcriptase, complementary DNA, copy DNA, cDNA, RNase H, polymorphism, somatostatin, *Mycoplasma mycoides*, concentration of termini, bimolecular recombinants, concatemers, homopolymer tailing, annealing, linkers, adapters, oligomers, methylase, gene bank, TA cloning, topoisomerase I, TOPO cloning, ligation-independent cloning (LIC), sequence- and ligation-independent cloning (SLIC), T4 DNA polymerase, primary library, amplify (library), promoter, strong promoter, weak promoter, upstream, consensus sequence, Pribnow box, TATA box, Hogness box, CAAT box, inducible, repressible, constitutive, Gateway cloning, site-specific recombination, entry vector, donor vector, destination vectors, Golden Gate cloning, type IIS restriction enzymes, Daniel Gibson, isothermal reaction, T5 DNA exonuclease.

Chapter 8

Cloning Strategies

The four stages in a gene cloning protocol are: *generation* of DNA fragments, *joining* the fragments to a suitable vector to produce recombinant DNA, *propagation* of the recombinants in a host cell and identification and *selection* of the required sequence (Fig. 1.1). In this chapter, we will look at some of the strategies that can be used to achieve the first two of these stages, focusing mainly on cloning eukaryotic sequences. The choice of the vector/host system for cloning will predetermine the method(s) available for getting the recombinant DNA into the host cell using one of the methods we looked at in Section 7.5. We will also see how the polymerase chain reaction (PCR, discussed in detail in Chapter 9) and the availability of near-complete genome sequence data for many organisms have changed the strategic approach to cloning. Selection of cloned sequences is covered in Chapter 10.

8.1 | Which Approach Is Best?

The complexity of any cloning strategy depends on a number of factors. Some of these are common to all protocols, and some are specific to particular applications. A strategy to isolate and sequence a relatively small DNA fragment from *Escherichia coli* will be different than a strategy to produce a recombinant protein in a transgenic eukaryotic organism. Each project will therefore be unique, and will present its own set of problems that have to be addressed by choosing the appropriate path through the maze of possibilities. Some key questions to consider are:

To complete the four key stages and achieve a successful outcome to a cloning procedure, a clear overall strategy is required at the outset.

- What is the overall aim of the procedure?
- How much is already known about the target sequence(s)?
- Is a cloning procedure required, or is there an alternative?
- What is the most appropriate source material – genomic DNA or mRNA?
- Is an appropriate vector/host system available?

Although these questions remain relevant to the design of an effective experimental system, the answers to them have changed as cloning technology has become more sophisticated. Let's add some historical context across the genomics timeline that we considered in Fig. 1.4.

8.1.1 Cloning in the Pre-genomic Era

The aim of a cloning procedure in the pre-genomic era was usually fairly well defined. In most cases, very little was known about the target gene sequence; interest might have been stimulated by the desire to better understand a particular disease or to investigate the structure and function of a protein. Some limited protein sequence data might have been available to help design molecular probes, and comparative information from other organisms could have been helpful if this was available. From the late-1970s, DNA sequencing began to have an impact, but this took a number of years to become established widely. Thus, most cloning applications in the pre-genomic period involved the generation of a collection of cloned sequences from which the target gene could be isolated. These became known as **clone banks, genomic libraries** or **cDNA libraries**.

Although the DNA represents the complete genome of the organism, it will contain non-coding DNA such as introns, control regions and repetitive sequences. This can sometimes present problems, particularly if the genome is large and the aim is to isolate a single-copy gene. However, if the primary interest is in the control of gene expression, it is obviously necessary to isolate the control sequences, so genomic DNA is needed.

Messenger RNA has two advantages over genomic DNA as a source material. Firstly, it is part of the **transcriptome**, *i.e.* the genetic information that is being expressed by the particular cell type from which it is prepared (mRNA is often defined as the **exome**, which is the set of exons in a genome; we will look at this in more detail in Chapter 13). This can be a powerful preliminary selection mechanism, as not all the genomic DNA will be represented in the mRNA population. Also, if the gene of interest is highly expressed, this may be reflected in the abundance of its mRNA, and this can make isolation of the clones easier. A second advantage of mRNA is that by definition, it represents the coding sequence of the gene, with any introns having been removed during RNA processing. Thus, production of recombinant protein is much more straightforward if a clone of the mRNA is available.

Some of the factors involved in genomic or cDNA library production are shown in Fig. 8.1. Very often the protocols and methods had to be developed and optimised in each laboratory, but as progress was made, the reliability of the methods became more settled and laboratories would often develop expertise in a particular area and were happy to share this with others. In many cases, individual clones or entire libraries might be shared with collaborators – this was sometimes known as the 'cloning by phoning' approach.

8.1.2 Cloning (or Not) in the Genomic and Post-genomic Eras

In addition to general improvements in the methodology, two things had a significant impact on cloning strategies from the early-1980s. Firstly, the use of DNA sequencing became firmly established, and the move towards genome-level sequencing projects meant that the amount of sequence data generated increased exponentially. Secondly, the development of the PCR in 1983 was a step-change event

Generation of libraries of cloned fragments was a key part of the pre-genomic cloning landscape.

Each cloning procedure is unique and presents a set of challenges that must be overcome by selection of the appropriate techniques – this is often made easier by using an optimised kit from a single supplier.

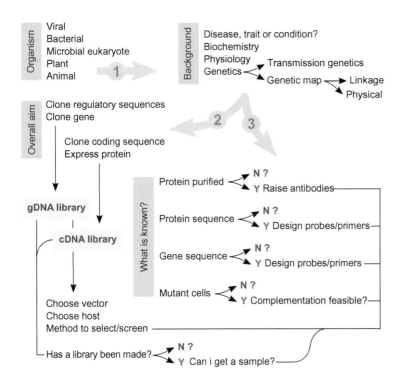

Organism
Viral
Bacterial
Microbial eukaryote
Plant
Animal

Background
Disease, trait or condition?
Biochemistry
Physiology
Genetics — Transmission genetics
Genetic map → Linkage
Physical

Overall aim
Clone regulatory sequences
Clone gene
Clone coding sequence
Express protein

gDNA library

cDNA library

What is known?
Protein purified — N ?
Y Raise antibodies
Protein sequence — N ?
Y Design probes/primers
Gene sequence — N ?
Y Design probes/primers
Mutant cells — N ?
Y Complementation feasible?

Choose vector
Choose host
Method to select/screen

Has a library been made? — N ?
Y Can i get a sample?

Fig. 8.1 Factors to consider when devising a cloning strategy in the pre-genomic era. The main areas to consider when designing a 'traditional' cloning strategy are the characteristics of the organism and what background knowledge exists (large arrow 1). From this, the overall aim of the protocol will determine whether genomic DNA (gDNA library) or mRNA (cDNA library) is the better option (large arrow 2). The detail of what is already known about the gene/protein needs to be considered (large arrow 3). This will inform both the cloning strategy and also the method that can be used to screen or select clones from the library. Where a 'no' appears (N ?), the options are either to choose a different route or to carry out investigation into the area that is lacking information, e.g. if the protein has not been purified, it may be necessary to do this to generate sequence data to design probes or primers (perhaps for DNA sequencing and/or PCR, which were becoming established in the 1980s). Although many of these questions and protocols have been superseded by more recent advances, if a novel organism or trait is being investigated, it may still be necessary to design a procedure to generate a clone bank and isolate a gene sequence directly.

that, in some cases, meant that a cloning protocol was not required. Development of the internet and World Wide Web, and the emergence of **bioinformatics** as a discipline (see Chapter 11), also played a critical part in moving to a much more informatics-based approach to DNA analysis. Yet another step-change occurred when next-generation sequencing methods were developed (see Section 6.8.1) and high-throughput genome sequencing became widely available and cost-effective. Over a fairly short time frame, the key question has therefore been 'what is already known about the sequence?' as the starting point for devising a cloning protocol.

Although genomic and cDNA libraries remain an important part of cloning technology, the range of informatics tools available today

Many cloning procedures can be designed and evaluated *in silico* using an informatics approach. This 'virtual cloning' can help avoid some of the early-stage potential errors that are likely to occur in a laboratory-based *de novo* approach.

means that a lot of the work in devising a cloning strategy is done in the computer (sometimes called the '*in silico*' approach; see Chapter 11). Most of the model organisms across the range of taxonomic levels have been sequenced to reference genome standard, opening up the possibility of evaluating and comparing genes and other sequences without having to carry out early-stage laboratory-based experiments. Sequences can be manipulated and the potential effects evaluated, primers for PCR or DNA sequencing can be devised, probe sequences can be generated and sub-clones constructed and assessed, so that when experiments are eventually carried out, a lot of the potential for error has already been eliminated. The strategic possibilities are therefore very different than was the case previously. Whilst certainly more complex, procedures are more efficient and precise, leading to effective use of resources. The post-genomic cloning strategy landscape is shown in Fig. 8.2.

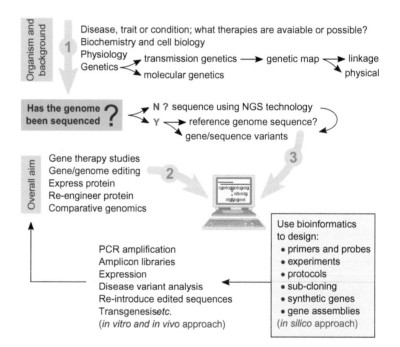

Fig. 8.2 Devising an investigative and experimental strategy in the post-genomic era. As in Fig. 8.1, questions about background knowledge remain relevant, but the key question now is 'has the genome been sequenced?' (large arrow 1). If 'Yes', check if a reference (consensus) version is available and/or how many variants have been assembled to enable comparative genomics. If the sequence is not available, it is now within reach of most laboratories to sequence (or have this done commercially) the genome of the organism in question using high-throughput next-generation sequencing technology. The aims of the investigation (large arrow 2) and the genome sequence data (large arrow 3) can now be brought together to carry out computer-based planning of experimental protocols. In many cases, this may not require primary cloning methods such as library construction, as the informatics approach can proceed straight to gene identification, modification and testing, and design of sub-cloning protocols to enable the *in vitro* and *in vivo* approaches to be much more efficient and focused.

8.2 | Generating DNA Fragments for Cloning

8.2.1 Genomic DNA

A representative genomic library should contain the entire genome of an organism as a set of overlapping cloned fragments. The fragments should ideally be generated by a sequence-independent procedure (such as sonication; see Fig. 6.13), thus avoiding bias towards any particular sequence. However, often a partial restriction digestion is used, with infrequent cutting of the DNA approximating to random fragmentation. This has the advantage of generating cohesive ends if an appropriate enzyme is used, so insertion into the vector is straightforward. A size selection procedure is sometimes used to produce a suitable range of fragment sizes for cloning. Finally, the cloned fragments should be maintained in a stable form with no misrepresentation of sequences due to recombination or differential replication of the cloned DNAs during propagation of the recombinants. Whilst these criteria may seem rather demanding, the systems available for producing genomic libraries enable these requirements to be met more or less completely.

> A genomic library is a rich resource for the scientist, as it represents the entire genome of an organism and (at least in theory) should contain all the genes and their control sequences.

The number of clones required for a genomic library depends on a variety of factors, the most obvious one being the size of the genome. The type of vector to be used also has to be considered, which will determine the size of fragments that can be cloned. In practice, library size can be calculated quite simply on the basis of the probability of a particular sequence being represented in the library. There is a formula that takes account of all the factors and produces a 'number of clones' value:

$$N = \ln(1 - P)/\ln(1 - a/b)$$

where N is the number of clones required, P is the desired probability of a particular sequence being represented (typically set at 0.95 or 0.99), a is the average size of the DNA fragments to be cloned and b is the size of the genome (expressed in the same units as a). Some genome library sizes are shown in Table 8.1.

> By linking genome size and the desired probability of a gene being isolated, the size of genomic library required can be calculated. However, this should only be used as a guide, as there is always the possibility of any given sequence not being present in the primary library.

8.2.2 Synthesis of cDNA

When cloning from mRNA, it has to be converted into DNA before being inserted into a suitable vector. This is achieved by using the enzyme reverse transcriptase (RTase; see Section 5.2.2) to produce complementary DNA (also known as **copy DNA** or **cDNA**). The classic early method of cDNA synthesis utilises the poly(A) tract at the 3' end of the mRNA to bind an oligo(dT) primer, which provides the 3'-OH group required by RTase (Fig. 8.3). Given the four dNTPs and suitable conditions, RTase will synthesise a copy of the mRNA to produce a cDNA·mRNA hybrid. The mRNA can be removed by alkaline hydrolysis and the single-stranded (ss) cDNA converted into double-stranded (ds) cDNA by using a DNA polymerase. In this second strand synthesis, the priming 3'-OH is generated by short hairpin loop regions that form at the end of the ss cDNA. After the second strand

> The first step in cloning from mRNA is to convert the mRNA into double-stranded complementary DNA (cDNA; also known as copy DNA) by using the enzymes reverse transcriptase and DNA polymerase.

Table 8.1 | Genomic library sizes for various organisms

Organism	Genome size (kbp)	Number of clones N, for P = 0.95 Insert size (kbp)		
		20 kbp	45 kbp	300 kbp
Escherichia coli (bacterium)	4.60×10^3	687	305	44
Saccharomyces cerevisiae (yeast)	1.22×10^4	1825	810	120
Caenorhabditis elegans (nematode worm)	1.00×10^5	1.5×10^4	6654	997
Arabidopsis thaliana (simple higher plant)	1.35×10^5	2.0×10^4	8984	1346
Drosophila melanogaster (fruit fly)	1.44×10^5	2.2×10^4	9583	1436
Mus musculus (mouse)	2.70×10^6	4.0×10^5	1.8×10^5	2.7×10^4
Homo sapiens (human)	3.1×10^6	4.6×10^5	2.1×10^5	3.1×10^4
Nicotiana tabacum (tobacco)	4.6×10^6	6.9×10^5	3.1×10^5	4.6×10^4
Triticum aestivum (hexaploid wheat)	1.7×10^7	2.5×10^6	1.1×10^6	1.7×10^5

Note: The number of clones (N) required for a probability (P) of 95% that a given sequence is represented in a genomic library is shown for a range of different organisms. Approximate genome sizes of the organisms are given (haploid genome size, if appropriate). Three values of N are shown, for 20 kb inserts (λ replacement vector size), 45 kb inserts (cosmid vectors) and 300 kbp inserts (BACs). Values less than 10 000 are written without exponents. The values should be considered as minimum estimates, as strictly speaking, the calculation assumes that: (1) the genome size is known accurately, (2) the DNA is fragmented in a totally random manner for cloning, (3) each recombinant DNA molecule will give rise to a single clone, (4) the efficiency of cloning is the same for all fragments, and (5) diploid organisms are homozygous for all loci. These assumptions are usually not all valid for a given experiment.

Synthesis of full-length cDNA from an mRNA template can be difficult to achieve, and is affected by a number of factors that need to be controlled carefully.

synthesis, the ds cDNA can be trimmed by S1 nuclease to give a flush-ended molecule, which can then be cloned in a suitable vector.

A range of problems can occur when synthesising cDNA using the method outlined above, two of which can be difficult to address. Firstly, synthesis of full-length cDNAs may be inefficient, particularly if the mRNA is relatively long. This is a serious problem if expression of the cDNA is required, as it may not contain all the coding sequence of the gene. Such inefficient full-length cDNA synthesis also means that the 3′ regions of the mRNA tend to be over-represented in the cDNA population. Secondly, problems can arise from the use of S1 nuclease, which may remove some important 5′ sequences when it is used to trim the ds cDNA.

Other methods for cDNA synthesis overcome the above problems to a great extent, and the original method is now rarely used. Using random oligonucleotide primers, as well as the oligo(dT) primer, can help generate longer cDNA transcripts, as the primers will bind at other sites within the mRNA, in addition to the 3′ poly(A) sequence. Another adaptation involves the use of oligo(dC) tailing to permit oligo(dG)-primed second-strand cDNA synthesis. The dC tails are added to the 3′ termini of the cDNA using terminal transferase. As this functions most efficiently on accessible 3′ termini, the tailing

(a)

(b)

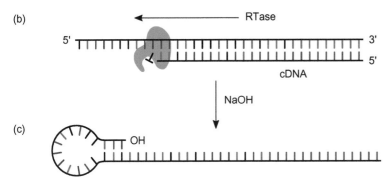

(c)

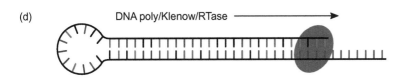

(d)

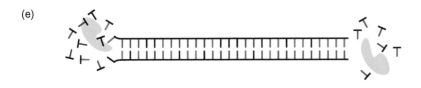

(e)

Fig. 8.3 Synthesis of cDNA. Poly(A)$^+$ RNA (mRNA) is used as the starting material. (a) A short oligo(dT) primer is annealed to the poly(A) tail on the mRNA. (b) The first cDNA strand is synthesised from the 3'-OH group using RTase. (c) The mRNA is removed by alkaline hydrolysis to give a single-stranded cDNA molecule. This has a short double-stranded hairpin loop structure that provides a 3'-OH terminus for second strand synthesis. (d) A DNA polymerase (e.g. T4 DNA polymerase, Klenow fragment or RTase) synthesises the second strand of the cDNA. (e) The double-stranded cDNA is trimmed with S1 nuclease to produce a blunt-ended double-stranded cDNA molecule. An alternative to the alkaline hydrolysis step is to use RNase H, which creates nicks in the mRNA strand of the mRNA·cDNA hybrid. By using this in conjunction with DNA polymerase I, a nick translation reaction synthesises the second cDNA strand.

reaction favours full-length cDNAs in which the 3' terminus is not 'hidden' by the mRNA template. The method also obviates the need for S1 nuclease treatment, and thus full-length cDNA production is enhanced further. Another method for synthesising the second cDNA strand is to use **RNase H** to create nicks in the mRNA template, which generates short RNA primers that can be used in what is effectively a nick translation reaction similar to that used with ds DNA (Fig. 6.3).

8.2.3   PCR Fragments

The PCR, which we will look at in more detail in Chapter 9, can be used to amplify defined DNA fragments if enough sequence information is available to design the pair of primers needed. In fact, if the aim of the procedure is to make enough of the target sequence for further analysis, cloning the sequence may not be required if the PCR product can be used directly. If cloning the target is the aim, using the PCR to amplify the sequence can make things much more straightforward than using a traditional clone library approach. The PCR approach essentially includes a selection and screening process that isolates the target sequence, which can be cloned into a plasmid for further use. Sub-cloning into a different vector can also be carried out using the PCR instead of the restriction digestion method.

> For some application, the amplification of fragments using the PCR removes the need for cloning using traditional methods.

Although using PCR to generate fragments for cloning is a major part of modern DNA technology, there are some constraints that can affect the use of the technique. If reliable sequence information is not available, it can be difficult to design suitable primers that are stable during the procedure. The PCR process can also generate errors in the sequence, which are of course amplified with each subsequent cycle. This can be a problem if the aim is to compare sequences across a range of samples, perhaps when investigating disease gene poly-morphism, where accuracy is important. To help avoid this problem, modern PCR technology very often uses high-fidelity polymerases that are less prone to error. Another potential issue is that not all sequences in a genome may be amplified with the same efficiency, and thus may be difficult to isolate.

> Base errors and differential efficiency of amplification can be problematic when using PCR-based methods. The impact of any errors increases with the number of amplification cycles.

8.2.4   Synthetic Biology: Making Genes from Scratch

In the early days of gene cloning, short stretches of DNA could be synthesised if required. This was often useful to fill in gaps in cloned sequences, or to generate oligonucleotides for use as probes or PCR primers. In 1972, a yeast tRNA gene was the first to be synthesised chemically *de novo*, and in 1977, the first expression of an artificial protein-coding gene was achieved with the short peptide hormone **somatostatin**.

> Synthesising gene sequences *de novo* is an option that was once restricted to short sequences. The technology is now much more widely available and can generate entire genes and even small genomes.

Despite the early advances, large-scale gene synthesis remained challenging. Improvements in oligonucleotide synthesis and assembly procedures from the mid-1990s led to advances in the technology, in some ways mirroring the development of high-throughput sequencing methods. However, the cost of gene synthesis did not fall to the same extent as sequencing, and the methods remain comparatively expensive. Despite the challenges, gene synthesis has developed into a widely available enabling technology, supported by a number of companies that produce genes on a commercial basis. It offers an additional way to generate fragments of DNA and will have a significant part to play in gene technology into the future. A major landmark in large-scale synthetic biology was the synthesis of an entire genome of 1.08 Mbp for the bacterium *Mycoplasma mycoides* in 2010, and the first synthetic yeast chromosome was made in 2014.

8.3 | Inserting DNA Fragments into Vectors

Having generated a population of DNA fragments for cloning, the next step is to insert these into the vector. The choice of vector/host combination will have largely determined the method used to prepare the fragments, and will have been considered at the start of the process as part of the overall cloning strategy. Fragments may be blunt-ended, have cohesive ends from restriction digestion or have end overhangs of various types due to processing. In some cases, fragment ends will have been filled in or 'polished' (as we saw for next-generation sequencing in Section 6.8.2). The methods available for joining fragment and vector to generate recombinant DNA are outlined below.

Joining the prepared DNA fragments to the vector is a critical step in cloning, with a number of possible outcomes, as well as the intended bimolecular or concatemeric recombinant DNA molecule.

8.3.1 Ligation of Blunt/Cohesive-Ended Fragments

The principles underpinning the ligation of fragments and vectors is similar for blunt- and cohesive-ended fragments. In essence, the ends of the fragments need to 'meet' to enable DNA ligase to form the phosphodiester bond (see Section 5.3), and the relative concentrations of fragment and vector molecules are important in optimising this interaction. The effective concentration of DNA molecules in cloning reactions is usually expressed as the **concentration of termini** (as these are the key functional elements in the reaction), often expressed as 'picomoles of ends', which can seem a rather strange terminology to the uninitiated.

The concept of the concentration of 'ends' of DNA molecules is important to ensure optimal conditions for the formation of recombinants.

The conditions for ligation of ends must be chosen carefully. In theory, when vector DNA and the target DNA fragments are mixed, there are several possible outcomes. For plasmid vectors, the desired result is for one target DNA molecule to join with one vector molecule, thus generating **bimolecular recombinants**, each with one insert. However, if concentrations are not optimal, the insert or vector DNAs may self-ligate to produce circular molecules, or the insert/vector DNAs may form **concatemers** instead of bimolecular recombinants. In practice, the vector is often treated with a phosphatase (either BAP or CIP; see Section 5.2.3) to prevent self-ligation, and the concentrations of the vector and insert DNAs are chosen to favour the production of recombinants. Where phage-based vectors are used, the aim is to generate recombinant concatemeric DNA that can act as the substrate for packaging *in vitro*, as we saw in Section 7.5.2 and Fig. 7.11.

When joining blunt-ended fragments, the main disadvantage is that it is an inefficient process, as there is no specific intermolecular

Blunt-end ligation is less efficient than cohesive-end ligation, as there are no 'sticky ends' to hold the molecules together, and the reaction is therefore dependent on random intermolecular associations between the ends of the target and vector molecules.

During recombinant formation, homopolymer tailing can generate more stable intermolecular associations than cohesive-end ligation, due to the longer sequence of paired bases in the A·T or G·C hybrids.

association to hold the DNA strands together whilst DNA ligase generates the phosphodiester linkages required to produce the recombinant DNA. Thus, high concentrations of the participating DNAs must be used, so that the chances of two ends coming together are increased. Another potential disadvantage of blunt-end ligation is that it may not generate restriction enzyme recognition sequences at the cloning site, thus hampering excision of the insert from the recombinant. This is usually not a major problem, as most vectors now have a polylinker (multiple cloning site). The set of restriction sites clustered around the target cloning site means that DNA inserted by blunt-end ligation can be excised by using one of the restriction sites in the cluster if the original site is not regenerated.

Ligation of fragments with cohesive ends is much more efficient than blunt-end ligation, as the molecules are held together by the complementary sequences generated by the restriction digest. We have already seen how this works in Chapter 5 (Table 5.1; Figs. 5.2 and 5.3).

8.3.2 Homopolymer Tailing

The use of **homopolymer** **tailing** has proved to be a popular and effective means of cloning cDNA. In this technique, terminal transferase is used to add homopolymers of dA, dT, dG or dC to a DNA molecule. Any pair of complementary bases can be used to add the appropriate homopolymer to the target and vector molecules, so we can have either dA·dT or dG·dC tailing. Homopolymers have two main advantages over other methods of joining DNAs from different sources. Firstly, they provide longer regions for **annealing** DNAs together than, for example, cohesive termini produced by restriction enzyme digestion. This means that ligation need not be carried out *in vitro*, as the DNA·vector hybrid is stable enough to survive introduction into the host cell, where it is ligated *in vivo*. A second advantage is specificity. As the vector and insert DNAs have different but complementary 'tails', there is little chance of self-annealing, and the generation of bimolecular recombinants is favoured over a wider range of effective concentrations that is the case for other annealing/ligation reactions.

An example of the use of homopolymer tailing for cloning cDNA in a plasmid vector is shown in Fig. 8.4. The vector is cut with *Pst* I and tailed by terminal transferase in the presence of dGTP. This produces dG tails. The insert DNA is tailed with dC in a similar way, and the two can then be annealed. This regenerates the original *Pst* I site, which enables the insert to be cut out of the recombinant using this enzyme.

8.3.3 Linkers and Adapters

Linkers and **adapters** (or adaptors) are two variations on a theme that we have already come across in Section 6.8 when discussing fragment

(a)

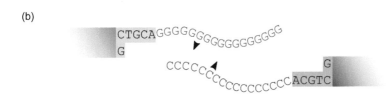

(b)

Fig. 8.4 Homopolymer tailing. (a) The vector and insert DNA are cut with *Pst* I, which generates protruding 3′-OH termini. (b) In this example, the vector is tailed with dG residues, using terminal transferase, which sequentially adds dG residues to the terminus. The insert DNA is tailed with dC in a similar way. The dC and dG tails are complementary and the insert can therefore be annealed with the vector to generate a recombinant. The *Pst* I sites can be regenerated at the ends of the insert DNA following transformation and repair by the host, which facilitates excision using the enzyme.

preparation for next-generation sequencing methods. Both linkers and adapters add sequence-specific features to the ends of DNA molecules, in the form of oligonucleotide constructs. Whilst the principles are similar, linkers and adapters differ slightly in the way that they achieve the aim of modifying the ends of molecules to generate cohesive termini on blunt-ended fragments.

Linkers are self-complementary oligomers that contain a recognition sequence for a particular restriction enzyme. One such sequence would be 5′-CCGAATTCGG-3′ that, in double-stranded form, will contain the recognition sequence for *Eco* RI (GAATTC). Linkers are synthesised chemically, and can be added to DNA by blunt-end ligation (Fig. 8.5(a)). When they have been added, the DNA/linker is cleaved with the linker-specific restriction enzyme, thus generating sticky ends prior to cloning. Although many copies of the linker are added to the ends of the fragments, digestion removes all but the one that is attached to the target fragment. One problem is if the DNA contains sites for the restriction enzyme used to cleave the linker, but this can be overcome by using a **methylase** to protect any internal recognition sites from digestion by the enzyme.

A second approach to cloning by addition of sequences to the ends of DNA molecules involves the use of adapters (Fig. 8.5(b)). These are synthesised as single-stranded oligomers that are different lengths, with partial complementary regions. When annealed together, an adapter with one blunt end and one sticky end is produced, which

Modifying the ends of DNA fragments using linkers and adapters is a useful technique to add cohesive ends to blunt-ended fragments.

(a)

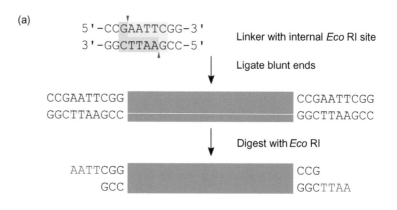

(b)

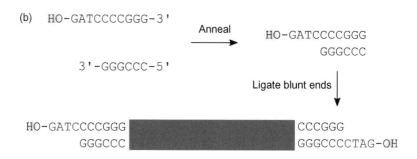

Fig. 8.5 Linkers and adapters. (a) Linkers. The 10-mer 5'-CCGAATTCGG-3' contains the recognition site for *Eco* RI. Linkers are added to blunt-ended DNA using DNA ligase. Only one linker addition to each end of the DNA is shown here (multiple copies are usually added, but the digestion with *Eco* RI leaves a single linker at each end). The construct is then digested with *Eco* RI, which cleaves the linker to generate protruding 5' termini. (b) Use of adaptors. In this example, a *Bam* HI adaptor (5'-GATCCCCGGG-3') is annealed with a single-stranded *Hpa* II linker (3'-GGGCCC-5') to generate a double-stranded sticky-ended molecule. This is added to blunt-ended DNA using DNA ligase. The DNA therefore gains protruding 5' termini without the need for digestion with a restriction enzyme. The 5' terminus of the adaptor can be dephosphorylated (leaving a 5'-OH) to prevent self-ligation.

can be added to the DNA to provide sticky-end cloning without digestion of the linkers. The 5' terminus of the overlap can be treated with phosphatase to generate a 5'-OH that prevents self-ligation. Adapters therefore have the advantage of avoiding the two issues that can affect linkers, *i.e.* internal restriction sites in the target DNA and multiple copies of the adapter being added to the fragment.

Linkers and adapters provide a flexible and efficient way of altering the ends of DNA molecules to incorporate a specific sequence-dependent function. They can be used in a variety of ways with many different types of vector. Although blunt-end ligation is most often used to add the linker to the DNA fragments, the low efficiency of this technique is not a problem as the concentration can be adjusted to give a high linker-to-fragment ratio that favours linker addition.

8.3.4 Other Methods for Joining DNA Fragments and Vectors

As cloning technology progressed, new methods to insert DNA into vectors were developed. Increased use of the PCR in analysis and cloning, and the advent of high-throughput sequencing and gene synthesis, meant that the sub-cloning technologies (that support the primary cloning methods of **gene bank** construction) had to be improved. In this section, we will look at four examples of how the technology has progressed to permit assembly of constructs without the need for restriction digestion.

One feature of the *Taq* polymerase used in the PCR reaction (see Section 9.2.1) is that it usually generates a single base extension, most often an adenine, at the 3′ ends of the amplified fragment – effectively the smallest version of a cohesive end. Thus, if a vector is generated with a single T at its 3′ ends, the construct can be generated by the pairing between the A and T residues (Fig. 8.6). Although not as thermodynamically stable as the four-base overhangs commonly found in cohesive ends generated by restriction enzymes, the inter-action is sufficient to hold the fragment and vector together to enable ligase to seal the nicks. This type of cloning is usually called TA cloning.

A novel way of cloning blunt-ended fragments uses the properties of the enzyme **topoisomerase I**, which acts as both an endonuclease and a ligase. The enzyme facilitates unwinding of DNA *in vivo* to relieve torsional stresses generated during DNA replication.

Even single base-pair overlaps can function as cohesive ends for some sub-cloning protocols.

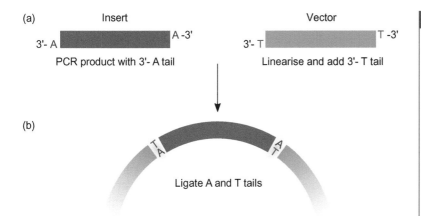

(a)   Insert   Vector

3′- A   A -3′   3′- T   T -3′

PCR product with 3′- A tail   Linearise and add 3′- T tail

(b)

Ligate A and T tails

Fig. 8.6 The TA cloning method. (a) Insert DNA generated by PCR using *Taq* polymerase usually has single-base (A) overhangs at the 3′ ends. The vector is tailed with a single T residue (often ddTTP is used to prevent sequential additions by terminal transferase, as the dideoxy form lacks the 3′-OH group necessary for elongation). (b) The complementary A·T base-pair is essentially a very short cohesive end that can be ligated to seal the nicks and generate a recombinant.

Fig. 8.7 TOPO blunt-end cloning. Topoisomerase I acts as both a nuclease and a ligase. (a) DNA for cloning is prepared as PCR-generated blunt-ended fragments. The vector, usually pre-prepared, is cut at the topoisomerase I recognition site. Topoisomerase remains covalently attached to the 3′ termini. (b) Mixing the fragments and vector results in the blunt-ended fragments being ligated into the vector by topoisomerase (acting as a ligase), which is then released.

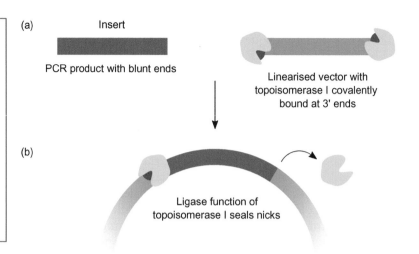

(a) Insert

PCR product with blunt ends

Linearised vector with topoisomerase I covalently bound at 3' ends

(b)

Ligase function of topoisomerase I seals nicks

The method is commonly called TOPO cloning, and can be used with the TA cloning process as well as with blunt-ended fragments. The technique uses a topoisomerase from vaccinia virus that recognises a unique site in the vector with the sequence 5′-(C/T)CCTT-3′. On cutting at this site, the topoisomerase remains covalently attached to the accepting ends. The fragment is then inserted at the cut site using the ligase function of the enzyme. The process is outlined in Fig. 8.7.

Other methods for restriction-free cloning include ligation-independent cloning (LIC) and a development of this called **sequence- and ligation-independent cloning (SLIC)**. As the names suggest, the process does not require ligase-catalysed sealing of the nicks in the recombinant. Instead, the stability of the annealed sequences is increased by ensuring that the regions of overlap are long enough to keep the hybrids together during the transformation process (thermodynamically, this is similar to homopolymer tailing). The complementary regions are added to the target fragments during PCR amplification. For LIC, the overlaps are around 10–12 bp and are generated by the 3′ → 5′ exonuclease activity of **T4 DNA polymerase** in a 'chew back' process that produces 5′ overhangs.

SLIC is similar to LIC in concept, but takes the process a stage further by generating longer overlapping regions of around 20–60 bp. Because of the increased length, the sequence homology does not need to be 100 per cent; thus, fragments generated by imprecise T4 exonuclease activity, or mixed/incomplete PCR, can be annealed in a stable hybrid. The homologous recombination mechanism of *E. coli* then repairs the fragments *in vivo* following transformation. The basis of the LIC/SLIC method is shown in Fig. 8.8.

Methods are now available that enable cloning of any fragment into any vector at any position, not dependent on restriction enzyme digestion, cohesive ends or even full sequence homology.

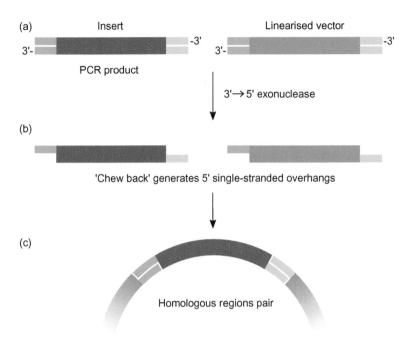

(a) Insert   Linearised vector

PCR product

$3' \rightarrow 5'$ exonuclease

(b)

'Chew back' generates 5' single-stranded overhangs

(c)

Homologous regions pair

Fig. 8.8 The concept of sequence/ligation-independent cloning. Ligation-independent cloning (LIC) and sequence- and ligation-independent cloning (SLIC) are similar in concept. (a) Fragments and vector are prepared with regions of sequence homology at each end. (b) The fragments are 'chewed back' using the $3' \rightarrow 5'$ exonuclease activity of T4 DNA polymerase to generate single-stranded 5' overhangs. In LIC, limited digestion generates overhangs of around 12 bp. In SLIC, different methods can be used to generate longer regions of sequence homology that can accommodate some level of sequence variation. (c) The homologous regions enable insertion of the target DNA into the vector.

8.4 | Putting It All Together

In this section, we will integrate some of the concepts we have discussed in the chapter, to show how the characteristics of the cloning system can be used to construct functional recombinant DNA molecules. These systems are now available commercially as cloning kits from various suppliers. This has made the technology widely accessible, and is a variation on the 'cloning by phoning' concept – although today the phone has probably been replaced by a website interface and the cost is likely to be considerably more than was the case when sharing resources with colleagues. The sophisticated assembly-based cloning methods outlined in Sections 8.4.4, 8.4.5 and 8.4.6 represent the culmination of years of painstaking development work by many people, and we should see the wider picture and reflect on the elegance of these cloning systems and how they are used.

There is a bewildering range of different vectors and cloning kits available from commercial suppliers. Even if a particular application is not already catered for, many companies now offer a custom design service that can construct vectors to the customer's specification.

8.4.1 Cloning in a λ Replacement Vector

Replacement vectors derived from bacteriophage λ, such as EMBL4 (Fig. 7.8(b)), have for many years been the method of choice for the production of genomic libraries. Fragments for cloning are usually generated by partial restriction digestion (see Section 5.1.2), and fragments in the range of 17–23 kbp selected for ligation. If *Sau* 3A (or *Mbo* I, which has the same recognition sequence) has been used as the digesting enzyme, the fragments can be inserted into the *Bam* HI site of the vector, as the ends generated by these enzymes are complementary. The insert DNA can be treated with phosphatase to reduce self-ligation or concatemer formation. If *Sal* I is used with *Bam* HI, the *Bam* cohesive ends are produced on the left and right arms of the vector, and the stuffer fragment is cleaved into three pieces at the internal *Sal* I sites (plus the two very small *Sal/Bam* fragments), which helps to prevent it from re-annealing. Ligation of DNA into EMBL4 is summarised in Fig. 8.9.

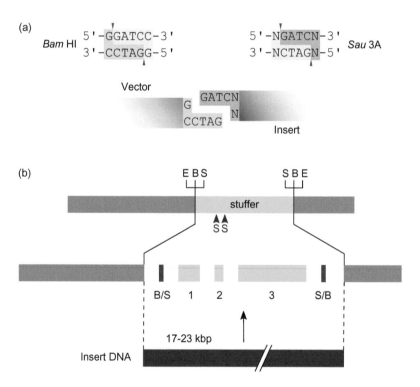

Fig. 8.9 Cloning *Sau* 3A fragments into a *Bam* HI site. (a) The recognition sequences and cutting sites for *Bam* HI and *Sau* 3A. In the *Sau* 3A site, N is any base. Cutting with these enzymes generates complementary 5′ cohesive ends with the sequence 5′-GATC-3′. (b) Ligation of *Sau* 3A-cut DNA into the λ replacement vector EMBL4. Sites on the vector are *Eco* RI (E), *Bam* HI (B) and *Sal* I (S). The vector is cut with *Bam* HI and *Sal* I, which generates three fragments from the stuffer (1–3, shown in yellow) and two very short *Sal* I/*Bam* HI fragments (B/S and S/B, shown in blue), and prevents the stuffer fragment from re-annealing. On removal of the stuffer, the *Sau* 3A-digested insert DNA can be ligated into the *Bam* HI site of the vector. Fragments between 17 and 23 kbp can be cloned in this vector.

Ligation, followed by packaging *in vitro,* produces what is known as a **primary library** of individual recombinant phage particles. Whilst this is theoretically the most useful type of library in terms of isolation of a specific sequence, it is a finite resource. Thus, a primary library is produced, screened and then discarded. If the sequence of interest has not been isolated, more recombinant DNA will have to be produced and packaged. Whilst this may not be a problem, there are occasions where a genomic library is to be used several times, and it is then necessary to amplify the library. This is achieved by plating the packaged phage on a suitable host strain of *E. coli,* and then resuspending the plaques by gently washing the plates with a buffer solution. The resulting phage suspension can be stored almost indefinitely, and will provide enough material for many screening and isolation procedures.

8.4.2 Expression of Cloned cDNA Molecules

Many of the routine manipulations in gene cloning experiments do not require expression of the cloned DNA. However, there are certain situations in which some degree of genetic expression is needed. A transcript of the cloned sequence may be required for use as a probe, or a protein product (requiring transcription and translation) might be needed as part of the screening process used to identify the cloned gene. Another common biotechnological application is where the recombinant DNA is used to produce a protein of commercial value. If eukaryotic DNA sequences are cloned, post-transcriptional and post-translational modifications may be required, and the type of host/vector system that is used is therefore very important in determining whether or not such sequences will be expressed effectively. The problem of RNA processing in prokaryotic host organisms may be obviated by cloning cDNA sequences, and this is the most common approach where expression of eukaryotic sequences is desired. In this section, we will consider some aspects of cloning cDNAs for expression, concentrating mainly on the characteristics of the vector/insert combination that enable expression to be achieved.

> Expression of cloned sequences, to generate RNA transcripts or protein products, presents a more demanding set of requirements than more basic cloning methods.

Assuming that a functional cDNA sequence is available, a suitable host/vector combination must be chosen. The host cell type will usually have been selected by considering aspects such as ease of use, fermentation characteristics or the ability to secrete proteins derived from cloned DNA. However, for a given host cell, there may be several types of expression vector, including both plasmid and (for bacteria) phage-based examples. In addition to the normal requirements such as restriction site availability and genetic selection mechanisms, a key feature of expression vectors is the type of promoter that is used to direct expression of the cloned sequence. Often the aim will be to maximise the expression of the cloned sequence, so a vector with a highly efficient promoter is chosen. Such a promoter is often termed a **strong promoter**. However, if the product of the cloned gene is toxic to the cell, a **weak promoter** may be required to avoid cell death due to over-expression of the toxic product.

> The critical part of an expression vector is the promoter sequence that is used to drive transcription of the cloned gene.

Promoters are regions with a specific base sequence to which RNA polymerase will bind. By examining the base sequence lying on the 5′

(**upstream**) side of the coding regions of many different genes, the types of sequences that are important have been identified. Although there are variations, these sequences all have some similarities. The 'best fit' sequence for a region such as a promoter is known as the **consensus sequence**. In prokaryotes, there are two main regions that are important. Some ten base-pairs upstream from the transcription start site (the -10 region, as the T_C start site is numbered $+1$), there is a region known as the **Pribnow box**, which has the consensus sequence 5'-TATAAT-3'. A second important region is located at around position -35, and has the consensus sequence 5'-TTGACA-3'. These two regions form the basis of the promoter structure in prokaryotic cells, with the precise sequences found in each region determining the strength of the promoter.

Sequences important for transcription initiation in eukaryotes have been identified in much the same way as for prokaryotes. Eukaryotic promoter structure is generally more complex than that found in prokaryotes, and control of initiation of transcription can involve sequences (*e.g.* enhancers) that may be several hundreds or thousands of base-pairs upstream from the T_C start site. However, there are important motifs closer to the start site. These are a region centred at around position -25 with the consensus sequence 5'-TATAAAT-3' (the **TATA box** or **Hogness box**) and a sequence in the -75 region with the consensus 5'-GG(T/C)CAATCT-3', known as the **CAAT box**.

In addition to the strength of the promoter, it may be desirable to regulate the expression of the cloned cDNA by using promoters from genes that are either **inducible** or **repressible**. Thus, some degree of control can be exerted over the transcriptional activity of the promoter; when the cDNA product is required, transcription can be 'switched on' by manipulating the system using an appropriate metabolite. Genes that are expressed at all times are known as **constitutive**. Some examples of promoters used in the construction of expression vectors are given in Table 8.2.

8.4.3 Cloning Large DNA Fragments in BAC and YAC Vectors

Bacterial artificial chromosomes (BACs) and yeast artificial chromosomes (YACs; see Fig. 7.10) can be used to clone very long pieces of DNA. The use of a BAC or YAC vector can reduce dramatically the number of clones needed to produce a representative genomic library for a particular organism, and this is a desirable outcome in itself. A consequence of cloning large pieces of DNA is that physical mapping of genomes is made simpler, as there are obviously fewer sequences to fit together in the correct order.

In practice, cloning in YAC vectors is similar to other protocols (Fig. 8.10). The vector is prepared by a double restriction digest, which releases the vector sequence between the telomeres and cleaves the vector at the cloning site. Thus, two arms are produced, as is the case with phage vectors. Insert DNA is prepared as very long fragments (a partial digest with a six-cutter may be used) and ligated into the

It is useful to be able to regulate the expression of cloned genes by using promoters that can be induced or repressed by certain conditions or metabolites.

Artificial chromosomes can accommodate very large pieces of cloned DNA and therefore reduce the number of clones needed to generate a representative genomic or cDNA library.

Table 8.2 Some promoters that can be used in expression vectors

Organism	Gene promoter	Regulation
Escherichia coli	*lac* operon	IPTG
	λP$_L$	λ cI protein
	Phage T7	T7 polymerase
Aspergillus nidulans	Glucoamylase	Starch
Tricoderma reesei	Cellobiohydrolase	Cellulose
Saccharomyces cerevisiae	Acid phosphatase	Phosphate depletion
	Alcohol dehydrogenase	Glucose depletion
	Galactose utilisation	Galactose
	Metallothionein	Heavy metals
	TEF1 (elongation factor)	Constitutive
Drosophila melanogaster	ACT5C (actin)	Constitutive
Mammalian cells	SV40 (Simian virus)	Constitutive
	CMV (cytomegalovirus)	Constitutive
	EF1A (elongation factor)	Constitutive
Various	Metallothionein	Heavy metals
	Heat-shock protein	Temperature >40°C

Note: Regulation of promoters can be inducible, repressible or constitutive (expressed at all times). For expression vectors, inducible or constitutive promoters are most useful. Artificial promoters can also be constructed by combining elements of different promoters to achieve the desired characteristics.

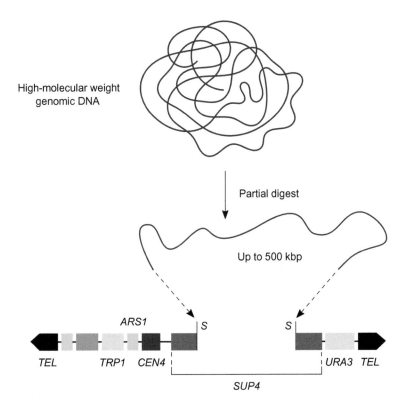

High-molecular weight genomic DNA

Partial digest

Up to 500 kbp

ARS1

S

S

TEL   TRP1   CEN4   URA3   TEL

SUP4

Fig. 8.10 Cloning in a YAC vector. Very large DNA fragments (up to 500 kb) are generated from high-molecular weight DNA, often by limited digestion with a restriction enzyme that cuts infrequently. The fragments are then ligated into a YAC vector (Fig. 7.10) that has been cut with *Bam* HI (to remove the inter-telomeric fragment) and the cloning site *Sma* I (S) that lies within the *SUP4* gene. The construct contains the cloned DNA and the essential requirements for a yeast chromosome, *i.e.* telomeres (*TEL*), an autonomous replication sequence (*ARS1*) and a centromere region (*CEN4*). The *TRP1* and *URA3* genes can be used as dual selectable markers to ensure that only complete artificial chromosomes are maintained.

cloning site to produce artificial chromosomes. Selectable markers on each of the two arms ensure that only correctly constructed chromosomes will be selected and propagated.

8.4.4 Gateway Cloning Technology

Gateway cloning technology was developed in the late-1990s by Invitrogen as one of the first 'modular' systems to enable rapid transfer of fragments from one vector to another. The procedure uses the recombination sites from bacteriophage λ (*att P* sites) and *E. coli* (*att B* sites) to enable insertion of fragments by the process of **site-specific recombination**. This converts the *att B* and *P* sites into *L* (left) and *R* (right) variants that can be used to reverse the integration, so fragments can be moved between vectors easily by using the appropriate mix of enzymes to drive the integration or excision processes. A selectable marker gene (*ccd B*, part of the F-plasmid toxin/antitoxin system) produces a protein that is toxic to the cell and is used as part of the exchange process to ensure that only cells with the desired constructs are viable. This results in very high cloning efficiency (approaching 100 per cent) that minimises the need for preliminary screening of clones.

> Gateway technology harnesses site-specific recombination to enable efficient insertion and excision of cloned fragments and simplify their transfer between vectors.

A variety of protocols can be derived from the Gateway system, which is now available on an open-access basis. The fragment of interest is cloned into an **entry vector** using standard techniques such as isolation from a library or PCR amplification and recombining with a **donor vector**. The entry clone can then be used to insert the fragment into a range of **destination vectors** for various purposes, including expression of the gene or transfer into a different host/vector system. In addition to cloning fragments *de novo*, collections of 'ready-made' entry vectors containing genes of interest from a range of organisms (*e.g.* human open-reading frame clones) can be obtained from a variety of sources, so that initial cloning and isolation of the target gene are not required. An example of the Gateway cloning procedure is shown in Fig. 8.11.

8.4.5 Golden Gate Cloning and Assembly

Although the enzymes were discovered in the mid-1990s, the term Golden Gate cloning was first used in 2008 to describe a method that used **type IIS restriction enzymes** (Table 5.1) to generate overlaps and enable seamless cloning of multiple fragments. The method of assembly and ordering of fragments is similar to other techniques in that a single reaction can be carried out to generate, assemble and ligate the fragments. Careful design of the cleavage sites is required to ensure that the overhangs enable assembly of fragments in the correct order; thus (as with many of the modern cloning techniques), time spent on the computer to design and check sequences is well spent before carrying out the actual experiment.

> Golden Gate assembly is based on using type IIS restriction enzymes to enable seamless assembly of multiple fragments in a single reaction.

The basis of Golden Gate cloning is that the cleavage site of the type IIS enzymes such as *Bsa* I is a short distance away from the recognition sequence and is not sequence-dependent. Thus, for an overhang of four nucleotides, there are 256 possible sequences (4^4) that can be used to

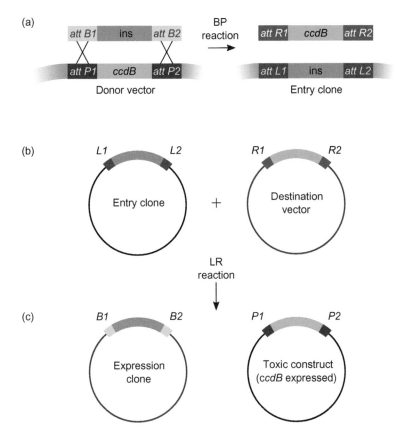

Fig. 8.11 The Gateway cloning system. In (a), one method of generating an entry clone is shown, with *att B* sites (B = bacterial) added to the target insert fragment to be cloned (ins). The *att P* sites (P = phage) from bacteriophage λ flank the *ccd B* gene in the donor vector plasmid. Recombination between the B and P sites (X) is catalysed by a BP Clonase mix (Invitrogen) and enables integration of the fragment into the donor vector, replacing the *ccd B* fragment. The recombination process converts the B and P sites into L and R variants. (b) The entry clone is mixed with a destination vector that has R sites flanking the *ccd B* gene, using the LR Clonase mix to catalyse the integration of the insert into the destination vector. (c) This generates an expression clone and a construct that expresses the *ccd B* gene to produce the Ccd B protein, which is toxic to Ccd B-sensitive host cells. Only clones with the desired inserts will be viable, which increases cloning efficiency to very high levels. A wide range of destination vectors is available for various applications. The vectors will also have selectable markers for antibiotic resistance (not shown for clarity).

generate recognition/cleavage site constructs that can be added to fragments for assembly (PCR primers can also be incorporated to enable amplification and incorporation of the cleavage sites). Once fragments and vector are prepared, the cleavage and annealing/ligation reactions can occur at the same time, as the restriction recognition site is not included in the ligated products. Thermal cycling at the optimum cleavage and ligation temperatures (37°C and 16°C, respectively) improves the efficiency of the reaction, and a final step at 55°C favours digestion to

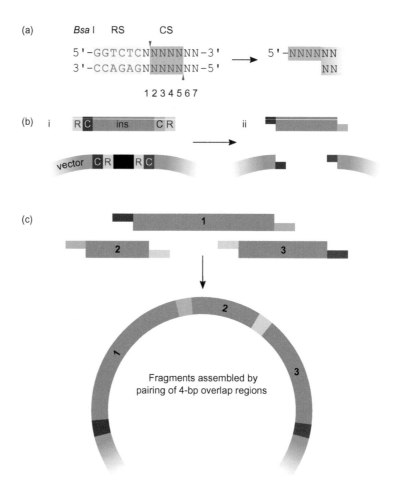

(a) Bsa I   RS   CS

5'-GGTCTCNNNNNNNN-3'   5'-NNNNNN
3'-CCAGAGNNNNNNN-5'   NN

1 2 3 4 5 6 7

(b) i

ii

(c)

1

2   3

2

1

3

Fragments assembled by
pairing of 4-bp overlap regions

Fig. 8.12 Golden Gate cloning and assembly. (a) The recognition and cutting sites for the type IIS restriction enzyme *Bsa* I. The recognition site (RS, highlighted in yellow) is a non-palindromic six base-pair sequence. The cutting site (CS, highlighted in red) is sequence-independent and separated from the RS by a single base-pair (position 1 in the 7-bp numbered sequence). The cutting site generates 4-bp 5′ overhangs by cutting between positions 1/2 and 5/6 in the diagram. (b) Preparation of vector and insert DNA. In (b)i, the insert fragments (ins) have *Bsa* I recognition and cutting sites (R and C) incorporated at each end, in opposite orientations. The vector has the R and C sites in the opposite orientation to those in the target fragments. On cutting with *Bsa* I ((b)ii), the overhangs are generated in the correct orientations to enable assembly. (c) Fragments are generated with overhangs that enable correct assembly into the vector, in a single-tube reaction with *Bsa* I and DNA ligase. Note that the *Bsa* I recognition sites are removed during assembly, resulting in a 'seamless' construct.

ensure that only complete, assembled and ligated constructs remain. As with the Gateway system, cloning efficiencies approaching 100 per cent can be achieved using this method. Successful assembly of over 50 individual fragments has been reported using the Golden Gate method, the basis of which is outlined in Fig. 8.12.

8.4.6   The Gibson Assembly Method

This technique, developed in 2009 by **Daniel Gibson**, is similar to the LIC and SLIC methods (see Section 8.3.4) in that it is a restriction-independent process that can join fragments by generating single-stranded regions of overlap. A major advantage of the method is that it can be carried out as a single **isothermal reaction** that contains the three enzymes needed (exonuclease, polymerase and ligase). The single-stranded regions (ranging from 15 to 40 bp) are generated by the 5′ → 3′ activity of **T5 DNA exonuclease** (note this is different from the 3′ → 5′ exonuclease activity of T4 DNA polymerase that is used in the LIC and SLIC methods). Following annealing of the overlap regions, the polymerase fills in any gaps and the ligase seals the nicks. As well as being a quick and efficient method of sub-cloning fragments from one vector to another, Gibson assembly can be used to join multiple fragments together, and has become a key method for synthetic biology applications. It was one of the methods used in the construction of the *Mycoplasma* synthetic genome that we noted in Section 8.2.4. The principles of Gibson assembly are outlined in Fig. 8.13.

Gibson assembly is a simple and powerful method that can be used to arrange several fragments in their correct relative order using sequence homology overlaps. Along with other assembly techniques, this is a key part of synthetic biology protocols.

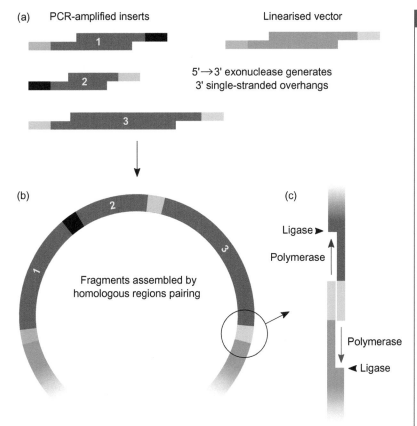

Fig. 8.13   Outline of the Gibson assembly method. This method is similar in concept to the LIC and SLIC methods (Fig. 8.8). The process can be completed in a single isothermal reaction. (a) DNA is prepared as PCR-amplified fragments with regions of sequence homology incorporated at each end. The vector DNA also has sequences that complement the target fragments' extensions. In this case, a three-fragment assembly is illustrated, with homologous regions indicated by green, black, turquoise and yellow. The regions are added asymmetrically to enable ordered assembly of the fragments. Vector sequences are complementary to the left and right regions of fragments 1 and 3, respectively (green and yellow for the vector in this illustration). The 5′ → 3′ exonuclease generates 3′ single-stranded overlapping regions. (b) The fragments anneal at the complementary regions, producing the assembly. DNA polymerase fills the gaps from the 3′-OH and the ligase seals the nicks. The yellow region is shown in detail in (c).

8.5 | Conclusion: Designing a Cloning Strategy

Selecting the most appropriate methods for a cloning procedure can seem daunting, but by asking the right questions, the options usually become clear. The overall aim of the procedure will often dictate the host/vector combination and the source of the fragments for cloning. We looked at how cloning strategies have changed from the *pre-genomic* to the *post-genomic* eras, with *bioinformatics-based* sequence manipulation *in silico* now often the first step in designing an experimental system.

The *PCR* has had an enormous impact on molecular biology, and is a common method for generating specific fragments (even sometimes removing the need for cloning altogether). However, there is still an important role for more 'traditional' methods that generate *clone banks* or *libraries*, from *genomic DNA* or *complementary DNA* (cDNA, synthesised from mRNA). More recently, making *synthetic genes* has become a viable option as technology has improved. The overall aim of the procedure will determine which method of fragment production is best.

We saw that joining fragments to the vector of choice can be accomplished in a number of ways, including *ligation* (blunt or cohesive ends), *homopolymer tailing*, using *linkers* or *adapters*, and *restriction-free* methods that are particularly useful for assembling cloned fragments. In the final part of the chapter, we looked at six examples, three from the early development of cloning technology and three that have emerged in response to the need for rapid cloning systems to keep pace with developments in sequencing and bioinformatics. These sophisticated *assembly-based cloning methods* illustrate how complex procedures can be made accessible by innovation and careful design.

Cloning has now matured and developed into a sophisticated technology that is supported by a vast infrastructure, both non-commercial and commercial. Although it is still sometimes the case that an entire strategy will be designed and enacted in the laboratory by making the vectors and cells, it is more common to use the kits, products and services that are available. Overall this approach is usually a cost-effective solution to any complex cloning task.

| Further Reading

Creager, A. (2020). Recipes for recombining DNA: a history of molecular cloning: a laboratory manual. *BJHS Themes*, 5, 225–43. URL [www.cambridge .org/core/journals/bjhs-themes/article/recipes-for-recombining-dna-a-history-of-molecular-cloning-a-laboratory-manual/ F4EE6A7FFA4991B714A33D474BC1CF76#]. DOI [https://doi.org/10.1017/bjt .2020.5].

El Karoui, M. *et al.* (2019). Future trends in synthetic biology – a report. *Front. Bioeng. Biotechnol.*, 7, 175. URL [www.frontiersin.org/article/10.3389/fbioe .2019.00175]. DOI [https://doi.org/10.3389/fbioe.2019.00175].

Gibson, D. G. (2011). Enzymatic assembly of overlapping DNA fragments. *Methods Enzymol.*, 498, 349–61. URL [www.ncbi.nlm.nih.gov/pmc/articles/PMC7149801/]. DOI [https://doi.org/10.1016/B978-0-12-385120-8.00015-2].

Marillonnet, S. and Grützner, R. (2020). Synthetic DNA assembly using Golden Gate cloning and the hierarchical modular cloning pipeline. *Curr. Prot. Mol. Biol.*, 130, e115. URL [https://currentprotocols.onlinelibrary.wiley.com/doi/10.1002/cpmb.115]. DOI [https://doi.org/10.1002/cpmb.115].

Reece-Hoyes, J. S. and Walhout, A. J. M. (2018). Gateway recombinational cloning. *Cold Spring Harb. Protoc.*, 2018 (1), pdb.top094912. URL [www.ncbi.nlm.nih.gov/pmc/articles/PMC5935001/]. DOI [https://doi.org/10.1101/pdb.top094912].

Websearch

People

In 1982, *Hiroto Okayama* and *Paul Berg* published a landmark paper. What did it describe? Find out as much as you can about the scientific careers of Okayama and Berg.

Places

What does '*EMBL*' stand for in the λ replacement vector EMBL4? Investigate the contribution this place has made to molecular biology.

Processes

One of the stages of restriction-free cloning processes like LIC, SLIC and Gibson assembly is the 'chew back' of part of the DNA fragment/vector to expose a single-stranded region. We saw that two different enzymes could be used for this – the $3' \rightarrow 5'$ exonuclease function of *T4 DNA polymerase* and the $5' \rightarrow 3'$ *T5 DNA exonuclease*. Find out a little more about how these enzymes work.

Reflections

What topics in this chapter have you found most challenging? Look for resources that help to illustrate the key points.

Cloning strategies

influenced by

overall aim — may be to

and if → **genome sequence is known** — use → **bioinformatics approach (in silico)**

genome sequence is known → and if → will be available in sequence database(s)

will be available in sequence database(s) — via → worldwide IT infrastructure — enables → bioinformatics approach (in silico)

bioinformatics approach (in silico):
- perform virtual experiments — to → test ideas before carrying out 'real' experiments
- design cloning protocols — to → make efficient use of time and resources

however → in vitro and in vivo experiments will still be required — to achieve aims →

e.g. → specialised vectors — to →
- amplicon cloning → assembly of contiguous fragments

and if → **genome sequence not known** → consider sequencing the genome using NGS

genome sequence not known → may need to → purify protein — to generate → amino acid sequence data — to → design and make oligonucleotides — for use as → probes / primers

primers — for → sequencing
primers — for → PCR
probes / primers → to screen → clone bank (library)

clone gene / clone cDNA — need → clone bank (library)

clone bank (library):
- genomic DNA → generate DNA fragments
- cDNA → make cDNA from mRNA

generate DNA fragments / make cDNA from mRNA — join to vector by →
- direct ligation
- homopolymer tailing
- linkers
- adapters

overall aim may be to:
- clone gene
- clone cDNA
- edit genes
- edit genomes
- amplify DNA
- make proteins

edit genes / edit genomes — to → specialised vectors

amplify DNA ← PCR

make proteins — related to →
- basic science — e.g. → gene structure, gene function, control, genomics
- biotechnology — e.g. → crop plants, transgenics, proteins, therapeutics
- medical — e.g. → diagnostics, screening, gene therapy, genome editing

PCR — to → amplify DNA

Concept Map 8

Chapter 9 Summary

Learning Objectives

When you have completed this chapter, you will be able to:

- Outline the history and principles underpinning the PCR
- Describe the methodology of PCR amplification
- Demonstrate the range of PCR variants, including nested, inverse, quantitative and digital
- Discuss the applications of the PCR and the processing of PCR products

Key Words

Polymerase chain reaction (PCR), Kary Mullis, oligonucleotides, DNA templates, DNA polymerase, Joshua Lederberg, end-point PCR, primer, thermostable DNA polymerases, thermal cycler, positive control, negative control, sequence (of primer), inosine, informatics-based, length (of primer), melting temperature (T_m), unique sequence, thermophilic, *Thermus aquaticus*, *Taq* polymerase, processivity, fidelity, half-life (of enzyme), *Thermus flavus*, *Thermus thermophilus*, *Pyrococcus furiosus*, hyperthermophile, hot-start PCR, reverse transcriptase PCR (RT-PCR), competitor RT-PCR, nested PCR, external primers, internal nested primers, inverse PCR (IPCR), quantitative PCR (qPCR), real-time PCR, real-time qPCR, quantification cycle (Cq), cycle threshold (Ct), calibration curve, absolute, relative, digital PCR (dPCR), limiting dilution PCR, chamber digital PCR (cdPCR), droplet digital PCR (ddPCR), Poisson distribution, RAPD-PCR (random amplification of polymorphic DNA), arbitrarily-primed PCR (AP-PCR), multiplex PCR, singleplex, amplicons, third-generation PCR, diagnostic testing.

Chapter 9

The Polymerase Chain Reaction

The topic of this chapter is the polymerase chain reaction (PCR), which was discovered by **Kary Mullis** and for which he was awarded the Nobel Prize in Chemistry in 1993. The PCR technique produces a similar result to DNA cloning – the selective amplification of a DNA sequence – and has become such an important part of the genetic engineer's toolkit that, for many applications, it has essentially replaced traditional cloning. In this chapter, we will look at some of the techniques and applications of PCR technology.

9.1 | History of the PCR

In addition to requiring someone to provide a spark of genius, major scientific breakthroughs depend on existing knowledge, the techniques available and often a little luck. In genetics and molecular biology, there have been many examples of scientists being in the right place, with the right 'mindset', investigating the right problem and coming up with a seminal discovery. Gregor Mendel, James Watson and Francis Crick are three names that stand out; Kary Mullis can legitimately be added to the list. In 1979, he joined the Cetus Corporation, based in Emeryville, California. By this time, the essential prerequisites for the development of the PCR had been established. Mullis was working on oligonucleotide synthesis, which by the early-1980s had become an automated and somewhat tedious process. Thus, his mind was free to investigate other avenues. In his own words, he found himself 'puttering around with oligonucleotides', and the main thrust of his puttering was to try to develop a modified version of the dideoxy sequencing procedure. His thoughts were therefore occupied with oligonucleotides, **DNA templates** and **DNA polymerase**.

Late one Friday night in April 1983, Mullis was driving to his cabin with a friend, and was thinking about his modified sequencing experiments. He was in fact trying to establish if extension of oligonucleotide primers by DNA polymerase could be used to 'mop up' unwanted

Major scientific breakthroughs require a number of things to come together at the right time, and the discovery of the PCR is a good example of this.

The basic premise of the PCR is quite simple – two primers are used so that each strand of the DNA serves as a template, and thus the number of strands of DNA doubles on each cycle of the PCR.

dNTPs in the solution, which would otherwise get in the way of his dideoxy experiment. Suddenly he realised that, if *two* primers were involved, and they served to enable extension of the DNA templates, the sequence would effectively be duplicated. Fortunately, he had also been writing computer programs that required reiterative loops – and realised that sequential repetition of his copying reaction (although not what was intended in his experimental system!) could provide many copies of the DNA sequence. Some hasty checking of the figures confirmed that the exponential increase achieved was indeed 2^n, where n is the number of cycles. The PCR had been discovered.

Subsequent work proved that the theory worked when applied to a variety of DNA templates. Mullis presented his work as a poster at the annual Cetus Scientific Meeting in spring of 1984. In his account of the discovery of the PCR in *Scientific American* (April 1990), he recalls how **Joshua Lederberg** discussed his results and appeared to react in a way that was to become familiar – the 'why didn't I think of that' acceptance of a discovery that is brilliant in its simplicity.

Over the past 40 years, the PCR technique has been adopted by scientists in a pattern similar to that for recombinant DNA technology itself. It became widely established in the early-1990s, with the number of papers published annually increasing tenfold from 1990 to 1994. Novel versions of the technique continue to be developed, with around 40 or so now available for a range of applications. We will describe only a few of these in this chapter; part of the *Websearch* activity will be to extend this to look at some of the more esoteric PCR-based methods.

As with many new technologies, over the past 40 years the PCR has developed into many variants, each with its own specific application.

9.2 | The Methodology of the PCR

The PCR is elegantly simple in theory. When a DNA duplex is heated, the strands separate or 'melt'. If the single-stranded sequences can be copied by a DNA polymerase, the original DNA sequence is duplicated. If the process is repeated many times, there is an exponential increase in the number of copies of the starting sequence. The length of the fragment is defined by the 5′ ends of the primers, which helps to ensure that a homogeneous population of DNA molecules is produced. Thus, after relatively few cycles (often around 30), the target sequence becomes greatly amplified, which generates enough of the sequence for identification and further processing. In the basic method, the products are analysed at the end of the process; thus, it is sometimes called **end-point PCR**.

9.2.1 Essential Features of the PCR

In addition to a DNA sequence for amplification, there are two requirements for the PCR. Firstly, a suitable **primer** is required. For most applications, *two* primers are necessary, one for each strand of the duplex. The primers should flank the target sequence, so some sequence information is required if selective amplification is to be achieved. The primers are synthesised as oligonucleotides, and added to the reaction in excess, so that each of the primers is always available following the denaturation step. A second requirement is a suitable form of DNA polymerase. Whilst it is possible to use a standard DNA polymerase (as was done in the early days of the PCR), this is inactivated by the heat denaturation step of the process and thus, fresh enzyme has to be added after each cycle of operation. The availability of **thermostable DNA polymerases** makes life much easier for the operator (see Section 9.2.3).

The use of a thermostable polymerase means that the PCR procedure can be automated, as there is no need to add fresh polymerase after each denaturation step. In addition to the DNA, primers and polymerase, the usual mix of the correct buffer composition and the availability of the four dNTPs are needed to ensure that copying of the DNA strands is not stalled due to inactivation of the enzyme or lack of monomers.

In operation, the PCR is straightforward. The target DNA and reaction components are usually mixed together at the start of the process, and the tube heated to around 90°C to denature the DNA. As the temperature drops, primers will anneal to their target sequences on the single-stranded DNA, and the polymerase will begin to copy the template strands. The cycle is completed (and restarted) by a further denaturation step. The operational sequence is shown in Fig. 9.1.

Automation of the PCR cycle of operations is achieved by using a programmable heating system known as a **thermal cycler**. This takes small microcentrifuge tubes (96-well plates or glass capillaries can also be used) in which the reactants are placed. Thin-walled tubes permit more rapid temperature changes than standard tubes or plates. Various thermal cycling patterns can be set according to the particular reaction conditions required for a given experiment, but in general the cycle of events shown in Fig. 9.1 forms the basis of the amplification stage of the PCR process. Although thermal cyclers are simple devices, they have to provide accurate control of temperature and similar rates of heating and cooling for tubes in different parts of the heating block. More sophisticated devices provide a greater range of control patterns than the simpler versions, such as variable rates of heating and cooling and heated lids to enclose the tubes in a sealed environment.

A final practical consideration in setting up a PCR protocol is general housekeeping and manipulation of samples. As the technique is designed to amplify small amounts of DNA, even trace contaminants (perhaps from a tube that has been lying open in a laboratory or from an ungloved finger) can sometimes ruin an experiment. Thus, the operator needs to be fastidious (or even a little paranoid!) about cleanliness when carrying out the PCR. Also, even the aerosols created by pipetting reagents can lead to cross-contamination, so good technique is essential. It is best if a sterile hood or flow cabinet can be set aside for setting up the PCR reactions, with a separate area used for post-reaction processing. Accurate labelling of tubes (primers,

The key requirements for the PCR are a DNA template, a pair of suitable oligonucleotide primers and a DNA polymerase.

The PCR is in essence a simple series of defined steps that is repeated many times. This means that the process can be automated easily with relatively simple technology.

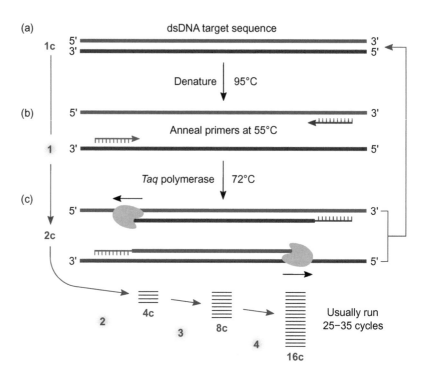

Fig. 9.1 The basis of the polymerase chain reaction (PCR). (a) Duplex DNA is heat-denatured at 95°C to give single strands. (b) The temperature is lowered to 55°C and two oligonucleotide primers anneal to their complementary sequences on the target DNA. Note that the optimum primer annealing temperature will have been determined by calculation of the T_m and experimental confirmation. (c) *Taq* polymerase (thermostable) synthesises complementary strands from the templates by primer extension at 72°C. The denature/prime/copy cycle is then repeated (blue arrows). The blue numbers/yellow circles show the cycles (4 in this diagram). Numbers in bold blue font are the number of copies of the target duplex at the end of each cycle. At the start, there is one copy (**1c**) of the target duplex. At the end of the first cycle, there are two copies (**2c**), then four (**4c**), *etc.* Usually around 30 cycles are completed for a typical PCR reaction. In theory, the number of fragments increases as 2^n, where n is the number of cycles. Thus a 30-cycle PCR will generate around 10^9 copies of a single target fragment.

Because the PCR is so efficient at making lots of copies of a DNA sequence, great care must be taken to avoid contamination with minute amounts of unwanted nucleic acid that would be amplified along with the target sequence.

target DNAs, nucleotide mixes, *etc.*) is also needed, and quality control procedures are critical. This is particularly important where analysis is being carried out for medical or forensic applications, and automated sample tracking using barcode-type technology is essential where large sample throughput is involved. Careful use of a positive control (a DNA sample known to generate a positive result) and a negative control (no DNA) is also important to ensure that the reaction has performed as expected.

9.2.2 Designing Primers for the PCR
Oligonucleotide primers are available from many commercial sources and can be synthesised to order in a few days. In designing primers,

(a)

| a.a. sequence | Phe - Leu - Pro - Ser - Ala - Lys - Trp - Ala - Tyr - Asp - Pro |

| Codons per amino acid | 2 | 6 | 4 | 6 | 4 | 2 | 1 | 4 | 2 | 2 | 4 |

Avoid   Better sequence

(b)

Ala - Lys - Trp - Ala - Tyr - Asp - Pro

```
GCAAAATGGGCATACGACCC-
  G   G       G   T  T
  C   C       C
  T   T       T
```

Number of possibilities   4 x 2 x 1 x 4 x 2 x 2   = 128

(c)

Ala - Lys - Trp - Ala - Tyr - Asp - Pro

```
GCIAAATGGGCITACGACCC-
   G          T  T
```

Number of possibilities   1 x 2 x 1 x 1 x 2 x 2   = 8

Fig. 9.2 Designing primers for the PCR. One method for designing primers to amplify coding sequences is to use part of the amino acid sequence of a protein. In (a), the amino acid sequence and number of codons per amino acid are shown. Amino acids with six codons (circled in yellow) are best avoided. The shaded run of amino acids is therefore a better starting point, as this has less degeneracy and thus fewer possible sequences. In (b), a mixed probe is synthesised by incorporating the appropriate mix of dNTPs at each degenerate position. Note that in this example, the final degenerate position for proline is not included, giving an oligonucleotide with 20 bases (a 20-mer). There are 128 possible permutations of sequence in the mixture. In (c), inosine (I, highlighted green) is used to replace the fourfold degenerate bases, giving eight possible sequences. Note that because two primers are needed for the PCR, a second region of the protein will have to be selected to enable a pair of primers to be designed that will generate a suitable size of amplified fragment to meet the aims of the experiment.

there are several aspects that have to be considered. Perhaps most obvious is the **sequence** of the primer – more specifically, where does the sequence information come from? It may be derived from amino acid sequence data, in which case the degeneracy of the genetic code has to be considered, as shown in Fig. 9.2. In synthesising the primer, two approaches can be taken. By incorporating a mixture of bases at the wobble position, a mixed probe can be made, with the 'correct' sequence represented as a small proportion of the mixture. Alternatively, the base **inosine** (which pairs equally well with any of the other bases) can be incorporated as the third base in degenerate codons.

The expansion of genome sequencing has had a major impact on all aspects of **informatics-based** applications, primer design for the PCR being one notable example. Analysis of genome sequence databases now enables primers to be identified easily, either from the target organism's genome (if the sequence is available) or perhaps by looking at the same gene from a different organism. Software tools are available to enable primers to be designed quickly and accurately.

Regardless of the source of the sequence information for the primers, there are some general considerations that should be addressed. The **length** of the primer is important. It should be long enough to ensure stable hybridisation to the target sequence at the required temperature. The **melting temperature** (T_m) is also critical, as this will determine what temperature is used for annealing the primers. This is usually one or two degrees below the T_m in the range of 52–58°C. The T_m depends on primer length and base composition, with primers with a higher G·C:A·T content being more thermo-dynamically stable. The T_m can be calculated easily, and is often checked empirically to determine the optimal binding temperature. The primer must also be long enough to ensure that it is a **unique sequence** in the genome from which the target DNA is taken. Primer lengths of around 20–30 nucleotides are usually sufficient for most applications. With regard to the base composition and sequence of primers, repetitive sequences should be avoided, and also regions of single-base sequence. Primers should obviously not contain regions of internal complementary sequence or regions of sequence overlap with other primers.

As extension of PCR products occurs from the 3′ termini of the primer, it is this region that is critical with respect to fidelity and stability of pairing with the target sequence. Some 'looseness' of primer design can be accommodated at the 5′ end, and this can sometimes be used to incorporate design features such as restriction sites at the 5′ end of the primer.

> When designing primers for the PCR, there are many aspects to be considered. The stringency of primer binding can have a great effect on the number of PCR products in any one amplification.

9.2.3 DNA Polymerases for the PCR

In theory, any DNA polymerase would be suitable for use in a PCR procedure. Originally the Klenow fragment of DNA polymerase was used, but this is thermolabile and requires addition of fresh enzyme for each extension phase of the cycle. This was inefficient in that the operator had to be present at the machine for the duration of the process and a lot of enzyme was needed. Also, as extension was carried out at 37°C, primers could bind to non-target regions, generating a high background of non-specific amplified products. The availability of a thermostable polymerase solved these problems, as the enzyme not only survived the denaturation temperature, but also could be used at a much higher temperature (72°C) for the synthesis part of the cycle. The first thermostable enzyme to be used for PCR was isolated from the **thermophilic** bacterium *Thermus aquaticus* and was called *Taq* polymerase (using the same naming convention as restriction enzymes, *i.e.* T and *aq* from the generic and specific parts of the organism's name).

The key features required for a DNA polymerase include **processivity** (affinity for the template, which determines the number of bases incorporated before dissociation), **fidelity** of incorporation, *rate of synthesis* and **half-life** of the enzyme at different temperatures.

> Thermostable DNA polymerases such as *Taq* polymerase have enabled automation of the PCR process without the need for fresh polymerase to be added after the denaturation stage of each cycle.

In theory, variations in these aspects shown by different enzymes should make the choice of a polymerase a difficult one; in reality, a particular source is chosen and conditions adjusted empirically to optimise the activity of the enzyme.

Of the features mentioned above, fidelity of incorporation of nucleotides is perhaps the most critical. Obviously, an error-prone enzyme will generate mutated versions of the target sequence out of proportion to the basal rate of mis-incorporation, given the repetitive cycling nature of the reaction. In theory, an error rate of 1 in 10^4 in a millionfold amplification would produce mutant sequences in around a third of the products. Thus, steps often need to be taken to identify and avoid such mutated sequences in cases where high-fidelity copying is essential.

> Any errors introduced during polymerase copying of DNA strands will become amplified to significant levels over a typical number of PCR cycles.

The development of modified and improved polymerases has addressed many of the issues that affected the early development of PCR, and today a wide variety of thermostable enzymes is available. These include several versions of recombinant *Taq* polymerase, as well as enzymes from *Thermus flavus* (*Tfl* polymerase) and *Thermus thermophilus* (*Tth* polymerase). In addition to *Thermus*-derived polymerases, a thermostable DNA polymerase from *Pyrococcus furiosus* is available. This organism is classed as a **hyperthermophile**, with an optimal growth temperature of 100°C, and is found in deep-sea vents.

One other interesting development of polymerase technology is the 'hot-start' PCR method. This was developed to help reduce non-specific amplification during initial stages of the reaction before the operating temperature cycle is stabilised. The polymerase is complexed with an antibody that inhibits enzyme activity until the temperature reaches a certain level, when the antibody is denatured. This acts as an 'on' switch and the polymerase begins synthesis from this point, which avoids any synthesis that might have occurred as the reaction temperature increases to operational level.

9.3 | More Exotic PCR Techniques

As the PCR became established, variations of the basic procedure were devised. This is still an area of active development and new modifications to, and applications for, PCR appear regularly. In this section, we will outline a few of these variations on the basic PCR process, and list the others to provide some 'primers' for searching for information about these techniques. Be prepared for a bit of 'acronym overload' when reading Table 9.1!

9.3.1 PCR Using mRNA Templates

An important and widely used variant of the basic PCR is **reverse transcriptase PCR** (RT-PCR). It can be useful in determining low levels of gene expression by analysing the PCR product of a cDNA prepared from the mRNA transcript. The process involves copying the mRNA

> Of the many variants of the PCR protocol, one of the most important is RT-PCR. This enables amplification of cDNA derived from mRNA samples.

Table 9.1 | Abbreviations, acronyms, descriptors and initialisms for variants of the PCR

AFLP-PCR (amplified fragment length polymorphism)
Allele-specific PCR
Alu-PCR
Arbitrarily-primed PCR[a]
Assembly PCR
Asymmetric PCR
CAPS-PCR (cleaved amplified polymorphic sequence)
cdPCR (chamber digital)
COLD-PCR (co-amplification lower denaturation temperature)
Colony PCR
COMPAS-PCR (complementary primer asymmetric)
Competitor RT-PCR
ddPCR (droplet digital)
Direct PCR
dPCR (digital)
ePCR/emPCR (emulsion)
epPCR (error-prone)
End-point PCR[b]
Fast-cycling PCR
GAWTS-PCR (gene amplification with transcript sequencing)
GC-rich PCR
High-fidelity PCR
HRM-PCR (high-resolution melt)
Hot-start PCR
In situ PCR
ISSR-PCR (intersequence-specific)
IPCR/inverse PCR
LAMP-PCR (loop-mediated isothermal amplification)
LATE-PCR (linear after the exponential)
Ligation-mediated PCR

Long-range PCR
Methylation-specific PCR
Miniprimer PCR
mrPCR (multiplex restriction)
Multiplex PCR
nanoPCR (nanoparticle-assisted)
Nested PCR
Overlap extension PCR
PCR[b] (standard PCR method)
PCR-RFLP (restriction fragment length polymorphism)
polony PCR (polymerase colony)
qPCR[c] (quantitative PCR)
RACE-PCR (rapid amplification of cDNA ends)
RAPD-PCR[a] (random amplified polymorphic DNA)
RAWTS-PCR (RNA amplification with transcript sequencing)
Real-time PCR[c]
Repetitive sequence-based PCR
RT-PCR (reverse transcriptase PCR)
RT-qPCR (reverse transcriptase quantitative PCR)
rhPCR (RNase H-dependent)
Single-cell PCR
Singleplex PCR
SSP-PCR (single specific primer)
Solid phase PCR
Suicide PCR
TAIL-PCR (thermal asymmetric interlaced)
TD-PCR (touchdown)
Two-tailed PCR
VNTR-PCR (variable number tandem repeat)

Note: Terms are listed alphabetically; those in blue are mentioned and/or described in the text, and are also in the glossary. Superscript letters indicate different terms for the same process. Where an abbreviation, acronym or initialism is listed, the term is defined in brackets.

using reverse transcriptase, as in a standard cDNA synthesis. Oligo(dT)-primed synthesis is often used to generate the first strand of cDNA if the RNA is a polyadenylated eukaryotic mRNA. Gene-specific PCR primers are then used as normal, although the first few cycles may be biased in favour of copying the cDNA single-stranded product until enough copies of the second strand have been generated

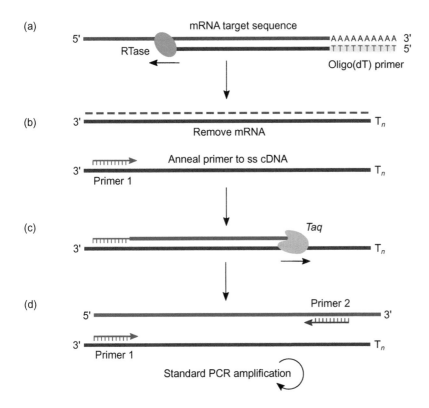

(a) mRNA target sequence

5' ————————————————AAAAAAAAA 3'
RTase —————————TTTTTTTTT 5'
Oligo(dT) primer

(b) 3' — — — — — — — — — — — — — — Tₙ
Remove mRNA

Anneal primer to ss cDNA
3' ———————————————————— Tₙ
Primer 1

(c) Taq
3' ———————————————————— Tₙ

(d) Primer 2
5' ———————————————————— 3'
3' ———————————————————— Tₙ
Primer 1

Standard PCR amplification

Fig. 9.3 RT-PCR. (a) Reverse transcriptase (RTase) is used to synthesise a cDNA copy of the mRNA. In this example, oligo(dT)-primed synthesis from the poly(A) tail of the mRNA is shown, although random or sequence-specific primers may also be used. In some cases, optimal results are obtained by using a mixture of oligo(dT) and random primers. (b) The mRNA is removed to leave a single-stranded (ss) cDNA product, and gene-specific primer is added (primer 1). (c) The second DNA strand is synthesised from primer 1 using Taq polymerase to generate double-stranded (ds) cDNA. (d) The ds cDNA is now used as the start point for a standard PCR reaction, with a second primer enabling amplification of the target fragment in a series of denature/prime/extend cycles (as detailed in Fig. 9.1). There are many kits available for RT-PCR, which can be carried out as a one-step procedure in a single tube or as a two-step process. The choice of method will depend on the aim of the procedure and the relative importance of the simplicity of the one-step reaction, or the flexibility and higher degree of control over reaction conditions that is possible in the two-step process.

to allow exponential amplification from both primers. This has no effect on the final outcome of the process. An overview of RT-PCR is shown in Fig. 9.3.

One use of RT-PCR is in determining the amount of mRNA in a sample (competitor RT-PCR). A differing, but known, amount of competitor RNA is added to a series of reactions, and the target and competitor amplified using the same primer pair. If the target and competitor products are of different sizes, they can be separated on a gel and the amount of target estimated by comparing with the amount of competitor product. When the two bands are of equal

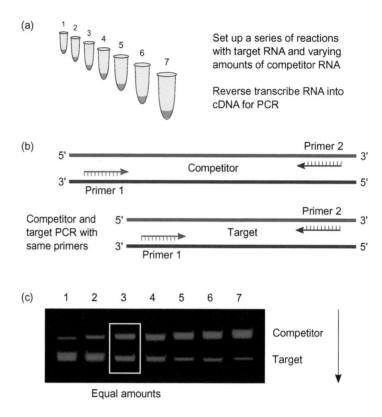

Fig. 9.4 Competitor RT-PCR. (a) A series of samples is spiked with increasing amounts of a competitor RNA that will produce a PCR product that is slightly different in size to the prospective target fragment. (b) The competitor RNA 'competes' with the target RNA for the primers and other resources in the reaction during the PCR. (c) When analysed on an electrophoresis gel, the products can be distinguished on the basis of size. Direction of movement in the gel is shown by the arrow. The equal band intensities observed in sample 3 (boxed) enables the amount of target mRNA in the original sample to be determined, as in this reaction, the competitor and target were present at equal concentrations at the start of the PCR process.

intensity, the amount of target sequence in the original sample is the same as the amount of competitor added. This approach is shown in Fig. 9.4.

9.3.2 Nested PCR

Nested PCR is a useful way of overcoming some of the problems associated with a large number of PCR cycles, which can lead to error-prone synthesis. The technique essentially increases both the sensitivity and fidelity of the basic PCR protocol. It involves using two sets of primers. The first **external** set generates a normal PCR product. Primers that lie inside the first set are then used for a second PCR reaction. These **internal** or **nested** primers generate a shorter product, as shown in Fig. 9.5.

(a)
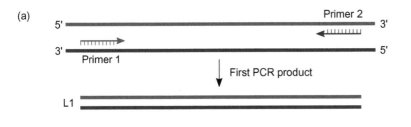

(b)
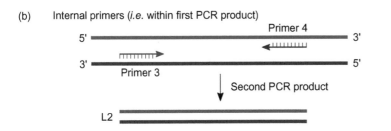

Fig. 9.5 Nested PCR. (a) Standard PCR amplification of a fragment using the first set of primers (primers 1 and 2). This generates a PCR product from this primer pair of length L1. (b) In the second PCR, a set of primers that lie inside the first pair is used to prime the synthesis of a shorter fragment (primers 3 and 4, product length L2). Nested PCR can be used to increase the specificity and fidelity of the PCR procedure.

9.3.3 Inverse PCR

Sometimes a stretch of DNA sequence is known, but the desired target sequence lies outside this region. This causes problems with the primer design, as there may be no way of determining a suitable primer sequence for the unknown region. **Inverse PCR (IPCR)** involves isolating a restriction fragment that contains the known sequence plus flanking sequences. By circularising the fragment, and then cutting inside the known sequence, the fragment is effectively inverted. Primers can then be synthesised using the known sequence data and used to amplify the fragment, which will contain the flanking regions. Primers that face away from each other (with respect to the direction of product synthesis) in the original known sequence are required, so that on circularisation, they are in the correct orientation. The technique can also be used with sets of primers for nested PCR. Deciphering the result usually requires DNA sequencing to determine the areas of interest. An illustration of inverse PCR is shown in Fig. 9.6.

9.3.4 Quantitative and Digital PCR

Although the basic PCR (end-point PCR) allows some level of quantification of products by comparing band intensities on a gel, more sensitive methods have been developed to measure how much of a DNA fragment (or mRNA transcript) is present in a sample. Two versions of this technique are now used routinely. Firstly, **quantitative PCR (qPCR)**, which is also known as real-time PCR or **real-time qPCR**. [*Confusion alert*: some authors may use the abbreviation RT to note *real time*, but this is conventionally taken to mean reverse transcriptase PCR, so should be avoided for the real-time version. Even

Quantitative and digital PCR methods use sophisticated technology to enable accurate quantification of the amount of target sequence in a sample.

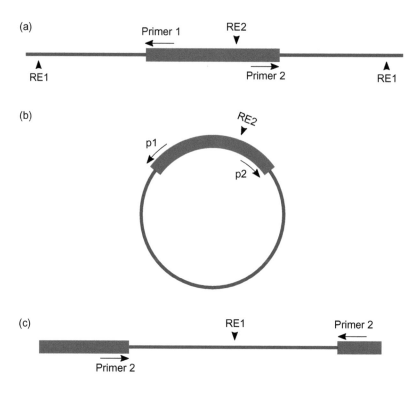

Fig. 9.6 Inverse PCR. (a) A region of DNA in which part of the sequence is known (thick red line). If the areas of interest lie outside the known region, inverse PCR can be used to amplify these flanking regions. Primers are selected to bind within the known sequence in the opposite orientation to that normally used (outward-facing instead of inward-facing). Restriction sites are noted as RE1 and RE2. If RE1 is used to cut the DNA, the resulting fragment can be circularised to give the construct shown in (b). If this is cut with RE2, the linear fragment that results has the unknown sequence in the middle of two known sequence fragments. Note that the unknown region is non-contiguous, being made up of two flanking regions joined at RE1. The PCR product will therefore contain these two regions plus the regions of known sequence.

more messy, mRNA analysis using qPCR can sometimes be called *real-time RT-qPCR*. It is best to stick to RT-PCR for the amplification of mRNA, qPCR for the quantitative version and RT-qPCR for mRNA quantification.]

The qPCR method monitors the reaction as it proceeds, by detection of a signal generated by fluorescent labels after each cycle. A **quantification cycle (Cq)** (often still known as the **cycle threshold (Ct)**) is set to avoid background 'noise' and ensure that the signal is strong enough to enable unambiguous detection. The basis of the technique is that the time (*i.e.* number of cycles) taken to reach the Cq depends on the amount of DNA at the start of the reaction – very low amounts of DNA need more cycles to be amplified to the Cq than higher concentrations of the target sequence. A **calibration curve** can be generated to enable estimation of the amount of the target

(a)

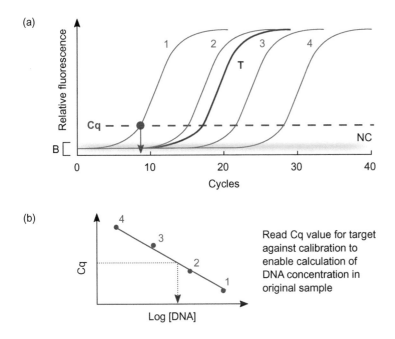

(b)

Fig. 9.7 Quantitative PCR. (a) Real-time or qPCR monitors the signal generated by fluorophores to follow the reaction progress through the early, exponential and plateau phases (usually around 40 cycles). The plot shows four known amounts of input DNA (blue lines 1–4) and an unknown sample (red line, T). Low levels of fluorescence (in the area marked B) are generated by initial amplification cycles plus background 'noise', and this sets the baseline for the reactions. A detection threshold is set that is sufficiently greater than the baseline to ensure that the signal is strong enough for unambiguous detection of amplified products. This sets the Cq value (quantification cycle, blue dashed line) that is used to determine the cycle at which the threshold is reached for each sample. (b) The Cq values determined for the standards are plotted against the concentration of DNA (log scale), and regression analysis used to plot the line of best fit. This can be used as a calibration graph to determine the concentration of the unknown by using the Cq value for this sample.

sequence (an **absolute** measurement, albeit dependent on the accuracy of the calibration), or the **relative** amount can be assessed by comparing the reaction against amplification of a known reference sequence. The qPCR method is shown in Fig. 9.7.

An extension of the qPCR method in recent years has led to the development of digital PCR (dPCR). Although the basis of the technique was first developed in the early-1990s under the name **limiting dilution PCR**, the technological improvements in miniaturisation and fabrication have made the method more accessible and reliable (in a sense, it parallels the development of next-generation sequencing techniques for DNA sequencing that we looked at in Section 6.8.1). Instead of all the sample fragments being in a single reaction, dPCR partitions the reaction mixture into thousands of sub-nanolitre scale reactions using two methods: (1) physical separation into nanowells in chamber digital PCR (cdPCR), or (2) a water-in-oil emulsion process can be used to generate droplets in **droplet digital PCR (ddPCR)**. Both techniques achieve the aim of separation of the target fragments into discrete units, such that each unit will have either zero fragments, or one or more than one fragment. Thus, we have a 'digital' situation where a well or droplet will score either positive or negative at the end-point of a PCR amplification. This lends itself to statistical analysis of the fit to a **Poisson distribution**, by which the concentration of the target fragment can be determined. The technique is particularly useful where there is a very small amount of the target fragment present (and thus a complex background due to the high level of non-target DNA), as the partitioning increases the effective concentration of the target in the partitioned reaction volume that contains a target fragment. A second advantage is that any contaminants that might inhibit the PCR reaction are diluted to the point where they are not an issue. The principle behind digital PCR is shown in Fig. 9.8.

9.3.5 RAPD and Several Other Acronyms

Normally the aim of the PCR is to generate defined fragments from highly specific primers. However, there are some techniques based on low-stringency annealing of primers. One of these is RAPD-PCR. This stands for **random amplification of polymorphic DNA**. The technique is also known as **arbitrarily primed PCR (AP-PCR)**. It is a useful method of genomic fingerprinting, and involves using a single primer at low stringency. The primer binds to many sites in the genome, and fragments are amplified from these. The stringency of primer binding can be increased after a few cycles, which effectively means that only the 'best mismatches' are fully amplified. By careful design, the protocol can yield genome-specific band patterns that can be used for comparative analysis. However, there can be problems associated with the reproducibility of the technique, as it is difficult to obtain similar levels of primer binding in different experiments. This is largely due to the mismatched binding of primers at low stringency that is the basis of the technique.

Although high-fidelity copying is usually the desired outcome of a PCR experiment, there are some procedures such as RAPD where low stringency of primer binding enables a DNA profile (made up of a number of amplified fragments) to be generated.

A final word on PCR variants. **Multiplex PCR** enables two or more fragments to be amplified in the same reaction, using different primer pairs. Usually 2–5 sets of primers are used. Primers can be unlabelled, labelled with different fluorophores or used in conjunction with labelled probes to enable identification. Different amplified fragment lengths can also help identification of the specific fragments on the basis of their mobility on a gel. There are several advantages that multiplex PCR has over single-fragment PCR (which is often called a **singleplex** reaction when comparing to multiplex). For example, internal standardisation can be achieved if one of the primers amplifies a known reference gene, the multiple targets can identify different genes or gene variants within a sample or the primers can identify multiple species/strains of pathogens. There is also the saving on reagent costs, and less variation in reaction conditions for a particular multiplex sample compared to individual reactions. The downside is

Although multiplex PCR presents some challenges in primer design and reaction optimisation, it offers a number of advantages over singleplex methods.

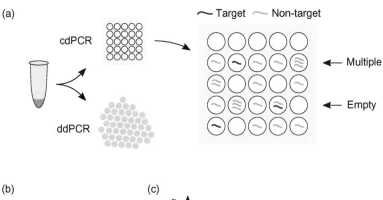

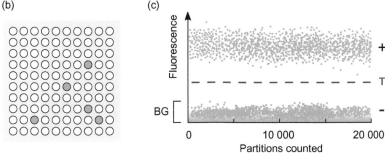

Fig. 9.8 Digital PCR. (a) The PCR reaction mix is partitioned into thousands of units; two methods use either sub-nanolitre volume chambers (cdPCR) or generate water-in-oil droplets (ddPCR). Each well or droplet will contain either no fragments or a single fragment, or more than one fragment. Target fragments will be distributed across the background of non-target or empty wells/droplets. (b) On amplification, the target fragments are scored as positive, and non-target and empty wells as negative, hence the 'digital' classification as either one or zero. (c) Wells or droplets are monitored, and the output processed. In this example, 20 000 wells are scanned and the relative fluorescence plotted. The non-target and empty well background (BG) generates a strong signal overall due to the larger number of wells in this category. A threshold detection level (blue dashed line, T) is set to ensure only positive wells are counted. The data are processed using statistical 'goodness of fit' to the Poisson distribution, and the amount of DNA can be calculated from this.

that multiplex PCRs can be difficult to set up, primer design can be tricky and differential amplification can be an issue due to primer or sequence length bias. However, these issues can usually be controlled successfully. The basis of multiplex PCR is outlined in Fig. 9.9.

In addition to the few PCR variants that we have looked at so far, there are many others. This has generated a whole new set of PCR abbreviations, acronyms, descriptors and initialisms. These are listed in Table 9.1. To avoid 'variant fatigue', we won't look at these in detail! However, it is worth having a look at information on the web (at least for some of them) to appreciate the impact that these increasingly novel and complex procedures have on the range of PCR-based applications.

(a)

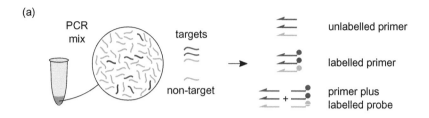

(b)

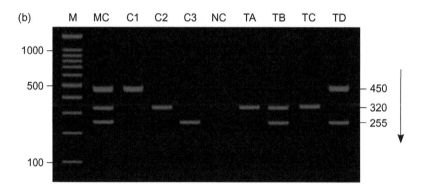

(c)

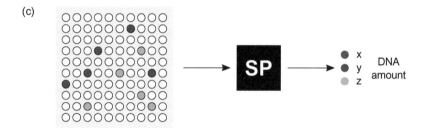

Fig. 9.9 Multiplex PCR. Multiplexing can be used with standard PCR, qPCR and dPCR. In (a), the PCR sample contains non-target fragments and three targets of interest. These may be from a number of sources. The technique is flexible and can detect, for example, different pathogens in clinical samples, different species in environmental samples or different genes in the same sample. Primer pairs can be unlabelled if standard qualitative electrophoresis is the detection method, or can be labelled with fluorophores either directly or in conjunction with a probe. (b) Agarose gel electrophoresis can be used to determine multiple fragments in multiplex PCR if they are different sizes. Here the example shows identification of three pathogens, using primer pairs that generate target amplicons of 450, 320 and 255 bp in length for pathogens 1, 2 and 3, respectively. Lanes are: M = 100 bp marker ladder; MC = multiplex positive control with DNA from all three pathogens; C1, C2 and C3 = positive controls for each pathogen separately; and NC = negative control (no DNA). Lanes TA–TD are four unknown test samples. The arrow shows the direction of movement of the fragments in the gel. Results indicate that samples TA and TC contain DNA from pathogen 2, TB from pathogens 2 and 3 and TD from pathogens 1 and 3. (c) For multiplex cdPCR, the fluorophore labels identify wells (or droplets if ddPCR) that contain the target fragments. A signal processor (SP) is used to analyse and interpret the output and calculate the amount of each of the target DNA fragments present in the sample.

9.4 | Processing and Analysing PCR Products

Once the PCR process has been completed, the DNA fragments that have been amplified (amplicons) can be analysed. For a standard end-point PCR, electrophoresis of the products is used to verify the amplicon size, and can also give some idea of purity, relative amount and any unexpected fragments. Blotting and hybridisation techniques may be used to identify specific regions of the sequence. If the PCR product is to be cloned, one of the methods we looked at in Section 8.3 can be used to insert the amplicon into a suitable vector for applications such as DNA sequencing or expression of the cloned gene.

As illustrated in Figs. 9.7, 9.8 and 9.9, variants of the standard PCR often have specific methods for visualising, evaluating and quantifying the results of the reaction. As the PCR protocols become more complex and technologically advanced, the detection system(s) also need to be developed to ensure that the output is captured and evaluated in an appropriate way.

> Analysis of PCR products can range from simple gel electrophoresis to complex technology that generates tens of thousands of data points that require statistical analysis. PCR products may be cloned in a suitable vector for further analysis and manipulation.

9.5 | Conclusion: The Game-Changing Impact of the PCR

Now and again a scientific discovery is made that changes the whole course of the development of a subject, and the PCR certainly has had this effect. In this chapter, we have seen that the basic theory of how the PCR works is in fact fairly simple, with *repetitive cycles* of denaturation, primer binding and primer extension acting as a copying mechanism that doubles the number of target fragment copies in each cycle. Automation of the process using a *thermal cycler*, and the availability of *thermostable DNA polymerases*, enabled the technique to become established widely. The standard PCR reaction proceeds for a number of cycles (30 or so), and the products analysed; for this reason, it is often called *end-point PCR*.

Oligonucleotide *primers* are the critical discriminating elements in the PCR, as they specify the DNA fragments to be amplified. Primer design is now most often carried by computation using sequence database information.

Somewhere around 50 or so different variants of the basic PCR exist, so we were only able to consider a few of these. *RT-PCR* enables mRNA to be amplified by generating cDNA prior to the PCR. *Nested* and *inverse* PCR are two conceptually elegant methods that illustrate how careful design can extend the specificity of the procedure. *Quantitative* and *digital PCR* methods enable accurate determination of the amount of nucleic acid in a sample, with the latest versions of digital PCR taking the procedure into technologically advanced areas. This is similar to the development of next-generation DNA sequencing, and dPCR is sometimes referred to as **third-generation PCR**.

Although *multiplex PCR* was our last look at a variant in detail, a list of the acronyms and names of PCR variants was presented in Table 9.1, with an encouragement to look for some more detailed information about these variants.

The applications of PCR technology are many and diverse, and new procedures are developed on a regular basis. Sometimes these involve the development of a more sophisticated form of PCR-based technology that has wide application potential. At other times, it may be that an existing technique is used in a novel way or to investigate the genome of an organism that has not been studied in detail. We will consider applications of the PCR more fully in Part 4 of this book.

Finally, **diagnostic testing** is a major use of PCR technology that entered the public consciousness (in an unwelcome way) during the COVID-19 pandemic, with the PCR constantly in the news as it was used extensively worldwide for detection of the SARS-CoV-2 virus. The public was exposed to some of the uncertainties of science during debates on PCR testing, reflecting some of the issues we considered in Sections 1.4 and 3.2. In years to come, when we look back at the period 2020–2022, might we perhaps note that this was the time when the most recognised triplet in biology changed from 'DNA' to 'PCR'?

Further Reading

Kralik, P. and Ricchi, M. (2017). A basic guide to real time PCR in microbial diagnostics: definitions, parameters, and everything. *Front. Microbiol.*, 8, 108. URL [www.frontiersin.org/articles/10.3389/fmicb.2017.00108/full]. DOI [https://doi.org/10.3389/fmicb.2017.00108].

Mullis, K. B. (1990). The unusual origin of the polymerase chain reaction. *Sci. Am.*, 262 (4), 56–65. DOI [https://doi.org/10.1038/scientificamerican0490-56].

Taylor, S. C. *et al.* (2017). Droplet digital PCR versus qPCR for gene expression analysis with low abundant targets: from variable nonsense to publication quality data. *Sci. Rep.*, 7, 2409. URL [www.nature.com/articles/s41598-017-02217-x#citeas]. DOI [https://doi.org/10.1038/s41598-017-02217-x].

Wong, M. L. and Medrano, J. F. (2005). Real-time PCR for mRNA quantitation. *BioTechniques*, 39 (1), 75–85. URL [www.future-science.com/doi/epub/10.2144/05391RV01]. DOI [https://doi.org/10.2144/05391RV01].

Xie, N. G. *et al.* (2022). Designing highly multiplex PCR primer sets with Simulated Annealing Design using Dimer Likelihood Estimation (SADDLE). *Nat. Commun.*, 13, 1881. URL [www.nature.com/articles/s41467-022-29500-4#article-info]. DOI [https://doi.org/10.1038/s41467-022-29500-4].

Websearch

People

Mention PCR and most who know a little about it will recognise *Kary Mullis* as the person who invented the process. Fill in the details of how Mullis developed the PCR, and have a wider look at some of the ideas and controversies associated with the man who changed the way in which modern molecular biology developed.

Places

Two odd little diversions here – the first place we are visiting is the web. Using Table 9.1, have a look at some of the terms listed that are not covered in the text. Use search terms based around the descriptor or abbreviation for the PCR variant. Try to build up a picture of how the basic PCR process has been modified to develop the new method. Our second place is Highway 128 in Northern California just where it branches off from Highway 101. Using the street view function of the well-known mapping application, put yourself on Highway 128 going North from Cloverdale. Somewhere along this road as it approaches Anderson Valley, the fundamentals of the PCR were put together by Kary Mullis one April night in 1983. It's just an ordinary road that was the site of an extraordinary discovery.

Processes

From a theoretical perspective, PCR is an elegant and conceptually straightforward process. In practice, like many protocols in biological research, it is a little bit less clear-cut. Have a look at two aspects of the PCR that need a little bit of puzzling to grasp fully – I suggest you get a few large pieces of paper to use a 'sketch and think' approach! Firstly, although PCR amplifies a defined fragment, the first few cycles generate products that are a little ragged in terms of fragment length. Why is this? Secondly, when working out 'how many fragments after how many cycles', we usually refer to amplification of a single target fragment. Why is this not likely to be the case in most applications, what difference would this make to the outcome, and is it likely to be a problem? Good luck!

Reflections

What topics in this chapter have you found most challenging? Look for resources that help to illustrate the key points.

The polymerase chain reaction

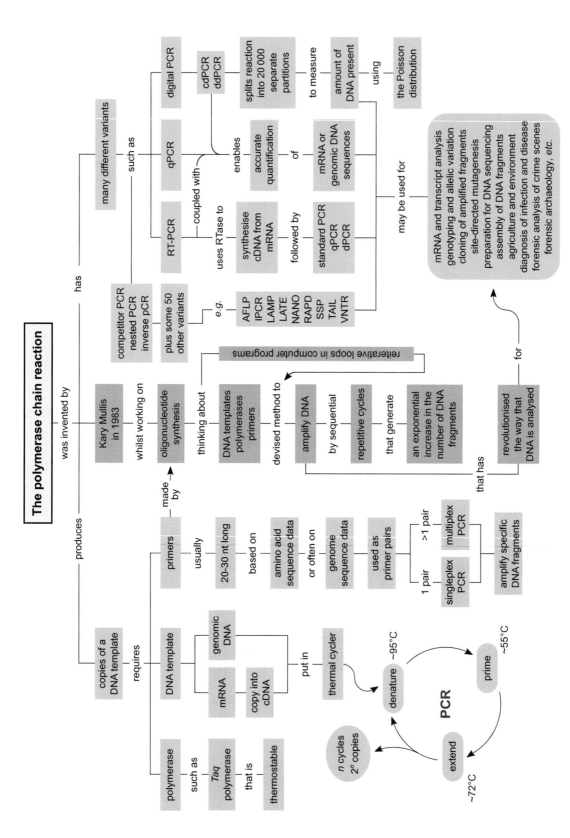

Concept Map 9

Chapter 10 Summary

Learning Objectives

When you have completed this chapter, you will be able to:

- Define the terms 'selection' and 'screening' with reference to clone analysis
- Evaluate a range of genetic selection and screening methods, including the use of chromogenic substrates and PCR
- Explain the use of nucleic acid probes in screening clone banks
- Outline the techniques used for analysis of cloned DNA fragments, including the Southern blotting technique
- Discuss the principles of sub-cloning and DNA sequencing for clone analysis

Key Words

Selection, screening, positive selection, negative selection, automation, antibiotic, ampicillin, tetracycline, chromogenic substrates, X-gal, β-galactosidase, IPTG, α-peptide, insertional inactivation, plaque morphology, cI repressor, complementation, auxotrophic, auxotroph, *hfl*, Spi$^+$, Ccd B protein, nucleic acid hybridisation, probe, homology, heterologous, plus/minus screening, chromosome walking, chromosome jumping, stringency, combinatorial screening, antibodies, antigen, polyclonal antibodies, monoclonal antibodies, immunological screening, rabbit anti-X IgG, goat anti-rabbit IgG, restriction map, Sir Edwin Southern, Southern blotting, Northern blotting, Western blotting, SDS-PAGE, Eastern blotting, Southwestern blotting, Northwestern blotting, dot-blotting, slot-blotting, microarrays, sub-cloning, massively parallel DNA sequencing, SMRT sequencing.

Chapter 10

Selection, Screening and Analysis of Recombinants

Success in any cloning experiment depends on being able to identify the desired sequence among the many different recombinants that may be produced. This may be straightforward for a simple sub-cloning procedure where the desired construct is present as a high proportion of the clones, or may be more difficult if working with a clone bank to isolate an unknown gene. In this chapter, we will look at some of the techniques that can be used to identify and characterise cloned sequences. As has been the case in previous chapters, we will include some examples that have played an important role in development of the technology, although they are not used widely today.

There are two terms that require definition before we proceed, these being selection and screening. Selection is where some sort of condition (*e.g.* the presence of an antibiotic) is applied during the growth of host cells containing recombinant DNA. The cells with the desired characteristics are therefore *selected* by their ability to survive. This approach ranges in sophistication, from simple selection for the presence of a vector, up to direct selection of cloned genes by complementation of defined mutations. Selection can be either **positive** (selecting for the presence of a gene product or function) or **negative** (selecting for the absence of a product or function). Screening a population of viable cells involves using a procedure that enables identification of the target sequence(s). Because only a small proportion of the large number of bacterial colonies or bacteriophage plaques being screened will contain the DNA sequence(s) of interest, screening requires methods that are highly sensitive and specific. In practice, both selection and screening methods may be required in any single experiment, and may even be used at the same time if the procedure is designed carefully.

In recent years, **automation** of various procedures in gene manipulation has become more widespread. In smaller laboratories, with perhaps only a few research projects at different stages of development, there may not be the same requirement for high-efficiency throughput as there might be in a commercially based company or a diagnostic laboratory. Robotic systems can be used to carry out many routine tasks, often in 96-well microtitre plates. In many cases, the technological advances in sample preparation, processing and

Identifying a clone can involve using some sort of selective mechanism (e.g. antibiotic or β-galactosidase status) and/or screening a population of clones with a specific probe (e.g. a labelled cDNA or oligonucleotide).

handling mean that very large numbers of clones can be handled easily, and many protocols are now developed specifically for use with automated procedures.

10.1 | Genetic Selection and Screening Methods

Antibiotic resistance marker genes provide a simple and reliable way to select for the presence of vectors in cells.

One of the simplest genetic selection methods involves the use of an **antibiotic** to select for the presence of vector molecules. For example, the plasmid pBR322 contains genes for resistance to **ampicillin** (Ap^r) and **tetracycline** (Tc^r). Thus, the presence of the plasmid in cells can be detected by plating potential transformants on an agar medium that contains either (or both) of these antibiotics. Only cells that have taken up the plasmid will be resistant, and these cells will therefore grow in the presence of the antibiotic. The technique can also be used to identify mammalian cells containing vectors with selectable markers.

Genetic selection methods can be simple (as above) or complex, depending on the characteristics of the vector/insert combination and on the type of host strain used. Such methods are extremely powerful, and there is a wide variety of genetic selection and screening techniques available for many diverse applications. Some of these are described below.

10.1.1 Use of Chromogenic Substrates

The use of **chromogenic substrates** in genetic screening methods has been an important aspect of the development of the technology. The most popular system uses the compound **X-gal** (5-bromo-4-chloro-3-indolyl-β-D-galactopyranoside), which is a colourless substrate for β-galactosidase. The enzyme is normally synthesised by *Escherichia coli* cells when lactose becomes available. However, induction can also occur if a lactose analogue such as **IPTG** (*iso*-propyl-thiogalactoside) is used. This has the advantage of being an inducer without being a substrate for β-galactosidase. On cleavage of X-gal, a blue-coloured product is formed (Fig. 10.1); thus, the expression of the *lac Z* (β-galactosidase) gene can be detected easily. This can be used either as a screening method for cells or plaques, or as a system for the detection of tissue-specific gene expression in transgenics.

The β-galactosidase/X-gal system has become a popular method of identifying cloned DNA fragments on the basis of an easy-to-read blue/white colour difference.

The X-gal detection system can be used where a functional β-galactosidase gene is present in the host/vector system. This can occur in two ways. Firstly, an intact β-galactosidase gene (*lac Z*) may be present in a vector such as a λ insertion vector. Host cells that are Lac$^-$ are used for propagation of the phage, so that the Lac$^+$ phenotype will only arise when the vector is present. A second approach is to employ the α-complementation system, in which only part of the *lac Z* gene (encoding a peptide called the **α-peptide**) is carried by the vector. The remaining part of the gene sequence is on the chromosome of the host cell. The region coding for the smaller, vector-encoded α-peptide is designated *lac Z'*. Host cells are therefore designated *lac Z'⁻*.

The α-peptide of β-galactosidase can be used in a vector to enable a more sophisticated form of X-gal blue/white identification to be achieved by complementation.

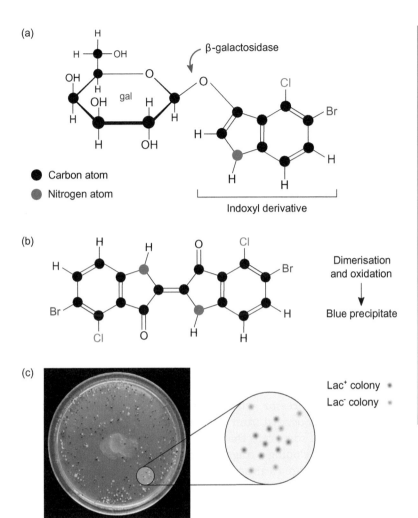

(a)

Carbon atom

Nitrogen atom

Indoxyl derivative

(b)

Dimerisation
and oxidation

Blue precipitate

(c)

Lac+ colony

Lac- colony

Fig. 10.1 Structure of X-gal and cleavage by β-galactosidase. (a) The colourless compound X-gal (5-bromo-4-chloro-3-indolyl-β-D-galactopyranoside) is cleaved by β-galactosidase to give galactose and an indoxyl derivative. (b) This forms dimers and is oxidised in air to generate the dibromo-dichloro derivative (5,5′-dibromo-4,4′-dichloroindigo), which forms a dense blue precipitate. (c) Colony screening using X-gal. Bacterial colonies are grown on a medium that contains X-gal and IPTG. Colonies that are phenotypically Lac+ (functional β-galactosidase) hydrolyse X-gal and are blue. Colonies that are Lac− (e.g. due to insertional inactivation of the *lac Z* gene in the cloning vector) are normal colour (usually called white colonies or plaques). *Source:* Photograph by Stefan Walkowski. Used under Licence CC-BY-SA-4.0 [www.creativecommons.org/licenses/by-sa/4.0/].

Blue colonies or plaques will only be produced when the host and vector fragments complement each other to produce functional β-galactosidase.

10.1.2 Insertional Inactivation

The presence of cloned DNA fragments can be detected if the insert interrupts the coding sequence of a gene. This approach is known as **insertional inactivation** and can be used with any suitable genetic system. Three systems will be described to illustrate the use of the technique.

Antibiotic resistance can be used as an insertional inactivation system if DNA fragments are cloned into a restriction site within an antibiotic resistance gene. For example, cloning DNA into the *Pst* I site of pBR322 (which lies within the Ap^r gene) interrupts the coding sequence of the gene and renders it non-functional. Thus, cells that

Antibiotic resistance status can be used to identify a vector carrying a cloned insert. If the insert is cloned into the coding sequence of an antibiotic resistance gene, the gene will be inactivated and the cell will be sensitive to the antibiotic. This is a very powerful method of selecting cells for further analysis.

harbour a recombinant plasmid will be Ap^sTc^r. This can be used to identify recombinants as follows – if transformants are plated firstly onto a tetracycline-containing medium, all cells that contain the plasmid will survive and form colonies. If a replica of the plate is then taken and grown on an ampicillin-containing medium, the recombinants (Ap^sTc^r) will not grow, but any non-recombinant transformants (Ap^rTc^r) will. Thus, recombinants are identified by their absence from the replica plate, and can be picked from the original plate and used for further analysis.

The X-gal system can also be used as a screen for cloned sequences. If a DNA fragment is cloned into a functional β-galactosidase gene in a λ insertion vector, any recombinants will be genotypically *lac Z⁻* and will therefore not produce β-galactosidase in the presence of IPTG and X-gal. Plaques containing such phage will therefore remain colourless. Non-recombinant phage will retain a functional *lac Z* gene, and therefore give rise to blue plaques. This approach can also be used with the α-complementation system; in this case, the insert DNA inactivates the *lac Z'* region in vectors such as the M13 phage and pUC plasmid series. Thus, complementation will not occur in recombinants, which will be phenotypically Lac⁻ and will therefore give rise to colourless plaques or colonies (Fig. 10.2).

Plaque morphology can also be used as a screening method for certain λ vectors such as λgt10, which contain the *cI* gene. This gene encodes the **cI repressor**, which is responsible for the formation of lysogens. Plaques derived from *cI⁺* vectors will be slightly turbid, due to the survival of some cells that have become lysogens. If the *cI* gene is inactivated by cloning a fragment into a restriction site within the gene, the plaques are clear and can be distinguished from the turbid non-recombinants. This system can also be used as a selection method (see Section 10.1.4).

10.1.3   Complementation of Defined Mutations

Direct complementation of a defined mutation may be feasible if the gene system is well known and an appropriate mutant is available. In this case, the defective gene function would be 'supplied' by the cloned sequence.

Direct selection of cloned sequences is possible in some cases. An example is where antibiotic resistance genes are being cloned, as the presence of cloned sequences can be detected by plating cells on a medium that contains the antibiotic in question (assuming that the host strain is normally sensitive to the antibiotic). The method is also useful where specific mutant cells are available, as the technique of complementation can be employed, where the cloned DNA provides the function that is absent from the mutant. There are three requirements for this approach to be successful. Firstly, a mutant strain that is deficient in the particular gene that is being sought must be available. Secondly, a suitable selection medium is required, on which the specific recombinants will grow. The final requirement, which is often the limiting step as far as this method is concerned, is that the gene sequence must be expressed in the host cell to give a functional product that will complement the mutation. This is not a problem if, for example, an *E. coli* strain is used to select cloned *E. coli* genes, as the cloned sequences will obviously function in the host cells.

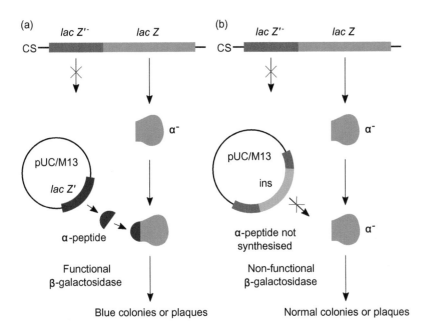

Fig. 10.2 Insertional inactivation in the α-complementation system. (a) The chromosome (CS) has a defective *lac Z* gene that does not encode the N-terminal α-peptide of β-galactosidase (specified by the *lac Z'* gene fragment). Thus, the product of the chromosomal *lac Z* region is an enzyme lacking the α-peptide (α⁻). If a non-recombinant pUC plasmid or M13 phage is present in the cell, the *lac Z'* gene fragment encodes the α-peptide, which enables functional β-galactosidase to be produced. In the presence of X-gal and IPTG, blue colonies or plaques will appear. (b) If a DNA fragment is inserted into the vector (ins, shown in green), the *lac Z'* gene is inactivated and no complementation occurs. Thus, colonies or plaques will not appear blue.

This approach has been used most often to select genes that specify nutritional requirements, such as enzymes of various biosynthetic pathways. Thus, genes of the tryptophan operon can be selected by plating recombinants on mutant cells that lack specific functions in this pathway (these are known as **auxotrophic** mutants or just **auxotrophs**). In some cases, complementation in *E. coli* can be used to select genes from other organisms such as yeast, if the enzymes are similar in terms of their function and they are expressed in the host cell. Complementation can also be used if mutants are available for other host cells, as is the case for yeast and other fungi.

Selection processes can also be used with higher eukaryotic cells. The gene for mouse dihydrofolate reductase (DHFR) has been cloned by selection in *E. coli* using the drug trimethoprim in the selection medium. Cells containing the mouse *DHFR* gene were resistant to the drug, and were therefore selected on this basis.

There are certain constraints on direct selection methods; some are more difficult to overcome than others.

10.1.4 Other Genetic Selection Methods

The use of the cI repressor system of λgt10 can be extended to provide a powerful selection system if the vector is plated on a mutant strain of *E. coli* that produces lysogens at a high frequency. Such strains are designated *hfl* (high frequency of lysogeny), and any phage that encodes a functional cI repressor will form lysogens on these hosts. These lysogens will be immune to further infection by phage. DNA fragments are inserted into the λgt10 vector at a restriction site in the *cI* gene. This inactivates the gene, and thus only recombinants (genotypically *cI*⁻) will form plaques.

A second example of genetic selection based on phage/host characteristics is the Spi selection system that can be used with vectors such as EMBL4. Wild-type λ will not grow on cells that already carry a phage, such as phage P2, in the lysogenic state. Thus, the λ phage is said to be **Spi⁺** (sensitive to *P2* inhibition). The Spi⁺ phenotype is dependent on the *red* and *gam* genes of λ, and these are arranged so that they are present on the stuffer fragment of EMBL4. Thus, recombinants, which lack the stuffer fragment, will be *red⁻ gam⁻* and will therefore be phenotypically Spi⁻. Such recombinants will form plaques on P2 lysogens, whereas non-recombinant phage that are *red⁺ gam⁺* will retain the Spi⁺ phenotype and will not form plaques.

A final example of a genetic selection system is the toxin/antitoxin system used in Gateway cloning vectors (see Section 8.4.4 and Fig. 8.11). The *ccd B* gene product (**Ccd B protein**) kills cells if present, and thus is a powerful selection system in that constructs that are *ccd B⁻* (destination vectors with inserts) will survive and any *ccd B⁺* constructs (no inserts) will not grow. This is a powerful system that can generate cloning efficiency approaching 100 per cent.

10.2 | Screening Using Nucleic Acid Hybridisation

General aspects of **nucleic acid hybridisation** are described in Section 6.5. It is a very powerful method of screening clone banks, and is one of the key techniques in gene manipulation. The production of a cDNA or genomic DNA library is often termed the 'shotgun' approach, as a large number of essentially random recombinants is generated. By using a defined nucleic acid **probe**, such libraries can be screened and the clone(s) of interest identified. As we have seen for other applications, the availability of genome sequences enables the design of probes or polymerase chain reaction (PCR) primers to be achieved easily by computation, and this has enabled a much more precise and targeted approach to screening clone banks.

10.2.1 Nucleic Acid Probes

The power of nucleic acid hybridisation lies in the fact that complementary sequences will bind to each other with a very high degree of fidelity. In practice, this depends on the degree of **homology** between the hybridising sequences, and usually the aim is to use a probe that has been derived from the same source as the target DNA. However, under certain conditions, sequences that are not 100 per cent homologous can be used to screen for a particular gene, as may be the case if a probe from one organism is used to detect clones prepared using DNA from a second organism. Such heterologous probes have been extremely useful in identifying many genes from different sources.

There are three main types of DNA probe, these being: (1) cDNA, (2) genomic DNA, and (3) oligonucleotides. Alternatively, RNA probes can be used if these are suitable. The availability of a particular probe will depend on what is known about the target gene sequence. If a cDNA clone has already been obtained and identified, the cDNA can be used to screen a genomic library and isolate the gene sequence itself. Alternatively, cDNA may be made from mRNA populations and used without cloning the cDNAs. This is often used in what is known as the 'plus/minus' method of screening. If the clone of interest contains a sequence that is expressed only under certain conditions, probes may be made from mRNA populations from cells that are expressing the gene (the plus probe) and from cells that are not expressing the gene (the minus probe). By carrying out duplicate hybridisations, the clones can be identified by their different patterns of hybridisation with the plus and minus probes. Although this method cannot usually provide a definitive identification of a particular sequence, it can be useful in narrowing down the range of candidate clones. The principle of the plus/minus method is shown in Fig. 10.3.

Genomic DNA probes are usually fragments of cloned sequences that are used either as heterologous probes or to identify other clones that contain additional parts of the gene in question. This is an important part of the techniques known as chromosome walking and chromosome jumping, and can enable the identification of

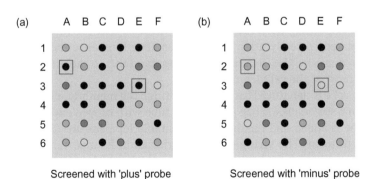

(a) Screened with 'plus' probe

(b) Screened with 'minus' probe

Fig. 10.3 Screening by the 'plus/minus' method. In this example, clones have been spotted on duplicate filters in a 6 × 6 array. Much larger arrays are usually prepared, or direct 'lifts' from agar plates can be used. In (a), one copy of the filter has been hybridised with a radioactive cDNA probe prepared from cells expressing the gene of interest (the 'plus' probe). In (b), the other copy of the filter has been hybridised with a probe prepared from cells that are not expressing the gene (the 'minus' probe). Some clones hybridise weakly with both probes (e.g. B1), and some strongly with both (e.g. C1). Some show a strong signal with the minus probe and a weaker signal with the plus probe (e.g. C5). Clones A2 (red square) and E3 (blue square) show a strong signal with the plus probe and a weaker signal with the minus probe. These would be selected for further analysis.

overlapping sequences which, when pieced together, enable long stretches of DNA to be characterised. We will consider chromosome walking and jumping in more detail in Chapter 15 when we look at medical applications of gene manipulation.

The use of oligonucleotide probes is possible where some amino acid sequence data are available for the protein encoded by the target gene. As we saw when looking at primer design for the PCR (see Section 9.2.2), a mixed probe can be synthesised that covers all the possible sequences by varying the base combinations at degenerate wobble positions. Alternatively, inosine can be used in highly degenerate parts of the sequence. The great advantage of oligonucleotide probes is that only a short stretch of sequence is required for the probe to be useful, and thus genes for which clones are not already available can be identified by sequencing peptide fragments and constructing probes accordingly.

When a suitable probe has been obtained, it can be labelled using one of the methods we looked at in Section 6.4. If a high-energy radioactive isotope such as ^{32}P is used, a probe of high specific activity can be generated. Alternatively, non-radioactive labelling methods such as fluorescent tags may be used if desired.

10.2.2   Screening Clone Banks

Colonies or plaques are not suitable for direct screening, so a replica is made on either nitrocellulose or nylon filters. This can be done either by growing cells directly on the filter on an agar plate (colonies) or by 'lifting' a replica from a plate (colonies or plaques). To do this, the recombinants are grown and a filter is placed on the surface of the agar plate. Some of the cells/plaques will stick to the filter, which therefore becomes a mirror image of the pattern of recombinants on the plate (Fig. 10.4). Reference marks are made so that the filters can be orientated correctly after hybridisation. The filters are then processed to denature the DNA in the samples, bind this to the filter and remove most of the cell debris.

The probe is denatured (usually by heating); the membrane filter is placed in a sealed plastic bag (or a plastic tube with capped ends), and the probe is added and incubated at a suitable temperature to allow hybrids to form. The **stringency** of hybridisation is important, and depends on conditions such as salt concentration and temperature. For homologous probes under standard conditions, incubation is usually at around 65–68°C. Time of incubation may be up to 48 hours in some cases, depending on the predicted kinetics of hybridisation. After hybridisation, the filters are washed (again the stringency of washing is important) and allowed to dry. They are then exposed to X-ray film to produce an autoradiograph, which can be compared with the original plates to enable identification of the desired recombinant.

An important factor in screening genomic libraries by nucleic acid hybridisation is the number of plaques that can be screened on each filter. Often an initial high-density screen is performed, and the plaques picked from the plate. Because of the high plaque density, it is often not possible to avoid contamination by surrounding plaques.

Oligonucleotides can be used as probes if designed carefully, taking the degeneracy of the genetic code into account.

Clone banks can be screened by producing a copy of the colonies (or plaques) on a nitrocellulose or nylon filter. This is then used to identify cloned sequences of interest using a specific nucleic acid probe.

The conditions for nucleic acid hybridisation have to be controlled carefully if optimum results are to be obtained. Key aspects include salt concentration and temperature.

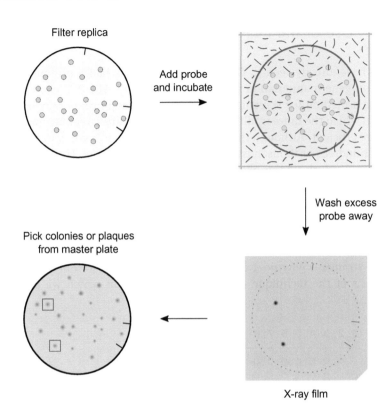

Filter replica

Add probe and incubate

Wash excess probe away

Pick colonies or plaques from master plate

X-ray film

Fig. 10.4 Screening clone banks by nucleic acid hybridisation. A nitrocellulose or nylon filter replica of the master Petri dish containing colonies or plaques is made. Reference marks are made on the filter and plate to assist with correct orientation. The filter is incubated with a labelled probe, which hybridises to the target sequences. Excess or non-specifically bound probe is washed off and the filter is exposed to X-ray film to produce an autoradiograph. Positive colonies (boxed) are identified and can be picked from the master plate.

Thus, the mixture is re-screened at a much lower plaque density, which enables isolation of a single recombinant (Fig. 10.5). This approach can be important if a large number of plaques has to be screened, as it cuts down the number of filters (and hence the amount of radioactive probe) required.

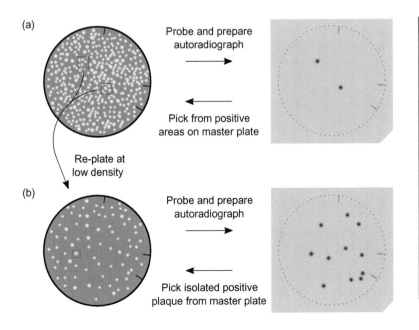

(a)

Probe and prepare autoradiograph

Pick from positive areas on master plate

Re-plate at low density

(b)

Probe and prepare autoradiograph

Pick isolated positive plaque from master plate

Fig. 10.5 Screening plaques at high and low densities. A radiolabelled probe was used to screen a genomic library in the λ vector EMBL3. Diagrams of the master plates and the autoradiographs are shown. (a) Initial screening was at a high plaque density, which identified two positive plaques on this plate. Phage particles were picked from the positive areas (red squares) and plated at a lower density. (b) The plaques were re-screened to enable isolation of individual plaques (e.g. green square). Many more positives are obtained due to the high proportion of 'target' plaques in the re-screened sample.

10.3 | Use of the PCR in Screening Protocols

As we have already seen in Chapter 9, the PCR has had a profound effect on gene manipulation and molecular biology. The PCR can be used as a method for screening clone banks, although there may be cases in which the more traditional approach to library screening may be more efficient. Although the PCR is easy to set up, it is still a major task if hundreds or thousands of clones have to be screened. There are, however, ways round this problem, and because the PCR is so specific (assuming that a suitable primer pair is available), it can be a useful way to identify clones.

One way of reducing the number of individual PCRs that have to be performed is to carry out PCRs on pooled samples of clones. This is known as **combinatorial screening**, and at first, the idea can seem a little strange – why pool clones if the aim is to isolate only one? The basis of the technique is shown in Fig. 10.6, in which we are

PCR can be used for screening clone libraries. Often clones are pooled from the rows and columns of microtitre plates, and a PCR carried out on the pooled samples.

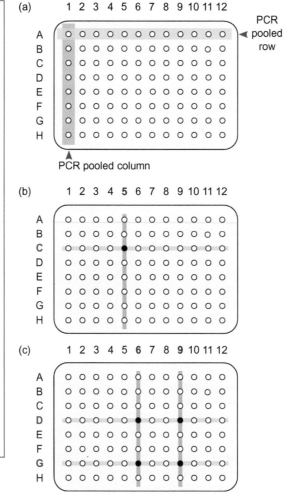

Fig. 10.6 The principle of combinatorial screening using the PCR. This shows a single 96-well microtitre plate. Clones are distributed in the wells and pooled across rows and columns, as shown in (a), highlighted in blue (row A pool) and red (column I pool). Row A clones (A1–A12) are pooled, and a single PCR performed. This is also done for column I clones (A1–H1). A single PCR is carried out for each pooled row and column; thus, for a 96-well plate, 20 PCRs are needed (8 row and 12 column pools). In (b), a result is shown where column 5 and row C PCRs show a positive result; thus, a positive clone must be in well C5. In (c), a different result is shown, with positives in rows D and G and columns 6 and 9. In this case, four clones (D6, D9, G6 and G9) would be retested to determine which were positives. The technique can be extended 'vertically' by adding more plates in a stacked arrangement and pooling in all three dimensions (row, column and vertical). The overall aim is to reduce the number of PCRs that need to be performed to achieve a specific identification.

considering clones that have been grown up in a single 96-well microtitre tray. As a library is likely to be much larger than 96 clones, the method can be used with many trays, with a vertical dimension added to the horizontal pooling of rows and columns in a single tray. Clones in the rows and columns of the plate are pooled, and a PCR performed on the mixed samples. If a clone is present in a 'row' pool, it must be in one of the 12 wells in that row. Similarly, if in a 'column' pool, the clone must be in that column. If a single positive is obtained, the clone is identified at the intersection of the row and column PCRs. If more than one positive is obtained, additional PCRs will need to be carried out to enable unequivocal identification of the clone.

10.4 | Immunological Screening for Expressed Genes

An alternative to screening with nucleic acid probes is to identify the protein product of a cloned gene by immunological methods. The technique requires that the protein is expressed in recombinants, and is often used for screening cDNA expression libraries that have been constructed in vectors such as λgt11 or plasmid-based expression vectors. Instead of a nucleic acid probe, a specific antibody is used.

Antibodies are produced by animals in response to challenge with an antigen, which is normally a purified protein. There are two main types of antibody preparation that can be used. The most common are **polyclonal antibodies**, which are usually raised in rabbits by injecting the antigen and removing a blood sample after the immune response has occurred. The immunoglobulin fraction of the serum is purified and used as the antibody preparation (antiserum). Polyclonal antisera contain antibodies that recognise all the antigenic determinants of the antigen. A more specific antibody can be obtained by preparing **monoclonal antibodies**, which recognise a single antigenic determinant. However, this can be a disadvantage in some cases. In addition, monoclonal antibody production is a complicated technique in its own right, and good-quality polyclonal antisera are often sufficient for screening purposes.

There are various methods available for **immunological screening**, but the technique is most often used in a similar way to 'plaque lift' screening with nucleic acid probes. Recombinant λgt11 cDNA clones will express cloned sequences as β-galactosidase fusion proteins, assuming that the sequence is present in the correct orientation and reading frame. The proteins can be picked up onto nitrocellulose filters and probed with the antibody. Detection can be carried out by a variety of methods, including

Antibody screening enables the specific identification of proteins that are synthesised from expression libraries. This technique can be used when a specific antibody is available.

the use of a non-specific second binding molecule such as protein A from bacteria. Alternatively, a second antibody can be used. This is raised in a different animal from which the primary antibody was prepared, and recognises the specifically bound primary antibody. Thus, if we are looking for protein X, and raise the antibody specific for protein X in a rabbit, the primary antibody would be **rabbit anti-X IgG**. The secondary antibody might be raised in goat and thus would be **goat anti-rabbit IgG**. Detection may be by radioactive label or by non-radioactive methods, including enzyme-linked and fluorophore-based methods similar to those outlined in Fig. 6.5.

10.5 | Analysis of Cloned Genes

Once clones have been identified by techniques such as hybridisation or immunological screening, more detailed characterisation of the DNA can begin. There are many ways of tackling this, and the choice of approach will depend on what is already known about the gene in question and on the ultimate aims of the experiment. In this section, we will look at some of the techniques that can be used for the initial analysis and manipulation of cloned genes.

10.5.1 Restriction Mapping

If the gene (or genome) sequence of the target fragment is known, many post-cloning manipulations can be designed and tested using computer software, with experimental confirmation. However, it is still sometimes necessary to construct an experimentally derived **restriction map** for a cloned fragment. Where phage or cosmid vectors have been used to clone relatively large pieces of DNA, a restriction map is helpful in isolating smaller fragments for subcloning into other vectors, the preparation of probes for chromosome walking and DNA sequencing. If a cloning protocol has involved the use of BAC or YAC vectors, with inserts in the size range of 300–500 kb, then so-called 'long-range' restriction mapping may be needed initially.

Generating a restriction map of a cloned DNA fragment is often one of the first tasks in analysing the clone.

We looked at the principle of restriction mapping in Section 5.1.3. In practice, the cloned DNA is usually cut with a variety of restriction enzymes to determine the number of fragments produced by each enzyme. If an enzyme cuts the fragment at frequent intervals, it will be difficult to decipher the restriction map, so enzymes with multiple cutting sites are best avoided. Enzymes that cut the DNA into 2–4 pieces are usually chosen for initial experiments. By performing a series of single and multiple digests with a range of enzymes, the complete restriction map can be pieced together. This provides the essential information required for more detailed characterisation of the cloned fragment.

10.5.2   Blotting Techniques

In many experiments, it is necessary to determine which parts of the original clone contain the regions of interest. This can be done by using a variety of methods based on blotting nucleic acid molecules onto membranes and hybridising with specific probes. Such an approach is, in some ways, an extension of clone identification by colony or plaque hybridisation, with the refinement that information about the structure of the clone is obtained.

The first blotting technique was developed by **Sir Edwin Southern** in 1975 and is eponymously known as Southern blotting. In this method, fragments of DNA, generated by restriction digestion, are subjected to agarose gel electrophoresis. The separated fragments are then transferred to a nitrocellulose or nylon membrane by a 'blotting' technique. The original method used capillary blotting, as shown in Fig. 10.7. Although other methods such as vacuum blotting and electroblotting have been devised, the original method is still used as it is simple and inexpensive.

When the fragments have been transferred from the gel and bound to the filter, it becomes a replica of the gel. The filter can then be hybridised with a labelled probe in a similar way to colony or plaque filters. As with all hybridisation, the key is the availability of a suitable probe, and an appropriate radioactive or fluorescent labelling method. After hybridisation and washing, the filter is exposed to X-ray film and an autoradiograph prepared (for radioactive probes) or scanned in an imaging device (fluorescent probes). An example of the use of Southern blotting in clone characterisation is shown in Fig. 10.8.

Although Southern blotting is a very simple technique, it has many applications and has been an invaluable method in gene analysis. The

> Blotting techniques, when used with restriction enzyme digestion and gel electrophoresis, can be used to identify particular regions of a gene in a cloned fragment of DNA.

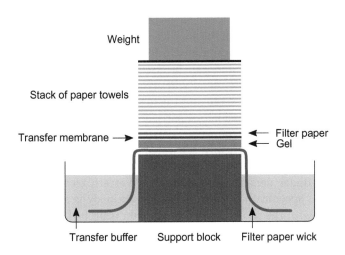

Fig. 10.7   Blotting apparatus. The gel is placed on a filter paper wick, and a nitrocellulose or nylon filter placed on top. Further sheets of filter paper and paper tissues complete the setup. Transfer buffer is drawn through the gel by capillary action, and the nucleic acid fragments are transferred out of the gel and onto the membrane.

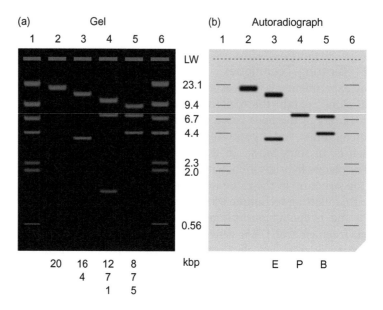

Fig. 10.8 Southern blotting. In this example, a 20 kbp fragment from a genomic clone is under investigation, and a radiolabelled cDNA copy of the mRNA is available for use as a probe. (a) The gel pattern of fragments produced by digestion with various restriction enzymes. (b) The autoradiograph resulting from the blot transfer, hybridisation and X-ray film detection. Lanes 1 and 6 have λ Hin d III markers, with sizes in kbp. These have been marked on the autoradiograph for reference. The position of the loading wells is marked LW. The intact 20 kbp fragment (lane 2) runs as a single band to which the probe hybridises. Lanes 3, 4 and 5 were digested with *Eco* RI (E), *Pst* I (P) and *Bam* HI (B). Fragment sizes are indicated under each lane in (a). The results of the autoradiography show that the probe hybridises to two bands in the *Eco* RI and *Bam* HI digests; therefore, the clone must have internal sites for these enzymes. The *Pst* I digest shows hybridisation to the 7 kbp fragment only. This would therefore be a good candidate for sub-cloning, as the gene may be located entirely on this fragment.

same technique can also be used with RNA, as opposed to DNA, and in this case is known as **Northern blotting** (by convention, capitalisation is usually retained for variants of the Southern technique, even though they are not eponymous). It is most useful in determining hybridisation patterns in mRNA samples, and can be used to determine which regions of a cloned DNA fragment will hybridise to a particular mRNA. However, it is more often used as a method of measuring transcript levels during expression of a particular gene.

Continuing round the compass, **Western blotting** involves the transfer of electrophoretically separated protein molecules to membranes. Often used with **SDS-PAGE** (polyacrylamide gel electrophoresis under denaturing conditions), Western blotting can identify proteins specifically if the appropriate antibody is available. The membrane with the proteins fixed in position is probed with the antibody to detect the target protein in a similar way to immunological screening of plaque lifts from expression libraries. Sometimes Western blots can be useful to measure the amount of a particular protein in the cell at any given time. By comparing with other data (such as amount of mRNA and/or enzyme activity), it is possible to build up a picture of how the expression and metabolic control of the protein are regulated.

There are yet further variations on the blotting theme. Predictably, Eastern blotting was a name that required a technique, and there was some jostling for use of the term. It is now used to refer to determination of post-translational modifications of proteins. A range of terms has also appeared to describe hybrid variants of blotting; thus, we have **Southwestern** and **Northwestern** blots for the investigation of DNA~protein and RNA~protein interactions, respectively.

If nucleic acid samples are not subjected to electrophoresis, but are spotted onto the filters, hybridisation can be carried out as for Northern and Southern blots. This technique is known as dotblotting, and is particularly useful in obtaining quantitative data in the study of gene expression. In some variants of the apparatus, the nucleic acid is applied in a slot rather than as a dot (not surprisingly, this is slot-blotting). As the available technology and fabrication methods developed, the principle of binding a sequence onto a substrate was extended to generate **microarrays** for a range of applications; we will look at this in more detail in Chapter 13.

Blotting without separation of sequences can be a useful technique for determining the amount of a specific sequence in a sample – often this is used to measure transcript levels in gene expression studies.

10.5.3   Sub-cloning

When a primary clone has been isolated, it is often necessary to move the fragment into another vector or vector/host system. This is known as **sub-cloning**, and can be used to generate constructs for expression, sequencing, probe generation, PCR amplification, assembly of complex constructs and more. The sophisticated cloning and assembly methods we looked at in Section 8.4 have made sub-cloning a much simpler and precise process than was the case just a few years ago, and usually a number of options is available for any sub-cloning procedure. Often a preferred supplier or well-tested laboratory protocol will determine which method is used.

10.5.4   DNA Sequencing

The development of rapid **massively parallel DNA sequencing** and **SMRT sequencing** (outlined in Section 6.8) has meant that determining the sequence of a clone (or clones) has become routine practice and is now often the first 'go to' step after isolation. Having the sequence enables an informatics approach to be taken to a whole range of post-isolation manipulations, and the ease of transferring cloned fragments between vectors opens up many

Determining the sequence of a cloned fragment is usually essential if gene structure is being studied. Sequencing can range from a small-scale project involving a single gene up to sequencing entire genomes.

different applications. Whilst not everything can be achieved by *in silico* methods alone, many of the design and checking steps can be achieved easily if the sequence is available. It is usually cost-effective (in terms of both time and money) to sequence early in the processing of cloned fragments.

10.6 | Conclusion: Needles in Haystacks

Isolating a specific sequence from a library of perhaps hundreds of thousands of clones is a bit like the proverbial 'needle in the haystack' (but with the added complication that the needle is made of the same material as the haystack). Thankfully a range of approaches is available to make the task possible. We looked at how *genetic methods* can be used to select clones by both *positive* and *negative* selection protocols, using the characteristics of vectors and host cells to enable desired clones to survive and make unwanted constructs non-viable. Genetic methods can also be used as a *screening* procedure, with processes such as *insertional inactivation* and the use of *chromogenic substrates*. Together, genetic selection and screening protocols provide a powerful way to isolate the clone of interest.

Screening clones using *nucleic acid hybridisation* is a highly specific and sensitive method that exploits the simple A·T and G·C base-pairing features of DNA. Probes can be used to screen a large number of clones in a single procedure, and variations such as the *plus/minus* method are useful where differential expression of a gene enables the preparation of different probes. The PCR can also be used to screen clones by designing primers specific for the gene of interest, and *combinatorial screening* reduces the number of reactions needed.

If a gene is cloned in an *expression vector*, the protein product can be detected using *immunological screening* if a suitable antibody is available. Like nucleic acid hybridisation, this exploits specificity of molecular recognition, although in this case, it is between antigen and antibody rather than *via* complementary base-pairs.

Preliminary analysis of clones requires a range of techniques that can include *restriction mapping*, either as a traditional experimental procedure or increasingly by *sequencing* the clone and manipulating the information *in silico*. Fragments are usually *sub-cloned* into other vectors for specific purposes, often using one of the sophisticated cloning and assembly methods available.

We also saw how *blotting techniques* can be used to characterise regions of cloned fragments. The original eponymous *Southern blotting* for DNA analysis was extended to Northern and Western blotting for RNA and proteins. Eastern, Southwestern and Northwestern variants extended the range further into the areas of *post-translational modification* of proteins and *nucleic acid~protein* interaction.

Further Reading

Benton, W. D. and Davis, R. W. (1977). Screening λgt recombinant clones by hybridization to single plaques *in situ*. *Science*, 196 (4286), 180–2. URL [www.science.org/doi/10.1126/science.322279]. DOI [https://doi.org/10.1126/science.322279].

Grunstein, M. and Hogness, D. S. (1975). Colony hybridization: a method for the isolation of cloned DNAs that contain a specific gene. *Proc. Natl. Acad. Sci. U. S. A.*, 72 (10), 3961–5. URL [www.pnas.org/doi/abs/10.1073/pnas.72.10.3961]. DOI [https://doi.org/10.1073/pnas.72.10.3961].

Prasanth, P. *et al.* (2016). High throughput gene complementation screening permits identification of a mammalian mitochondrial protein synthesis (ρ^-). *Biochim Biophys Acta*, 1857 (8), 1336–43. URL [www.sciencedirect.com/science/article/pii/S0005272816300469]. DOI [https://doi.org/10.1016/j.bbabio.2016.02.021].

Wang, W. *et al.* (2021). Recent advances in strategies for the cloning of natural product biosynthetic gene clusters. *Front. Bioeng. Biotech.*, 9, 692797. URL [www.frontiersin.org/article/10.3389/fbioe.2021.692797]. DOI [https://doi.org/10.3389/fbioe.2021.692797].

Zavala, A. G. *et al.* (2014). A dual color Southern blot to visualize two genomes or genic regions simultaneously. *J. Virol. Methods*, 198, 64–8. URL [www.ncbi.nlm.nih.gov/pmc/articles/PMC4010645/]. DOI [https://doi.org/10.1016/j.jviromet.2013.12.019].

Websearch

People

Sir Edwin Southern is best known for developing the blotting technique that bears his name. The importance of this is illustrated by the fact that many different variants of the technique were developed. Have a look at the career of this significant figure in DNA analysis.

Places

The availability of rapid and cost-effective DNA sequencing methods has changed the way in which we analyse DNA. One of the major centres for DNA sequencing is the *Wellcome Sanger Institute* (formerly the Sanger Centre). Examine the history of the institute from its inception in 1991, and trace how the type of science carried out has changed in line with technological developments in sequencing.

Processes

Two methods for clone analysis that we did not cover in the text are based on translation of mRNA *in vitro*. These are called **hybrid arrest translation (HART)** and **hybrid release translation** or *hybrid select* translation. Although not used widely today, they represent an important stage of the development of cloning technology. Have a look for information on these two techniques. Evaluate the contribution they made in the period when they were developed, and why they were replaced by other methods.

Reflections

What topics in this chapter have you found most challenging? Look for resources that help to illustrate the key points.

Selection, screening and analysis of recombinants

requires → **highly specific techniques**

highly specific techniques — **to enable** → **characterisation of the cloned gene**

highly specific techniques — **dependent on** → base-pairing in DNA and RNA / expression of cloned genes / antigen:antibody recognition / sequencing the cloned gene

characterisation of the cloned gene — **by**:

- **blotting techniques** — **to identify** → **important regions within cloned genes** — **includes** → **Southern Northern Western** — **to analyse** → **DNA RNA proteins**
  - Southern Northern Western — **also** — **Southwestern Northwestern Eastern** — **to analyse** → **DNA–protein interactions / RNA–protein interactions / post-translational modifications**

- **restriction mapping** — **to determine** → **restriction enzyme cutting sites**

- **DNA sequencing** — **enables** → **bioinformatics approach (in silico)** — **for** → **designing sub-cloning protocols** — **using** → **specialised vectors** — **for** → **production of proteins from cloned genes / assembly of contiguous fragments**
  - restriction enzyme cutting sites — **helpful for** → designing sub-cloning protocols

two approaches:

screening

using a → **clone bank (library)**

clone bank (library) — **and**:
- **PCR** — **by a** → **pooling method**
- **antibody** — **first** → **1° ab** — **then** → **2° ab** — **detect by, e.g.** → **enzyme reaction** — **using a** → **chromogenic substrate**
- **a probe** — **e.g.** → **RNA / cDNA / gDNA / oligo nt** — **labelled with** → **radioisotope** — **or a** → **fluorophore**

a probe — **uses** → base-pairing in DNA and RNA

selection

for → **presence of vector** — **e.g. by** → **antibiotic resistance** — **can use** → **insertional inactivation** — **to** → **isolate cloned fragments**

or by → **direct genetic selection** — **requires** → **mutant host cell** — **and** → **a functional cloned gene** — **to enable** → **complementation or marker rescue**

insertional inactivation / a functional cloned gene — **can use with** → chromogenic substrate

Concept Map 10

Chapter 11 Summary

Learning Objectives

When you have completed this chapter, you will be able to:

- Define and outline the scope of bioinformatics
- Explain the generation and use of biological data sets
- Identify a range of useful web resources for bioinformatics, including nucleic acid and protein databases
- Evaluate the impact that bioinformatics is having on the biosciences

Key Words

Bioinformatics, computer science, molecular biology, World Wide Web (www), uniform resource locators (URLs), suggested search terms (SSTs), information technology (IT), biological data sets, sequence data (*re.* nucleic acids and proteins), data warehouse, data mining, text mining, computer hardware, computer software, Φ-X174, lambda, coordinated database management, organise, collate, annotate, data curation, primary databases, secondary databases, European Molecular Biology Laboratory (EMBL), GenBank, DNA Data Bank of Japan (DDBJ), International Nucleotide Sequence Database Collaboration (INSDC), European Bioinformatics Institute (EBI), European Nucleotide Archive (ENA), National Center for Biotechnology Information (NCBI), National Institute of Genetics, open reading frames (ORFs), Universal Protein Resource (UniProt), Protein Information Resource (PIR), Swiss-Prot, Swiss Institute for Bioinformatics, UniProt Knowledgebase (UniProt KB), UniProt/Swiss-Prot, UniProt/TrEMBL, UniParc, UniRef, Worldwide Protein Data Bank (wwPDB), Protein Data Bank in Europe (PDBe), SnapGene, A plasmid Editor (ApE), M. Wayne Davis, bibliographic databases, Human Protein Atlas, protein~protein interactions (PPIs), interactome, human reference interactome (HuRI), the 'GIGO' effect, single-pass sequence data, finished sequence data, algorithm, basic local alignment search tool (BLAST), Ensembl, Genome Data Viewer (GDV), lactase gene (LCT).

Chapter 11

Bioinformatics

A good indication that a discipline has become firmly established and embedded across a range of applications is that it becomes more difficult to define in relatively simple terms. By this measure, bioinformatics is now a major discipline in its own right, and not simply a supporting technology that arose from computational biology and helped to frame bioscience as it entered the era of genome sequencing. Although bioinformatics primarily sits at the interface between **computer science** and **molecular biology**, it is also informed by many other subjects, including chemistry, physics, mathematics, statistics, medicine and biotechnology.

In writing an introduction to bioinformatics, there is an additional difficulty above the usual 'how much detail is needed to make sense of the topic?' question. Due to the nature of the discipline, bioinformatics is almost exclusively hosted within the World Wide Web (www), in terms of both the key data sets on which the subject is based and also the software tools that are needed to access and interrogate the data. It is a dynamic and actively evolving discipline, and the best way to appreciate how it all works is to actually *use* some of the tools available. In this chapter, we will therefore not attempt to explore all aspects of the topic in detail but will consider some of the most important central concepts. Inevitably there are some lists of uniform resource locators (URLs) and **suggested search terms (SSTs)**, but I have tried to keep these to some of the key sites that will help you to get started. The best way to learn more about the subject is to dive in and see the extent to which the use of data presentation and management tools has transformed biological databases in terms of level of complexity and ease of use (relatively speaking!). If you are using this book as a course or module text, your tutor will be able to guide you to the resources that are appropriate for your course, and some additional information on using bioinformatics sites can be found in the website that supports this book.

> Bioinformatics involves the use of computers to store, organise and interpret biological information, often in the form of sequence data. The World Wide Web is used extensively to share information with the global scientific community.

11.1 | What Is Bioinformatics?

Bioinformatics can be described in simple terms as the use of information technology (IT) for the analysis of biological data sets. It links the areas of bioscience and computer science, although it can be

---

Table 11.1 | Useful 'gateway' websites for bioinformatics

Site/page	URL	SST
European Bioinformatics Institute (EBI)	www.ebi.ac.uk/	ebi
EBI training site	www.ebi.ac.uk/training/on-demand	ebi on dem
The Wellcome Sanger Institute	www.sanger.ac.uk/	sanger
National Center for Biotechnology Information (NCBI)	www.ncbi.nlm.nih.gov/	ncbi
NCBI education site	www.ncbi.nlm.nih.gov/home/learn/	ncbi learn
Swiss Institute of Bioinformatics (SIB)	www.sib.swiss/	swiss bioinf
The Expasy resource portal of SIB	www.expasy.org/	expasy
The Bioinformatics Organisation	www.bioinformatics.org/	bioinform org

Note: The suggested search terms (SSTs) can be used to find the URLs easily. For those new to the topic, the education sites of EBI and NCBI are good places to start. The Bioinformatics Organisation is a user-group community site and offers some useful introductory information. All websites provide extensive links to other relevant sites.

The use of information technology to store and organise sequence data is coordinated by an international network of centres of excellence in bioinformatics, each of which has a specific role to play in the maintenance and development of the information.

difficult to define the limits of the subject. This is fairly typical of emerging interdisciplinary subjects, as they, by definition, have no long-established history and are generally pushing the boundaries of research in the topic. A good illustration of this is how the concept of the biological database itself has changed (and continues to change) to reflect the increasing sophistication of how the information is generated, stored, annotated and used. Some useful gateway sites for bioinformatics are listed in Table 11.1.

Although it can be difficult to define what bioinformatics *is*, it is relatively easy to define what bioinformatics is *not* – it is not simply using computers to look at sequence data. The IT requirements are of course central to, and critical for, the development of the discipline, but bioinformatics has broadened considerably over the past few years, with novel interactive and predictive applications emerging to supplement the original functions of data storage and analysis that marked the early development of the subject. The key elements of bioinformatics are summarised in Fig. 11.1.

11.1.1 Computing Technology

Computers are ideal for the task of analysing complex data sets, such as **sequence data** for **nucleic acids** and **proteins**. This often requires a series of computing operations, each of which may be relatively simple in isolation. However, the computation may need to be repeated thousands or millions of times, and it is therefore important that this is achieved quickly and accurately. Computers don't get tired, they generally don't make mistakes if programmed correctly and they can store large amounts of information in digital form. The emergence of genome sequencing as a realistic proposition in the early-1990s required the concomitant development of IT systems to deal with the output of large-scale sequencing projects, and thus

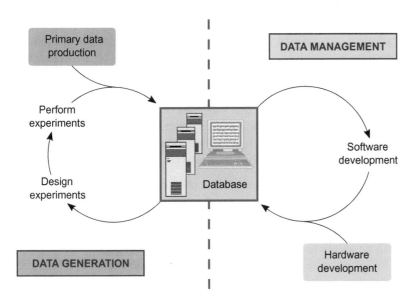

Fig. 11.1 Key elements in bioinformatics. The central core of bioinformatics is the database, of which there are many types. Data generation and data management are the two main activities that contribute to the development and use of the database. Critical elements are: (1) the availability of computer hardware and the development of software to manage the data, and (2) the generation of primary data to populate the database and enable derivative (secondary) databases to be developed. Genome sequencing projects have been the most significant contributors to bioinformatics database expansion over the past few years.

bioinformatics became fully established as a discipline in its own right. New terms such as data warehouse (a store of information) and data mining (interrogation of databases of various types) were coined to describe aspects of bioinformatics, with text mining (interrogation of bibliographic databases) providing an important supporting role.

Over the past 20 years or so, the staggering developments in desktop and server-based computer hardware have meant that bioinformatics is not the preserve of those with access to major computing facilities. Even a relatively low-cost desktop machine is powerful enough to access the various databases that hold the information, and all institutions and companies have IT infrastructure that provides secure and robust internet access. Thus, there is essentially no constraint on accessing and using the information from a hardware point of view. Computer software has, as might be expected, been developed in tandem with the computing power available and the requirements imposed by the nature of the data sets that are generated. Much of the software that is needed for the manipulation and interrogation of information is now made available free of charge by the various host organisations that run the bioinformatics network, although some packages are available from various companies on a

Computer hardware has developed to the point where desktop machines are powerful enough to act as the interface between the scientist and the various sequence databases and interrogation programs. Software and sequence data are usually available free of charge to the scientific community.

commercial basis. There are also some standalone packages, for applications such as vector and overlap design for assembly sub-cloning procedures, that can be used on a local network or an individual workstation basis.

11.1.2   The Impact of the Internet and World Wide Web

As noted in the introduction to this chapter, bioinformatics is essentially hosted within the web, which provides a number of benefits that could not otherwise be achieved easily:

- Access to a worldwide network of centres of expertise
- Easy data upload to the main databases
- No need to store large data sets locally
- Curated database services with shared aims and common standards
- Open access to most of the main sites and databases
- Web-hosted software (thus access to the most recent versions and updates)
- Regular updates to databases, thus no redundant information

In many ways, the impact of the web on bioinformatics is similar to that in other areas of life in general, and we are all perhaps a little guilty of taking this for granted. We should remind ourselves often that a lot of effort is needed to maintain and develop a resource of this magnitude.

11.2 | Biological Data Sets

11.2.1   Generation and Organisation of Information

It could be argued that bioinformatics really began to take off with the advent of rapid DNA sequencing methods in the late-1970s. Up until this point, the rate of acquisition of biological sequence data was limited, and thus there was relatively little pressure to develop common storage and analysis methods to cope with the information. Significant milestones occurred with the sequencing of the genomes of the bacteriophages **Φ-X174** (5 386 base-pairs, 1977) and lambda (48 502 base-pairs, 1982).

As sequencing techniques became established and used more extensively, the rate of data generation obviously increased, and thus the need for **coordinated database management** became greater. This need can be illustrated by looking at a simple example using some DNA sequence data (Fig. 11.2). This shows a short sequence of 50 bases, listed in various formats. Even with this length of sequence, it is clear that there are several requirements if sense is to be made of the information. These are:

- The need for accurate generation and capture of information
- A clear and logical presentation of the sequence in visual format
- Annotation of the sequence to enable orientation and identification of features
- Consistent use of annotation by different users/providers

Database management requires a high level of co-operation to establish common procedures for the storage, organisation and annotation of the information. The current state of development of the bioinformatics network represents an astonishing achievement that parallels the Human Genome Project in terms of significance.

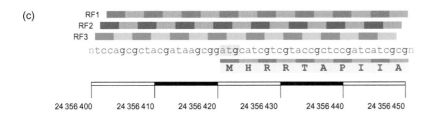

(a)
```
TCCAGCGCTACGATAAGCGGATGCATCGTCGTACCGCTCCGATCATCGCG
AGGTCGCGATGCTATTCGCCTACGTAGCAGCATGGCGAGGCTAGTAGCGC
```

(b)
```
           10        20        30        40        50
            .         .         .         .         .
 i  TCCAGCGCTACGATAAGCGGATGCATCGTCGTACCGCTCCGATCATCGCG

 ii TCCAGCGCTA CGATAAGCGG ATGCATCGTC GTACCGCTCC GATCATCGCG

iii tccagcgcta cgataagcgg atgcatcgtc gtaccgctcc gatcatcgcg
```

(c)
RF1
RF2
RF3

```
    ntccagcgctacgataagcggatgcatcgtcgtaccgctccgatcatcgcgn
                          M  H  R  R  T  A  P  I  I  A
```

24 356 400   24 356 410   24 356 420   24 356 430   24 356 440   24 356 450

Fig. 11.2 Some options for presenting sequence DNA information. (a) A short DNA sequence is shown in double-stranded format. By convention, the top strand is written 5′ → 3′ left to right. Bases are usually coloured green for A, red for T, black for G and blue for C. Three different ways of presenting the sequence are shown in (b)i, (b)ii and (b)iii. Usually only one strand is shown (the 'top' strand in the 5′ → 3′ direction). In (b)i, upper-case font is used with above-line numbering every ten bases. In (b)ii, spaces are used to separate groups of ten bases, and in (b)iii, lower-case font is used with separation. Lower-case avoids any confusion between G and C, as g and c are more easily distinguished. In (c), an annotated version of the sequence is shown, with the three reading frames identified as RF1, RF2 and RF3. A start codon (ATG in the DNA strand) that lies in reading frame 1 is highlighted in yellow, with a translated sequence of amino acids indicated by the single-letter amino acid codes (highlighted in light yellow). The numbering shows bases from the start of the chromosome – the region shown here is therefore some 24 million base-pairs from the end of the chromosome. The annotation will have other features indicated such as other open reading frames (ORFs), regions of sequence variation, patching of gaps or errors and so on.

As biological databases now deal with trillions of data points (*e.g.* bases in a sequence database), it is clearly a major task to **organise**, **collate** and **annotate** the information – the sequence itself is actually one of the simplest components in terms of organisational complexity. The process of managing all these aspects is often called **data curation**.

11.2.2 Primary and Secondary Databases

There are obviously many different ways to collate and store sequence information, and thus many different databases exist. It can be useful to think of databases as falling into two main categories. Repositories for experimentally derived sequence information are often known as **primary databases**. The main primary databases today are nucleic

Biological databases can be classified as either primary or secondary databases, although the distinction between these categories is becoming less important as database technology improves and annotation becomes more sophisticated.

acid sequence databases, although it is interesting to note that some protein sequences were collated in the early-1960s, and thus protein databases actually predate large-scale nucleic acid sequence databases.

Secondary databases are variants that have been derived from primary databases in some way. They may represent collections of sequence data arranged by organism or phylogenetic group, or may be based on genome projects. Alternatively, derived protein sequences, gene expression data or data restricted to a particular group of nucleic acids (e.g. ribosomal RNA sequences) can be used as the basis for establishing a secondary database. We can illustrate the development of sequence databases by looking briefly at two of the main types, dealing with nucleic acid and protein sequence data.

11.2.3 Nucleic Acid Databases

The first nucleic acid database to be established was set up by the **European Molecular Biology Laboratory (EMBL)** in 1980. Soon afterwards **GenBank** was established in the USA, and in 1986, the **DNA Data Bank of Japan** (DDBJ) began collating information. These three database providers now make up an international collaboration that is the major source of primary annotated nucleic acid sequence data, with the **International Nucleotide Sequence Database Collaboration (INSDC)** playing a key role in overseeing the collaboration. The sequence databases hosted and maintained by the **European Bioinformatics Institute (EBI)** are now part of the **European Nucleotide Archive (ENA)**. In the USA, the **National Center for Biotechnology Information (NCBI)** runs GenBank, and the DDBJ is hosted by the **National Institute of Genetics** in Japan.

The International Nucleotide Sequence Database Collaboration (INSDC) helps to oversee the activities of the three main primary sequence databases.

As would be expected as an information resource develops, nucleic acid databases have become organised in a variety of ways to reflect both the type of information and its intended use. Databases associated with different organisms is an obvious way to arrange information; thus, we have genome databases for all of the main model organisms. Things get a little more complex with further subdivision of database information, and one of the key aims of data curation is to ensure that data sets are annotated consistently and can be accessed and compared easily. It is a complex task. More information about nucleic acid databases can be found on the websites listed in Table 11.2.

One of the most striking aspects of nucleic acid databases is the increase in size since the first sequences were entered manually in the early-1980s. Trying to make sense of this growth was relatively straightforward over the first few years, with sequence data increasing exponentially and databases relatively well defined. As would be expected, as data acquisition progressed, the exponential increase continued and the data quickly became complex, with many database variants and repeatedly sequenced genomes. Thus, the simple question 'how many bases have been sequenced?' is no longer one that can

Growth of the various nucleic acid sequence databases has been exponential since the early-1980s, and it is now difficult to visualise the size of the data sets in a meaningful way.

Table 11.2	Nucleic acid sequence database websites	
Site/page	URL	SST
Nucleic acid sequence database sites		
International Nucleotide Sequence Database Collaboration (INSDC)	www.insdc.org/	insdc
EMBL-EBI	www.ebi.ac.uk/	emblebi
European Nucleotide Archive (ENA)	www.ebi.ac.uk/ena/	enaebi
GenBank	www.ncbi.nlm.nih.gov/Genbank/	genban
DNA Data Bank of Japan	www.ddbj.nig.ac.jp/	ddbj
Genome sequencing and other database sites		
The Wellcome Sanger Institute	www.sanger.ac.uk/	sanger
Databases at the European Bioinformatics Institute (EBI)	www.ebi.ac.uk/Databases/	ebiservices
Databases at the National Center for Biotechnology Information (NCBI)	www.ncbi.nlm.nih.gov/guide/all/	ncbidata

be answered meaningfully. To try to give a sense of how sequence data have proliferated, the growth in GenBank since 1982 is shown on a semi-log scale in Fig. 11.3. In the December 1982 release, GenBank had 680 338 bases from 606 sequences. In the February 2022 release, there were 1.17×10^{12} bases from over 236 million sequences. I'm not sure that any graph can represent this adequately – I certainly have difficulty trying to visualise what this means – but in a practical sense, it really doesn't matter, as the total number of bases sequenced is no longer a particularly useful concept.

11.2.4 Protein Databases

In looking at protein sequence databases, we can distinguish between database entries that have been determined by direct methods and protein sequences derived from nucleic acid database information by translation of the mRNA sequence. The direct method is similar to nucleic acid sequencing in that primary data are generated from the source protein, and additional biochemical and physical analysis is often possible. Thus, a reasonably complete characterisation of the protein, perhaps with biological activity fairly well established, is the standard to aim for. With derived sequence data, modifications to the amino acids may not be immediately obvious, and there is a risk that the derived sequence may not in fact be a functional protein in the cell if analysis of **open reading frames (ORFs)** is used to translate all potential mRNAs. Despite this potential problem, around 95 per cent of protein sequences are derived from nucleic acid translation data.

One of the key challenges for database developers is to ensure that databases can 'talk' to each other, so that complex data conversion is

Although protein sequencing has not produced primary data at the same rate as DNA sequencing, a wealth of information about protein form and function is available in the various protein databases.

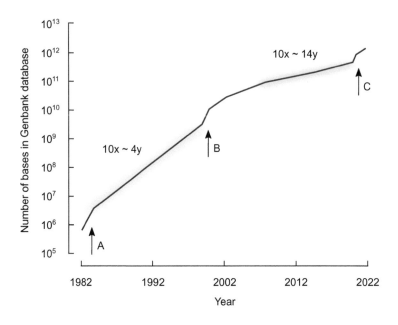

Fig. 11.3 Growth of the GenBank sequence database. Set up in 1982, the database has been a major source of sequence information. The exponential increase in sequence data means that a linear plot is essentially meaningless over this time frame, so a semi-logarithmic plot is shown here. The plot shows two main phases of steady growth, with tenfold increase times of around 4 years (from 1984 to 1999, yellow highlight) and 14 years (from 2008 to 2019, green highlight). Periods of a greater rate of increase are shown as A (early increase from a low baseline level), B (sequence data from genome sequencing contributing to the increase) and C (the latest period of increased growth, from 2020). It is difficult to grasp the scale involved; the number of bases in the database has increased some 1.8 millionfold since 1982. *Source:* Data from GenBank, NCBI, National Library of Medicine, US [www.ncbi.nlm.nih.gov/genbank/statistics/]. Reproduced with permission.

avoided where possible. As with nucleic acid databases, there is now a range of different protein database resources available. As proteins are structurally more complex than nucleic acids, how proteins fold and their three-dimensional (3-D) structure, function and location in cells and tissues have been a major emphasis as protein databases have developed.

The key umbrella resource for proteins is the **Universal Protein Resource** or **UniProt**, which represents the product of close collaboration between the major protein sequence repositories in Europe and the USA. UniProt was set up in 2002, with three major partners involved. The **Protein Information Resource (PIR)**, hosted by Georgetown University in the USA, joined forces with the **Swiss-Prot** database, developed and maintained by the **Swiss Institute for Bioinformatics**, and the EBI. The central core of UniProt is a set of four databases. The **UniProt Knowledgebase (UniProt KB)** has two

Table 11.3 Protein sequence database websites

Site/page	URL	SST
Database sites		
UniProt KB – the key gateway site for protein database information. You can access UniProt KB/Swiss-Prot (the SIB protein resource and part of UniProt) and UniProt KB/ TrEMBL (translated EMBL resource) from this site	www.uniprot.org/	uniprotkb
The Protein Information Resource (PIR, hosted by Georgetown University, and part of UniProt)	https://proteininformation resource.org/pirwww/	pir www
Analysis and proteomics sites		
The proteomics server (*Expert Protein Analysis System*) of SIB	www.expasy.org/	expasy
The bioinformatics toolbox of the EBI	www.ebi.ac.uk/services	ebiservices
Worldwide Protein Data Bank (wwPDB)	www.wwpdb.org/	wwpdb
Protein Data Bank in Europe (PDBe)	www.ebi.ac.uk/pdbe/	pdbe

sub-sections derived from the original Swiss-Prot database and the EMBL nucleotide sequence database. The difference is that the **UniProt/Swiss-Prot** version is manually curated to a high level of consistency and accuracy, a task that requires significant input from scientists skilled in the art of checking and annotating protein sequence information. As genome projects began to generate large amounts of data, the Swiss-Prot standard could not be maintained at the same pace, so an automated version was devised to enable relatively quick and accurate translation and annotation of sequences. This is now the complement to the Swiss-Prot version and is known as **UniProt/TrEMBL** ('translated EMBL'). The other two databases are **UniParc** (UniProt Archive, curated as non-redundant unique sequences) and **UniRef** (UniProt Reference Clusters, with data as clusters of identical sequences).

As with nucleic acid databases, the scale of currently available protein sequence data is very impressive – at the time of writing (April 2022), the UniParc database had some 486 million entries. Another useful database is the **Worldwide Protein Data Bank (wwPDB)**, which curates information about the 3-D structure of proteins and other biological molecules, and the **Protein Data Bank in Europe (PDBe)**. Further details about protein databases can be found in the websites listed in Table 11.3.

Bioinformatics is not restricted to the collation and management of DNA and protein sequence data, but is a much broader discipline that continues to generate novel applications and insights.

11.2.5 Other Bioinformatics Resources

Although we tend to associate bioinformatics primarily with sequence databases, the impact of the subject has expanded as the integration and evaluation of data sets have developed. We will look briefly at four areas that illustrate this.

Designing vectors for applications such as Gibson assembly, Golden Gate cloning and other sub-cloning procedures has become much simpler with access to software that enables plasmid design to be carried out in the computer. Usually small-scale in contrast to the large sequence databases, this is a good example of how subsets of information can be used with locally installed software (although web-based software can also be used) to generate and cross-check sequence overlaps and alignments before any experimental work is carried out. There are several versions of this type of application, including commercial products such as **SnapGene** (www.snapgene .com/), and free versions such as **A plasmid Editor (ApE)** (https:// jorgensen.biology.utah.edu/wayned/ape/) by **M. Wayne Davis** of the University of Utah. SnapGene has a free student version that can be used during an academic course, and the ApE program is certainly worth downloading and using to get a feel for how this type of software functions.

Using search terms to find information on the web is a form of informatics that most of us use frequently (with, of course, the attendant care and attention that we considered in Section 1.4). Another example is searching the scientific literature, although it is sometimes easy to underestimate the importance of **bibliographic databases**. However, access to the literature is a critical part of the scientific process, and IT-based searching has made life much easier than was the case in the pre-internet era.

The **Human Protein Atlas** (www.proteinatlas.org/) illustrates how existing resources can be integrated to derive a novel application – in this case, mapping human proteins in cells, tissues and organs. The project was set up in 2003, and provides a core resource to investigate the cellular locations and levels of expression of different gene products. This is providing new insights into how various processes function at the molecular level within cells and tissues. The website has a number of educational resources that illustrate this approach and are worth a detailed look.

Another aspect of protein analysis is the study of **protein~protein interactions** (PPIs). This is a key part of the study of the interactome, with the aim of trying to understand how the cell functions as a consequence of how its proteins interact with each other. As with the protein atlas, a derived resource called the **human reference interactome (HuRI)** (www.interactome-atlas.org/) is providing a focus for this type of research. We will look at the interactome in more detail in Section 13.4.

Many more applications of bioinformatics have been developed over the past few years, and there are few areas of modern bioscience that are not reliant on an informatics-based approach at some point. The range and scope of the subject are summarised in Fig. 11.4.

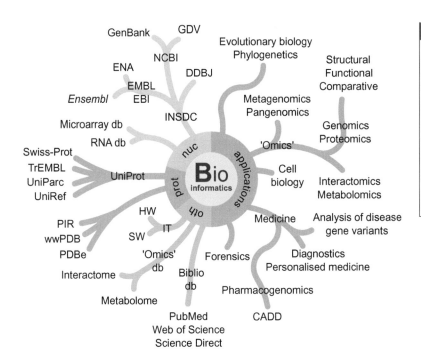

GenBank   GDV
Evolutionary biology
Phylogenetics
NCBI
ENA   DDBJ   Structural
EMBL   Functional
Ensembl   EBI   Comparative
INSDC   Metagenomics
Microarray db   Pangenomics
RNA db   Genomics
Swiss-Prot   'Omics'   Proteomics
TrEMBL
UniParc   UniProt   **B**io   Cell
UniRef   informatics   biology   Interactomics
                                            Metabolomics
HW
PIR   IT   Medicine   Analysis of disease
wwPDB   SW   gene variants
PDBe   'Omics'   Diagnostics
        db   Forensics   Personalised medicine
Interactome   Biblio
              db   Pharmacogenomics
Metabolome
        PubMed   CADD
        Web of Science
        Science Direct

Fig. 11.4 The scope of bioinformatics. Some of the key areas that make up bioinformatics are shown, with database organisation for nucleic acids (nuc, blue), proteins (prot, purple) and others (oth, green) on the left, and applications on the right (pink). Abbreviations not used in the text are db (database), IT (information technology), hardware (HW) and software (SW), and CADD (computer-aided drug design).

11.3 | Using Bioinformatics as a Tool

Having illustrated the scope of bioinformatics by looking at the basic organisation of nucleic acid and protein databases, the obvious question is 'what use can be made of such information?' At this point, things become a little difficult – not only due to the large range of topics, but also because the subject is IT-based and really requires active involvement to explore fully. In this section, we will look at some examples to illustrate the topic, but I would encourage the reader to have a look more widely at the websites listed in this chapter and build up a picture of how bioinformatics works. Some more detailed examples and tasks are given on the website that accompanies this book.

11.3.1 Avoiding the 'GIGO' Effect – Real Experiments

The '**Garbage In Garbage Out**' (**GIGO**) **effect** is a well-known phrase in computing, which was coined to stress the importance of valid and reliable data entry when programming or entering information. GIGO can apply to any situation where there is the possibility of data generation and/or entry errors, and is particularly appropriate when considering bioinformatics. Although many aspects of bioinformatics could give rise to data corruption, there are essentially three classes of problem, covering data generation, entry and processing.

The generation of primary data is obviously dependent on the availability of a reliable and reproducible experimental method. For example, DNA sequencing techniques are now well established and

Although the core information for bioinformatics is based on data sets of various sorts, it is vital that the information in these data sets is determined by careful experimentation to avoid propagation of errors. Thus, there is a high value placed on 'finished' sequence data that have been rigorously checked and confirmed.

generally demonstrate a very high degree of validity and reliability. Automation has reduced the incidence of human error, and with multiple 'reads' of both strands of DNA, the accuracy of the resulting data is getting close to 100 per cent for some technologies. Some human intervention to resolve anomalies may be required for tricky stretches of sequence, but this adds to the fidelity of the process rather than detracts from it.

Having generated the sequence, the transfer of data from the sequencing software to the database is the next step where data migration could present problems. Although the days of keyboard entry for the bulk of primary sequence data are now over, there can be issues due to the interface between different computer network systems. However, these are often of a technical nature and, if resolved, should not affect the stage of data collection.

Data processing is an area where there is a further risk of introducing errors. There is the possibility of incorrect information, derived experimentally, persisting through data generation and entry processes. Thus, any computation carried out can often compound the problem by 'cloning' the incorrect data and establishing it in the database. There is a real issue with the 'typewritten therefore must be correct' tendency; we tend to trust any data that appear to be part of a formal data set. Thus, the sequence could be 'wrong', but the incorrect data could be perfectly plausible and thus very difficult to identify. Any such incorrect data could be used for the production of derived sequence data sets or other purposes, thus making the problem worse. Quality control for sequencing requires careful design of techniques and procedures and rigorous cross-checking of sequence data. Distinction is usually made between single-pass sequence data, which will be inaccurate in places, and finished sequence data, where the inconsistencies have been removed by several sequencing runs and rigorous validation of the sequence.

The use of computer analysis and prediction to carry out 'experiments' *in silico* enables researchers to search, identify and modify gene and/or protein sequences easily. Results can then be tested by experiments *in vitro* or *in vivo* to determine biological significance.

11.3.2 Avoiding the Test Tube – Computational Experimentation

Assuming that the GIGO effect can be reduced to an acceptable level, the interrogation of databases provides many novel insights, including new avenues of research that emerge due to increased investigation of existing data. As already noted for a number of procedures, an *in silico* approach can be useful in preparing for experimental work, ironing out any anomalies and cross-checking sequence alignment, restriction site positions and so on. This does not replace 'real' practical work, but it can offer a very quick route to establishing what experiments might prove most incisive and informative. This cycle of experimental data generation and analysis, as outlined in Fig. 11.1, is a central pillar of bioinformatics.

Algorithms and search functions are critical features of database management. It is important that they work effectively and efficiently to enable large amounts of data to be analysed.

A critical part of searching databases is the algorithm used to carry out the operations of the search. One example is basic local alignment search tool (BLAST) that is used in most database search functions. As database size increases, searching for elements such as sequence homology, sequence variants and ORFs becomes more challenging. In addition to the search algorithms, curation of the data and the structure of the database assume an even greater importance. One obvious example of this approach is the translation of a DNA sequence into the predicted protein sequence. However, identification of a protein-sized ORF by computer analysis does not necessarily mean that the protein will be produced in the cell. To establish the presence of such a protein, identification and analysis will be required using a range of experimental techniques to characterise biological activity and function of the protein. We can trace this approach from nucleic acid sequence, through translated and annotated protein sequence, to cellular location and comparative expression, as illustrated by the HuRI project.

11.3.3 Presentation of Database Information

Our final consideration in this chapter is about data presentation, without which any database would be of very limited use. In particular, the availability of presentation interfaces that do not require high-level computer programming skills has been essential in ensuring widespread access to the large nucleic acid and protein databases. Some of the genome browsers available are listed in Table 11.4.

The human genome is perhaps the most interesting sequence to consider (largely because of what it represents, rather than for any inherent bioinformatics reason). As we are considering web-based presentation tools, the best way to read this part of the book is with the book propped up beside your computer, so that you can access the websites and see for yourself how the various features work. Two major genome browsers are **Ensembl** and **Genome Data Viewer (GDV)**. Ensembl has been developed by the European Molecular Biology Laboratory and the Wellcome Sanger Institute, and GDV is the NCBI version.

Let's consider presentation of the human genome using Ensembl (accessed October 2022). When you log on to the Ensembl site (Table 11.4), an index page shows the genomes available. The genome assembly used (GRCh38.p13) and Ensembl release (108 in this case) are shown on the page. Clicking on the *Homo sapiens* button brings up another index of the various features available, plus some very useful background information. Opening up the karyotype view, a graphic of

Although a basic data file may be useful, complex data sets are much more accessible with clear user interfaces, annotation and presentation. This aspect of database management is a major part of bioinformatics, and is perhaps the critical step that has transformed the subject into a universally accessible resource.

The best way to learn about biological databases is to explore their various features directly. A good place to start is by looking at genome data using a web-based genome browser such as Ensembl or Genome Data Viewer.

Table 11.4	Presentation interfaces for genome sequence databases		
Site/page		URL	SST
The EMBL-EBI automated annotation tool (Ensembl) for genome database presentation. Located at the Wellcome Genome Campus in Hinxton, UK		www.ensembl.org/	ensembl
A useful video/webinar introduction to Ensembl can be found on YouTube at the URL listed		www.youtube.com/ watch?v=JoJGg8xd0e0	
The National Center for Biotechnology Information (NCBI) site that lists genome sequences available and enables viewing through the Genome Data Viewer (GDV)		www.ncbi.nlm.nih.gov/ genome/gdv/	ncbigdv
The University of California at Santa Cruz (UCSC) genome browser.		https://genome.ucsc.edu/	ucscgen

the chromosomes (the karyotype) appears, along with some additional information about the genome. Have a look at the various features that can be accessed; you will have to try these in various combinations to see how the whole thing fits together. We will look at one specific example of searching for a gene to illustrate.

Let's look at the **lactase gene**, designated **LCT** (a helpful video tutorial is available for this, shown in Table 11.4). From the home or whole genome page, type LCT into the search box (top right). A list of the entries for LCT appears – select the top one that indicates 'LCT (Human gene)' to pull up a summary of the data and the graphic of the region on chromosome 2 where LCT is located. The graphic looks quite complex, but a key point to note here is the *size* of the region we are looking at. This is 69.33 kbp of chromosome 2, from positions 135 787 850–135 837 184. The LCT gene is located on the reverse strand, indicated by the arrowhead to the left of the gene abbreviation ($<$ LCT-201). Click on the 'Go to region in detail' link and 51.31 kbp comes up from the region, with more information and a zoom bar on the right-hand side of the page. Note that at this stage, no sequence information is presented, and also the karyotype and overview graphic remains at the top, covering 1 Mbp. Zoom in through the various levels and see how the view changes, noting how features go out of view as the 'magnification' increases.

Now, all this mouse-clicking and instant access to information seem straightforward – but we need to think a little about what is actually being presented to get some sort of perspective. Zoom in to maximum level for the LCT example. On my computer, this shows 104 bp, across 365 mm of the monitor's screen – that's 3.5 mm per basepair. Chromosome 2 is 242 193 529 bp long, and the LCT gene is just over half way (0.56) along from the left end of the chromosome. If we were to think of the chromosome as a piece of string, and we were looking at the 104 bp region, the string would extend 475 km to the left of the screen and 373 km to the right … yes, *kilo*metres. Remember that there are 22 other chromosomes, and we get some idea of the scale of presentation required.

As the information is so readily accessible, and the navigation relatively straightforward, it is easy to become a little blasé when looking at presentation tools such as Ensembl. If this happens, think of the 848 km of string needed to represent chromosome 2 (or the 10 800 km for the entire human genome) and reflect on the advances in sequence generation, collation, curation and annotation, and on the development of computing hardware, software and internet technology, that have made such a rich resource available to all.

11.4 | Conclusion: Bioscience and 'Big Data'

The phrase 'big data' is often used to indicate amounts of information that we cannot easily comprehend, and the expansion of genome sequencing in the 1990s pushed biological data sets into this category. *Bioinformatics* emerged from the more broadly based field of *computational biology* to help make sense of the data that were accumulating at ever increasing rates, and quickly became established as a discipline in its own right. The central element in bioinformatics is the database, of which the two most well-known are the *nucleic acid* and *protein* sequence databases. *Primary* databases are repositories of experimentally generated data, and *secondary* databases are derived by manipulating the information to provide new ways of presenting, analysing, annotating and interpreting the data set. Management of database information is called *data curation*.

Bioinformatics emerged through the concomitant development of *data generation*, computer *hardware* and computer *software* needed to analyse the data sets. For nucleic acid sequences, there are now many different variants of major databases, based on the core information that is managed by an international consortium (the *INSDC*). Protein databases have expanded into areas of structural analysis and mapping the cell and tissue locations and expression patterns of proteins, and areas such as the study of the *interactome* generate topic-specific variants of large data sets.

Paradoxically, the sophisticated development of the global database offering is now making simple questions like 'how many bases (or genomes) have been sequenced?' difficult to answer. In the annual database issue collated by the journal *Nucleic Acids Research* in early 2022, the paper from the *European Nucleotide Archive* noted that it contained 2.7×10^{15} base-pairs of raw data and 2.5×10^9 assembled and annotated sequences – this is undoubtedly in the realm of 'big data'.

Although bioinformatics enables interrogation of large amounts of information using a range of computational tools, experimental work is still a key element in generating data and confirming the biological function of genes and proteins. Most areas of research in bioscience and medicine now involve bioinformatics-based input at some point, and this approach will continue to provide new insights as data interpretation methods continue to develop. In addition to 'routine' sequence-based data analysis, *data mining*, *text mining* and search functions for *bibliographic databases* are important parts of bioinformatics.

Visualising and using biological databases require a *graphics-based interface* to present the information in a format that can be accessed without the need for advanced database management skills. This is one of the most significant aspects of bioinformatics and is sometimes taken for granted. It is a source of wonder to those of us who remember GenBank being provided on a set of 5.25-inch floppy disks for use in a text-based monochrome computer!

Further Reading

Diniz, W. J. S. and Canduri, F. (2017). Bioinformatics: an overview and its applications. *Genet. Mol. Res.,* 16 (1). URL [www.funpecrp.com.br/gmr/year2017/vol16-1/pdf/gmr-16-01-gmr.16019645.pdf]. DOI [https://doi.org/10.4238/gmr16019645].

EMBL–EBI Training. Bioinformatics for the terrified. URL [www.ebi.ac.uk/training/online/courses/bioinformatics-terrified/what-bioinformatics/].

EMBL–EBI Training. Introductory bioinformatics: a curated set of EMBL-EBI online courses. URL [www.ebi.ac.uk/training/online/courses/introductory-bioinformatics-pathway/]. DOI [https://doi.org/10.6019/TOL.IntroPathway-t.2021.00001.1].

NIH/NLM National Center for Biotechnology Information. Site guide for NCBI bioinformatics resources. URL [www.ncbi.nlm.nih.gov/guide/all/].

Roumpeka, D. D. *et al.* (2017). A review of bioinformatics tools for bioprospecting from metagenomic sequence data. *Front. Genetics*, 8, 23. URL [www.frontiersin.org/article/10.3389/fgene.2017.00023]. DOI [https://doi.org/10.3389/fgene.2017.00023].

Websearch

People

Modern bioinformatics is hosted on, and used through, the *internet* and *World Wide Web*. The web was invented by *Sir Tim Berners-Lee*, who also designed the first web browser.

Arguably we owe all the developments in bioinformatics (and, of course, lots of other things as well) to the concept that he developed into the www. Have a look at the career of Sir TimBL (as he is known), and his role in the standardisation of web-based protocols through the *World Wide Web Consortium (W3C)*.

Places

As is fitting for the concept of a cloud-based virtual system, we won't look at physical places for this chapter. Rather, we will have a look at the two main nucleotide sequence database interfaces – the EMBL-EBI *Ensembl* browser and NCBI's *Genome Data Viewer*. I suggest using a human gene – the *CFTR* gene (cystic fibrosis transmembrane conductance regulator) is an interesting one, but you may wish to choose something else. Have a look in both browsers and note their differences and similarities. Which one do you prefer, and why?

Processes

The *BLAST* search function is a mainstay of database management. What is BLAST, and what does it enable you to do?

Reflections

What topics in this chapter have you found most challenging? Look for resources that help to illustrate the key points.

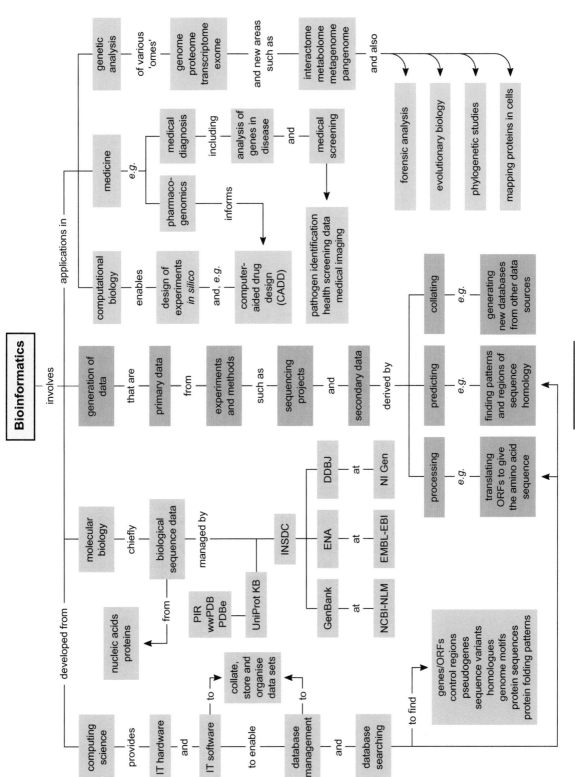

Bioinformatics

Concept Map II

Chapter 12 Summary

Learning Objectives

When you have completed this chapter, you will be able to:

- Demonstrate the scope and potential of genome editing
- Review the principles of genome editing
- Explain knockout and knockin technology
- Discuss the range of techniques available for editing genomes, including gene targeting, and the use of engineered nucleases, zinc-finger nucleases and TALENS
- Outline the techniques for editing RNA

Key Words

Mutant, meiosis, DNA repair, recombineering, genome editing, gene editing, genome engineering, engineered nucleases, RNA editing, prime editing, homologous recombination (HR), deleterious mutation, knockout (KO) mice, Mario Capecchi, Sir Martin Evans, Oliver Smithies, embryonic stem cells (ES cells), chimera, heterozygous, homozygous, knockin (KI), Jackson Laboratory, Knockout Mouse Programme (KOMP), International Mouse Phenotyping Consortium (IMPC), deletion cassette, yeast deletion collection, yeast knockout (YKO) set, double-strand breaks (DSBs), coordination centre, zinc-finger nucleases (ZFNs), *Fok* I, *Flavobacterium okeanokoites*, non-homologous end joining (NHEJ), homology-directed repair (HDR), error-prone, indels, transcription activator-like effectors (TALEs), repeat variable di-residue (RVD), TALEN, Yoshizumi Ishino, clustered regularly interspaced short palindromic repeats (CRISPR), CRISPR-associated (Cas) proteins, Cas9 endonuclease, CRISPR-Cas9, Emmanuelle Charpentier, Jennifer Doudna, guide RNAs (gRNAs), protospacer adjacent motif (PAM), nickase, deaminase, transcription factors, polygenic, David Liu, reverse transcriptase (RTase), prime editing guide RNA (pegRNA), FEN1, adenosine deaminase acting on RNA (ADAR), adenosine, inosine, Cas13, RNA-specific endonuclease, paradigm-shifting technology.

Chapter 12

Genome Editing

Since Mendel's work was rediscovered in 1900, the development of genetic analysis has been dependent on being able to 'tag' or label genes. The availability of mutant cells and organisms, and the realisation that mutants were a consequence of changes to the gene sequence, underpinned much of the development of microbial and molecular genetics. Studies on the process of **meiosis** and on **DNA repair** mechanisms showed how recombination could enable genetic crossing-over and the repair of breaks in DNA. With the development of techniques for gene cloning and sequencing DNA, introducing defined changes into particular genes became a possibility if a suitable mechanism could be devised. The use of recombination mechanisms in genetic engineering is sometimes called **recombineering**; the word mimics the action!

Although we tend to think of genome editing (also known as gene editing, and sometimes as **genome engineering**) as a relatively recent development, the technique has its origins in the 1970s and 1980s, when work with yeast and mice showed that genes could be inactivated. Later, the development of **engineered nucleases** provided a different approach that made the techniques more accessible. In addition, **RNA editing** is an option that has some potential advantages for therapeutic use, and the recent development of **prime editing** illustrates how the techniques for editing genes continue to develop.

> Genome editing developed from initial studies on the inactivation of genes using the natural process of homologous recombination.

All techniques for editing genes or genomes have a number of common aims, requirements and potential difficulties:

- The gene being targeted for editing needs to be identified within the genome
- The mechanism for introducing the change in the sequence must be able to introduce the change at a specified site within the target gene
- The process needs to be efficient
- 'Off-target' (*i.e.* unintended) effects need to be minimised

In addition to the technical challenges, editing genomes raises a number of significant ethical questions that have to be considered. In this chapter, we will look at some of the key methods that can be used to edit genes and genomes, and look at ethics in the context of medical applications in Section 15.4.

> As with many aspects of gene manipulation, genome editing raises significant ethical issues that have to be considered carefully.

12.1 | Gene Targeting

The first methods for altering genes were carried out using the process of **homologous recombination (HR)**. This enables DNA sequences to be exchanged at regions of sequence homology (as we saw for the Gateway cloning system in Fig. 8.11). Insertion of a DNA sequence into a chromosomal gene causes disruption of gene function (similar to the process of insertional inactivation in basic cloning methods) and acts in effect as a **deleterious mutation**. The principle of the process for gene targeting is shown in Fig. 12.1.

The generation of **knockout (KO) mice** was the first significant development using this procedure, and resulted in the 2007 Nobel Prize being awarded to **Mario Capecchi**, **Sir Martin Evans** and **Oliver Smithies**. In addition to the molecular biology needed to generate the gene disruption that produces the knockout, the work of Sir Martin Evans on **embryonic stem (ES) cells** in the early-1980s was critical to the success of the procedure. To produce alterations in the genes of a

> The first genome editing procedures generated mice in which gene function was disrupted; these were called 'knockout mice'.

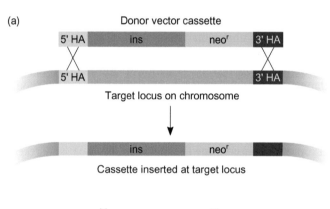

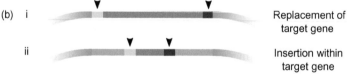

Fig. 12.1 Gene targeting using homologous recombination. (a) A donor vector is constructed with the sequence to be inserted (ins, highlighted purple) and a selectable marker appropriate for the cell being used. In this example, the neomycin phosphotransferase gene is inserted to confer resistance to neomycin (neo^r, highlighted green). Regions of homology with the target gene sequence are located at the 5′ and 3′ ends of the construct (homologous arms, HA). Vector constructs like this are sometimes called cassettes. The HA regions enable recombination and the cassette is integrated into the chromosomal locus. For some applications, the marker gene can be excised after selection of cells containing the construct. Two potential formats that depend on the location of the HA sequences are shown in (b). In (b)i, the insert sequence has replaced the target gene, whilst in (b)ii, it has been inserted within the target sequence. Careful design of the donor vector enables both knockout and knockin outcomes to be achieved.

multicellular organism, ideally all cells would be altered. In practice, this is impossible for an adult organism, so the target was altered in cultured ES cells that were then implanted in mouse blastocysts. The developing embryo will have two types of cell (the normal cells and the altered cells) and is known as a chimera. If the germ line cells carry the altered gene, crossing the adult chimeric mouse with a wild-type mouse results in some animals in which all cells have one copy of the altered gene (heterozygous for the knockout), and these can be further crossed to generate homozygous knockout mice.

As well as generating *loss* of gene function in knockout mice, the technique can also be used to introduce a functional gene into the genome. This gain of function technique is known as the knockin (KI) method. There are now many thousands of different KO and KI mice available, and the design and generation of specific KO/KI mice can be provided by a range of suppliers. One of the most significant contributors to the field is the **Jackson Laboratory** (known as the Jax laboratory), which is also involved with the **Knockout Mouse Programme (KOMP)** managed by the **International Mouse Phenotyping Consortium (IMPC)**. A lot of useful background information can be found in the websites of these organisations.

Knockout and knockin mutants have become essential tools for investigating gene function in mice, with many thousands of variants available.

Homologous recombination has also been used to generate loss of function knockouts in the yeast *Saccharomyces cerevisiae*. The techniques used initially involved transposon tagging to generate mutants, with later techniques using a **deletion cassette** approach to knockout some 6 000 ORFs in over 20 000 mutant strains to produce the **yeast deletion collection** or **yeast knockout (YKO) set**. As with the mouse, this has been an invaluable resource to study gene function using the yeast model.

The yeast deletion collection is an important resource for studying genes in this model organism.

12.2 | Genome Editing Using Engineered Nucleases

Although gene targeting continues to be an important technique, methods based on recombination have been developed into areas where more precise alteration of the target sequences can be achieved. The use of engineered nucleases is a good example of how innovative thinking can lead to new techniques that are often derived from naturally occurring systems and processes. In this section, we will look at three examples that can repair double-strand breaks (DSBs) generated by the altered nuclease. The possible outcomes of stimulating the repair of DSBs are summarised in Fig. 12.2. In the final part of this section, we will consider prime editing, a recent development that does not require generating DSBs as part of the process.

The generation of engineered nucleases is a good example of how adapting naturally occurring systems can enable new techniques to be developed.

12.2.1 Zinc-Finger Nucleases

The zinc-finger (ZF) protein domain was discovered in 1985 in transcription factor IIIA from *Xenopus laevis* (African clawed toad). The domain has a zinc ion as a **coordination centre** that binds two areas

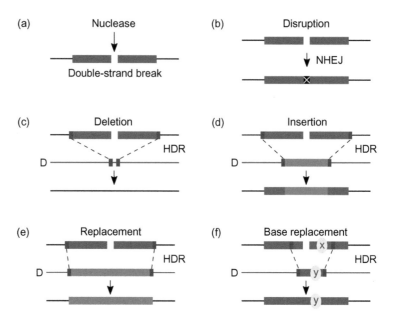

Fig. 12.2 Possible outcomes for the repair of double-strand breaks. (a) A double-strand break is generated in a gene sequence using a nuclease. (b) Repair by non-homologous end joining (NHEJ) is usually error-prone and introduces indels that disrupt the gene, inactivating it. (c)–(f) Introducing a donor vector sequence (D) to drive homology-directed repair (HDR) offers a number of options. In (c), a donor with adjacent regions homologous to the gene flanking regions (blue, homologous arms or HA regions) essentially deletes most of the gene sequence. In (d), a donor carrying a sequence flanked by the HA sites enables the sequence to be inserted (this usually inactivates the target gene, but a replacement functional gene may also be introduced in this way). In (e), the entire gene is replaced, and in (f), a single base replacement (x to y) can be achieved. There are many other potential configurations, in addition to these possible outcomes, depending on how the donor vector, donor insert and HA recombination sites are arranged during construction of the vector.

of the protein, often through two cysteine and two histidine residues (C2H2), to produce a finger-like interaction with the DNA major groove. A number of ZFs linked in tandem array can recognise a variety of DNA sequences (as triplets), and thus were of interest as possible sequence-specific recognition structures. In 1994, a three-finger engineered protein was used to block expression of an oncogene in a mouse cell line, and demonstrated the possibility of using the ZF with other functional domains to generate engineered enzymes.

Zinc-finger nucleases (ZFNs) combine the ZF domains with endonuclease function. The cleavage domain of the type IIS restriction endonuclease *Fok* I (from *Flavobacterium okeanokoites*) is commonly used, as this is distinct from the recognition domain and can be separated from this and joined to the ZF region. The enzyme functions as a dimer; thus, the two strands of the DNA bind a separate ZF-enzyme complex and generate a DSB. This break can be repaired by

Zinc-finger nucleases and TALENs use different mechanisms to recognise DNA sequences, and can be combined with a type IIS endonuclease function to enable cutting of DNA at defined sites.

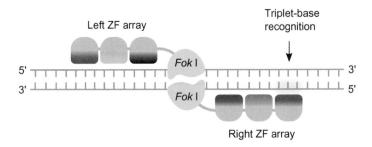

Fig. 12.3 Zinc-finger nucleases. The nuclease is generated by linking the DNA cleavage domain from the type IIS restriction enzyme *Fok* I with a number of zinc-finger (ZF) domains. The ZFs generally recognise triplet combinations of nucleotides (highlighted yellow), and an array of these can be constructed to recognise a specific DNA sequence. Two arrays of three ZFs are shown in this example; the *Fok*-derived cleavage domain functions as a dimer; thus, two ZFN constructs are needed to effect the double-strand cleavage.

either **non-homologous end joining (NHEJ)** or HR, also known as homology-directed repair (HDR). The NHEJ mechanism functions in the absence of any additional donor DNA sequence and generates **error-prone** repairs that tend to introduce insertions or deletions (indels) into the sequence, often shifting the reading frame and resulting in loss of function of the target gene. If a donor sequence is introduced, this can be incorporated into the DSB site and can produce a gene edit. The use of ZFNs is shown in Fig. 12.3.

12.2.2 TALENs

The use of *Fok* I with a different sequence recognition system was developed when **transcription activator-like effectors (TALEs)** were discovered in the plant pathogen *Xanthomonas* genus. The central feature of TALEs is a domain that has a highly conserved repeat region of 33–35 amino acids, within which is a variable region of two amino acids known as the **repeat variable di-residue (RVD)** region. It is this short sequence that generates the specificity for sequence recognition at the single base level. All possible variants of RVDs have been decoded, and by changing the arrangement of the sequences, a highly specific sequence recognition system can be generated. Combining the TALEs with the cleavage domain of *Fok* I generates the TALEN (transcription activator-like effector nuclease). TALENs were discovered in 2010 and were easier to generate and use than ZFNs, and thus became established more widely. As with ZFNs, they can be used to generate gene knockouts by NHEJ repair, or to introduce donor sequences using the HDR process. The use of TALENs is outlined in Fig. 12.4.

12.2.3 The CRISPR-Cas9 System

In 1987, **Yoshizumi Ishino** discovered unusual repetitive sequence elements in *Escherichia coli* whilst looking at the genes involved in phosphate metabolism. At the time, partly due to the lack of extensive

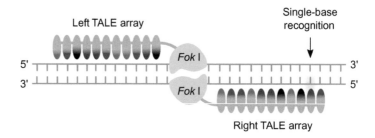

Single-base
recognition

5'

Fok I

3'

3'

Fok I

5'

Right TALE array

Fig. 12.4 TALENs. TALENs are similar to ZFNs in that two recognition–cleavage constructs are required to generate double-strand breaks. An array of TALEs is linked to the *Fok*-derived nuclease domain. A minimum of 11 TALEs is usually required to ensure specificity. Each TALE has a conserved region (purple) and a repeat variable di-residue (RVD) region that specifies single-base recognition (highlighted yellow; the TALE-specific recognition is shown in green, black, blue and red). Thus, a recognition sequence can be generated by assembling the appropriate array of TALEs.

sequence data for comparison, the significance of these repeats was unclear. It was not until the early-2000s that spacer regions in the repeats were found to be similar to bacteriophage sequences, and the repeats were proposed to be part of a bacterial defence system against bacteriophage, prophage and plasmids. The term **clustered regularly interspaced short palindromic repeats (CRISPR)** was proposed in 2002 to harmonise the variety of names that had been used, as more examples of the sequences were found in different prokaryotes. Subsequently, **CRISPR-associated (Cas) proteins** were found, covering a range of functions, including nuclease action as in the **Cas9 endonuclease**. It was not until 2007 that the functioning of the CRISPR-Cas system was confirmed experimentally as being a sort of 'adaptive memory-based immune system' for prokaryotes, where sequences from prior infections are stored in the repeat spacer regions and used as a library to activate the system when foreign sequences are detected.

The CRISPR-Cas9 editing system quickly became established and used in many thousands of laboratories around the world due to its specificity and ease of use.

Building on the earlier work, by 2012, the **CRISPR-Cas9** genome editing tool had been developed, and in 2015, the journal *Science* named CRISPR as the breakthrough of the year. In 2020, **Emmanuelle Charpentier** and **Jennifer Doudna** were awarded the Nobel Prize in recognition of their central roles in developing the CRISPR-Cas9 system.

The basis of using CRISPR as an editing system relies on the use of **guide RNAs (gRNAs)** that enable precise sequence-specific location of the complex. The Cas9 endonuclease then cuts the DNA to generate the DSB, which is repaired by the NHEJ or HDR processes (as for the ZFN/TALEN systems). A short 2–6 bp sequence called the **protospacer adjacent motif (PAM)** is present in the target DNA, but absent from the CRISPR sequence (this prevents the system from cleaving the

(a)

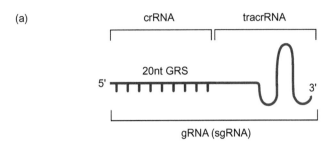

crRNA   tracrRNA

20nt GRS

5'   3'

gRNA (sgRNA)

(b)

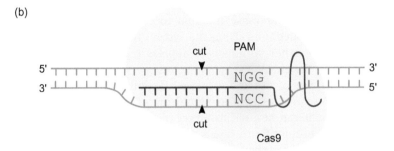

cut   PAM

5'   3'
                    NGG
3'   5'
                    NCC

cut

Cas9

Fig. 12.5 The CRISPR-Cas9 genome editing system. (a) In the naturally occurring system, two RNAs are produced – a CRISPR RNA (crRNA) and a *trans*-activating CRISPR RNA (tracrRNA) – that pair to produce the guide RNA (gRNA). For the engineered system, a single gRNA (sgRNA) is generated that contains the specific sequence required for target gene recognition (here this is labelled the gene recognition sequence or GRS). This is 20 nucleotides in length (shown shorter for clarity). (b) The CRISPR-Cas9 system functions by binding to the target sequence and generating a double-strand break. There are three levels of recognition/checking before cleavage of the DNA. The PAM sequence of 5'-NGG-3' enables rapid location of possible cutting sites in the genome by the Cas protein 'scanning' for PAM sites. The gRNA/GRS binding then unwinds the DNA helix in the binding region in preparation for cleavage. A final check of the fidelity of base recognition then enables cleavage to proceed.

bacterial CRISPR region). The PAM for Cas9 is 5'-NGG-3' (where N is any base) and lies 2–6 nt downstream of the gRNA recognition sequence. When the PAM is present with a suitable gRNA, the Cas nuclease cuts some 3–4 nt upstream of the PAM. By designing the gRNA and ensuring the correct location of the PAM sequence, it is relatively straightforward to design gRNA sequences that target specific regions of genes. A second function of the PAM sequence is that it enables rapid 'searching' for possible cut sites; when a PAM sequence is found, the gRNA can be checked against the target sequence and only if homology is found will the system cut the DNA target. The basis of using CRISPR-Cas9 for gene editing is shown in Fig. 12.5.

Cas9 can be modified to accommodate other proteins; this offers great potential for delivering specific functions to particular regions of the genome.

As the CRISPR-Cas9 system provides a means of locating any sequence in the genome, it is possible to harness this for other purposes, in addition to generating DSBs. The nuclease domains can be inactivated, which enables Cas9 to be modified in a number of ways. If one of the nuclease domains is inactivated and the other retained, the enzyme generates a single-strand nick in the DNA rather than a DSB (this is known as nickase activity). Using two nickases with different gRNAs enables the generation of staggered (overlapping) ends rather than blunt-end cuts, which can improve the efficiency and fidelity of introducing donor DNA sequences into the cut site. Other protein components that can be added to the Cas protein include a deaminase function to enable single-base mutations to be generated, and using the system to assemble **transcription factors** to either stimulate or repress transcription of genes targeted using the gRNA.

As well as ease of use and increased flexibility in designing very specific gRNA sequences, another significant advantage of the CRISPR system is that it can be used to target many genes at once. This is particularly important for the analysis of **polygenic** disease traits that are influenced by more than one gene, and there is great potential for devising therapies to repair defective genes that cause or contribute to severe disease conditions.

12.2.4 Prime Editing

The development of the CRISPR-Cas9 system was a major milestone in gene and genome editing, with two main advantages over previous methods being its specificity and relative ease of use. Following on from the single-base alteration capability developed in 2016, **David Liu's** laboratory pushed the technology further in 2019 with the prime editing technique. This is based on a nickase version of the Cas9 protein, coupled with a reverse transcriptase (RTase) enzyme. The nickase does not generate DSBs, so the potential problems associated with repairing these do not apply. The RTase is able to copy the gRNA, which includes the sequence containing the edit as well as the recognition sequence. This extended gRNA is known as a **prime editing guide RNA** (pegRNA). A cellular nuclease (FEN1, involved in DNA repair) trims the 5′ 'flap' overhang and favours the incorporation of the new sequence in a stable form. Prime editing is shown in Fig. 12.6.

Prime editing represents a further refinement of the CRISPR-Cas9 system, with some significant advantages, particularly for therapeutic intervention where a high degree of precision is required.

Further development of the prime editing system has improved the efficiency, reduced the number of indels generated (to a level some tenfold lower than is found in the HDR process) and also reduced off-target effects. The technique has therefore some significant advantages over the standard CRISPR-Cas9 system for potential therapeutic use, for which the process needs a level of precision that is not required in a non-clinical setting.

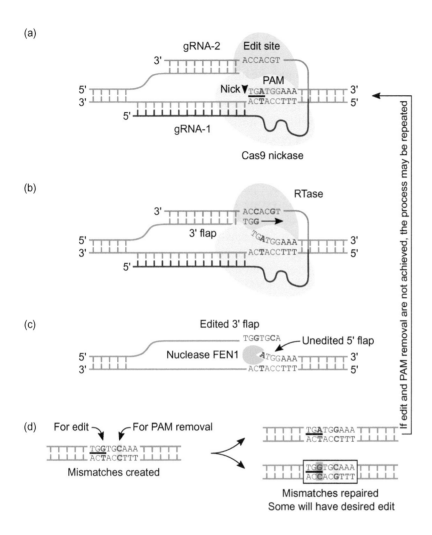

Fig. 12.6 The basis of prime editing. Prime editing uses a Cas9/RTase complex and an extended prime editing guide RNA (pegRNA). In this example, the aim is to alter a TGA stop codon. DNA/RNA strand orientations, and the position of the pegRNA, have been adjusted for clarity in places. (a) The Cas9 is modified to produce a nickase that cuts one strand only. The guide RNA has two areas, shown here as gRNA-1 (the target gene recognition sequence) and gRNA-2 (the RTase primer and edit site). The Cas9 nickase binds to the PAM site and cuts upstream of this (black arrowhead) to expose the target stop codon (TGA, black underline). (b) The target gene 3′ 'flap' sequence binds the 3′ end of the guide RNA (gRNA-2) to produce a 3′-OH primer for RTase, which copies the edit sequence (arrowed). The edit template sequence contains the target alterations, in this case a C in the stop codon (red bold) and a G in the PAM site (blue bold). (c) The edited 3′ flap is complete. The unedited 5′ flap on the other strand of the target gene is degraded by the cellular nuclease FEN1. (d) The edited region generates a mismatch with the unedited region (a G·T in the stop codon and a C·C in the PAM sequence). Mismatch repair either restores the original arrangement (top right) or makes the base change (s) (bottom right). Thus, in some cases, the desired edit (bottom right, box highlight) will be achieved – the TGA>TGG to deactivate the stop codon, and the TGG>TGC to remove the PAM site and prevent re-editing. If the edit is unsuccessful and the original configuration is restored, the process can repeat. *Source:* Modified from Scholefield, J. and Harrison, P. T. (2021). *Gene Therapy*, 28, 396–401. DOI [https://doi.org/10.1038/s41434-021-00263-9]. Used under Licence CC-BY-4.0 [https://creativecommons.org/licenses/by/4.0/].

12.3 | Editing RNA as an Option

Although methods for editing RNA transcripts are not as well developed as those for DNA, there are some potential advantages in editing the transient RNA molecules rather than the stable DNA of the genome.

Although the techniques we have considered so far have been directed at DNA, RNA is also a potential target for gene (or, more accurately, *transcript*) editing. RNA editing occurs naturally in cells, but it is a relatively rare event and not usually considered part of the post-transcriptional RNA processing mechanisms of polyadenylation, RNA splicing and 5'-capping. The process was first discovered in the late-1980s, and although its therapeutic potential was explored briefly in the early-1990s, not much progress was made. When the topic began to generate some interesting findings in 2012, the discoveries were somewhat overshadowed by the excitement that surrounded the use of the CRISPR system that was published at the same time. More recently, RNA editing has been more actively developed and is demonstrating something of a resurgence.

The mechanisms for RNA editing discovered and developed to date are rather more limited than those for DNA editing. One common process in cells involves an enzyme that is called adenosine deaminase acting on RNA (ADAR) that binds to double-stranded RNA (dsRNA) and causes deamination, changing the nucleoside adenosine to inosine (although not having one of the four canonical bases A, G, C or T, inosine acts functionally like guanine). Although limited in application, this does hold some promise for developing gRNAs that target parts of mRNAs that carry gene-derived mutations. As well as ADAR, in the Cas protein family, **Cas13** has RNA-specific endonuclease activity that may prove to be useful for editing RNA in cells.

One obvious question is 'why edit RNA when we can edit DNA?' – after all, RNA is more unstable and is a transient part of the cell's metabolic activity. In fact, these features in some ways make RNA editing a *more* attractive option, particularly for therapeutic intervention. Any off-target effects will disappear as the RNA is degraded over time, and the problem of stably inherited off-target changes to DNA is not an issue. Thus, RNA editing is more akin to a drug-based therapy than a permanent genetic change, which can be an advantage if the therapy can be designed and delivered effectively. Another advantage is that the system requires only gRNAs to be introduced into the cell, thus avoiding potential immune response problems that can accompany the introduction of proteins as in Cas-based systems. There is much as yet untapped potential in RNA editing, and the next few years should see some significant progress in this area.

12.4 | Where Can Genome Editing Take Us?

The short, trite, but possibly correct answer to this question may be 'as far as we want to go'. The development of engineered nucleases for genome editing illustrates how ingenuity and technical skill can enable major progress to be made in a short time frame. Genome editing is now used across the world in many thousands of laboratories, for a broad range of diverse applications. It certainly sits with gene cloning, the polymerase chain reaction and large-scale genome sequencing as being a **paradigm-shifting technology**. As we will see in Chapters 15 and 16, this has tremendous potential for medical science and clinical intervention, but also carries significant risk if not developed, regulated and implemented carefully.

Genome editing is arguably the most significant development of the first 50 years of gene manipulation; its potential is essentially unlimited, and how we handle this potent technology will pose challenges over the next 50 years.

12.5 | Conclusion: From Genome Read to Genome Write

In this fairly short chapter, we have covered what is potentially the single most important development of the past 20 years or so, namely the development of a suite of relatively accessible techniques that enable scientists to edit gene sequences in cells. There has been a marked transition across the period from when the first *knockout mice* were produced, using gene targeting by homologous recombination, to the present when we now have a range of *engineered nucleases* that target specific sequences with a high degree of precision. We looked at how *zinc-finger nucleases* and *TALENs* can be used to identify target sequences and generate double-strand breaks that are then repaired using either error-prone *non-homologous end joining* (generating *indels* that mostly knock out gene function) or *homology-directed repair* that can introduce donor DNA to replace parts of genes and generate loss-of-function or gain-of-function mutants.

Continuing the engineered nuclease story, we saw how adaptation of a bacterial defence system enabled the *CRISPR-Cas9* method to be established and adopted widely in a short time frame, and the further refinement of this into *prime editing* that does not require double-strand breaks. Together these two techniques have essentially unlimited potential across a wide range of areas in fundamental research, biotechnology and medicine. Finally, we had a brief look at the potential that RNA editing offers for future applications, particularly in the area of clinical intervention.

In a technical sense, genome editing mirrors the development of next-generation sequencing techniques for high-throughput DNA sequencing that enabled genomes to be deciphered. However, from the perspective of potential applications, genome editing offers the potential to *rewrite* genomes instead of simply reading the sequence. Coupled with synthetic biology approaches that can enable the design and synthesis of artificial genes, we now have a set of tools that are extremely powerful and have the potential to alter genome function in a significant way. The challenge, not unique to this area, will be to use this potential appropriately.

Further Reading

Carroll, D. (2011). Genome engineering with zinc-finger nucleases. *Genetics*, 188 (4), 773–82. URL [www.ncbi.nlm.nih.gov/pmc/articles/PMC3176093/]. DOI [https://doi.org/10.1534/genetics.111.131433].

Doudna, J. and Sternberg, S. (2018). *A Crack in Creation: The New Power to Control Evolution*. Penguin/Vintage, London. ISBN 978-1-784-70276-2.

Joung, J. K. and Sander, J. D. (2013). TALENs: a widely applicable technology for targeted genome editing. *Nat. Rev. Mol. Cell. Biol.*, 14 (1), 49–55. URL [www.ncbi.nlm.nih.gov/pmc/articles/PMC3547402/]. DOI [https://doi.org/10.1038/nrm3486].

Lander, E. S. (2016). The heroes of CRISPR. *Cell*, 164 (1–2), 18–28. URL [www.cell.com/fulltext/S0092-8674%2815%2901705-5]. DOI [https://doi.org/10.1016/j.cell.2015.12.041].

Scholefield, J. and Harrison, P. T. (2021). Prime editing – an update on the field. *Gene Ther.*, 28, 396–401. URL [www.nature.com/articles/s41434-021-00263-9#citeas]. DOI [https://doi.org/10.1038/s41434-021-00263-9].

Websearch

People

Development of the CRISPR-Cas9 method resulted in the award of the Nobel Prize to *Jennifer Doudna* and *Emmanuelle Charpentier* in 2020. Have a look at their careers and the collaboration that led to the development of this important genome editing technique.

Places

We mentioned the *Jackson Laboratory* as an important contributor to knockout/knockin work in mice. Find out about the laboratory – where it is located, what it does and how it is managed and funded. Also, see what the *International Mouse Phenotyping Consortium* does.

Processes

Have a look at how a *deletion cassette* is constructed and used to generate mutants in yeast. What similarities and differences are there when compared with other gene manipulation methods that we have looked at so far?

Reflections

What topics in this chapter have you found most challenging? Look for resources that help to illustrate the key points.

Genome editing

involves → **engineered nucleases**

used mostly to → **edit genomic DNA sequences**

edit genomic DNA sequences — **can be used to** → **edit RNA transcripts** — **but has** → **limited scope** — *e.g.* → ADAR Cas13

edit RNA transcripts — **and thus** → **regulate gene expression** — **some benefits** →
- effects are transient
- genome not altered

engineered nucleases → **hybrid enzymes**

engineered nucleases — **that have** →
- **catalytic cleavage domain** — **derived from** → *Fok* I (IIS RE) Cas9 nuclease
- **DNA sequence recognition domain** — **based on** → protein motifs guide RNAs — **such as** → zinc-fingers TALEs gRNA sgRNA pegRNA

catalytic cleavage domain / DNA sequence recognition domain → **generate double-strand breaks in the target sequence** → **Cas9 catalytic domains can be further engineered** →
- generate single-strand breaks
- deliver other protein functions

generate double-strand breaks in the target sequence →
- zinc-finger nucleases
- TALENs
- CRISPR-Cas9
- prime editing

zinc-finger nucleases / TALENs / CRISPR-Cas9 — **repaired by** →
- **non-homologous end joining** — error-prone and can cause indels
- **homology-directed repair** — more accurate repair of target

prime editing → **single-strand breaks** — **repaired by** → mismatch repair

editing genes and genomes raises significant ethical issues around how we will use the technique

gene targeting

gene targeting — **using** → **homologous recombination** — *i.e.* → **integration of DNA at regions of sequence homology** — **which is a** → **natural process** — **that** → **can be manipulated**

homologous recombination — **for** →
- **gain of function** — **or** → **knockin**
- **loss of function** — **or** → **knockout**

knockout — **first with** → **mice** — **and** → **yeast**

mice — **in** → **embryonic stem cells** → **chimeric mice** — **breed to give** → **heterozygous adults** — **then** → **homozygous adults**

Concept Map 12

Part 4

Genetic Engineering
in Action

Chapter 13 Summary

Learning Objectives

When you have completed this chapter, you will be able to:

- Outline the methods available for the analysis of gene structure and function
- Explain the development and use of DNA microarrays
- Discuss the range of genome sequencing projects, including mapping whole genomes
- Describe the other '-omes' that are now defined and studied: the transcriptome, proteome, metabolome and interactome
- Identify some of the current and likely future developments in genomics

Key Words

Open reading frames (ORFs), bait, prey, gel retardation, electrophoretic mobility shift assay (EMSA), supershift EMSA, David Galas, Albert Schmitz, DNA footprinting, DNase protection, chromatin immunoprecipitation (ChIP) assay, pull-down assay, yeast hybrid system, auxotrophic marker, *GAL 4* promoter, one-hybrid system (Y1H), two-hybrid system (Y2H), three-hybrid system (Y3H), primer extension, S1 mapping, deletion analysis, DNA microarray, genomics, renaturation kinetics, hyperchromic effect, abundance classes, repetitive sequence DNA, genetic map, physical map, recombination frequency, linkage mapping, genetic markers, physical markers, sequence-tagged site, Robert Sinsheimer, Renato Dulbecco, Charles DeLisi, Paul Berg, Walter Gilbert, Human Genome Organisation (HUGO), Human Genome Project (HGP), Telomere-to-Telomere (T2T) consortium, monogenic, Online Mendelian Inheritance in Man (OMIM), pedigree analysis, neutral molecular polymorphisms, restriction fragment length polymorphism (RFLP), Huntington disease (HD), cystic fibrosis (CF), sickle-cell disease, retinoblastoma, Alzheimer disease, minisatellites, microsatellites, variable number tandem repeats (VNTRs), genetic fingerprinting, DNA profiling, short tandem repeats (STRs), clone contig, Craig Venter, whole genome shotgun, Francis Collins, biome, Hans Winkler, transcriptome, DNA chip, exome, exons, expressome, proteome, two-dimensional electrophoresis, metabolome, interactome, protein~protein interactions (PPIs), whole genome sequencing (WGS), exome sequencing, whole exome sequencing (WES), transcriptome sequencing (RNA-Seq), structural genomics, functional genomics, comparative genomics, epigenomics, comparative genome analysis, personalised medicine, single nucleotide polymorphisms (SNPs), copy number variants (CNVs), trinucleotide repeat, genome-wide association study (GWAS), 1000 Genomes Project, David Cameron, 100 000 Genomes Project.

Chapter 13

Investigating Genes, Genomes and 'Otheromes'

In this chapter, we will look at some of the methods that can be used to investigate gene structure and function, and see how gene manipulation technology has opened up the world of the genome. This has led to the development of bioinformatics-based techniques for analysing genomes and gene expression in ways that were not thought possible a few years ago.

13.1 | Analysis of Gene Structure and Function

Although the contribution of classical genetic analysis should not be underestimated, much of the fine detail regarding gene structure and expression remained a mystery until the techniques of gene cloning enabled the isolation of individual genes. Many of the techniques used to characterise cloned DNA sequences provide information about gene structure, with one of the aims of most experiments being the determination of the gene sequence. However, even when the sequence is available, there is still much work to be done to interpret the various structural features of the sequence in the context of their function *in vivo*. Whilst the advent of the powerful techniques of bioinformatics has moved gene analysis on to a different level, there is still a need for characterisation of gene function by observation and experiment to provide a complete picture of gene structure and expression.

13.1.1 A Closer Look at Sequences
The availability of next-generation sequencing (NGS) has changed the way that genes are studied, with high-quality genome sequences now available for all the model organisms and thousands of other related strains and species. Even if the organism you are working on has not had its genome sequenced, it is likely that a related genome can be used to reduce some of the initial isolation and cloning procedures, although current sequencing technology means that it is often cost-effective to go straight to sequencing the genome anyway.

The key advantage of having a genome sequence is that a lot of the analysis can be carried out *in silico*. Searches can be made for regions of interest such as **promoters**, **enhancers**, **coding sequences**, *etc.* Coding regions can be translated to give the amino acid sequence of the protein, and **open reading frames (ORFs)** can be examined to see if they are in fact protein-coding sequences or not. Accurate restriction maps can be generated easily, and the sequence can be compared with others from different organisms, samples or patients. This comparative analysis can help in the study of phylogenetic relationships between groups of organisms, and in the analysis of the type and prevalence of disease alleles in a clinical setting.

Although computer analysis of a sequence is a very useful tool, it usually needs to be backed up with experimental evidence of structure or function. For example, if a previously unknown gene is being characterised, it will be necessary to carry out experiments to confirm where the important regions of the gene are, and what the function of the gene product is. Often the data produced will confirm the function inferred from the sequence analysis, although experiments sometimes generate novel findings that may not have been predicted. Thus, it is important that the computational and experimental sides of sequence analysis are used in concert.

> It is important that any computer-based analysis of sequence data is linked with experimentally derived information where possible, particularly where functional aspects of genes and proteins are being investigated.

13.1.2 Finding Important Regions of Genes

One of the key aspects in the control of gene expression is how proteins interact with DNA. Thus, it is important to find the regions of a sequence to which the various types of regulatory proteins will bind, and today a range of methods is available to do this. The terms **bait** and **prey** are sometimes used in describing the various assays, with the bait being the target sequence and the prey the protein of interest.

A relatively simple way to analyse DNA/protein binding is to generate a set of fragments from the region of interest. The protein under investigation (perhaps RNA polymerase, a repressor protein or some other regulatory molecule like a transcription factor) is added and allowed to bind to its site. If the fragments are then subjected to electrophoresis, the DNA/protein hybrid will run more slowly than a control fragment without protein, and can be detected by its reduced mobility. This technique is known as **gel retardation** or the **electrophoretic mobility shift assay (EMSA)**. There are now several variants of this technique, including identification of the bound protein using an antibody that is specific for the target protein. This is called a **supershift EMSA** in that the retardation is usually greater due to the additional effect of the antibody. The basis of the EMSA technique is shown in Fig. 13.1.

> Techniques based on gel electrophoresis can be useful in determining which regions of genes are involved in binding regulatory proteins.

An early method for determining protein binding sequences in a DNA fragment was devised by **David Galas** and **Albert Schmitz** in 1978. This is called **DNA footprinting** (or the **DNase protection** method). The technique is elegantly simple, and relies on the fact that a region of DNA that is complexed with a protein will not be susceptible to attack by DNase I (Fig. 13.2). The DNA fragment under investigation is radiolabelled and mixed with a suspected regulatory protein.

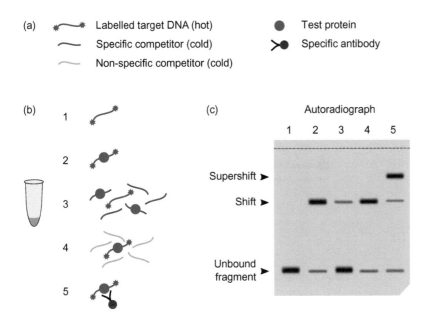

Fig. 13.1 The electrophoretic mobility shift assay (EMSA). (a) Key reaction components are the labelled target DNA fragment ('hot'), other test sequences that are similar to the target (specific competitors) or different (non-specific), the protein (or proteins) to be tested and an antibody specific for the regulatory test protein. (b) A series of reactions is set up with the target DNA fragment only (1), target plus test protein (2), target, protein and specific competitor fragment (3), target, protein and non-specific competitor fragment (4) and target, protein and antibody (5). The reactions are run on a polyacrylamide gel; unbound DNA fragments migrate fastest and move towards the bottom of the gel. Where protein binds to the target, mobility is slower and results in a shift in the position of the band. Often a radioactive label is used and detected by autoradiography. A typical result is shown in (c). Lane 1 is the fragment only control, and lane 2 shows the shift due to the bound test protein. There is usually residual unbound fragment present, so bands appear in all lanes at the unbound fragment position. Lanes 3 and 4 show the effect of competitor fragments, which are present in excess. A specific competitor (lane 3) will bind the test protein and thus reduce the strength of the signal in the shift position. A non-specific competitor (lane 4) will produce a result similar to the target fragment only (lane 2). Finally, if an antibody is used to bind the test protein, mobility will be slowed further (a supershift, lane 5). Note that there is often some residual signal in the shift position due to bound proteins that have not complexed with the antibody.

DNase I is then added, so that limited digestion occurs; on average, one DNase cut per molecule is achieved. Thus, a set of nested fragments will be generated, and these can be run on a sequencing gel. The region that is protected from DNase digestion gives a 'footprint' of the binding site within the molecule.

A more recent technique known as the **chromatin immunoprecipitation (ChIP) assay** uses an alternative approach to examining protein binding sequences. In this technique, the protein and DNA to which it is bound are cross-linked using formaldehyde, and the DNA sheared into fragments. Specific proteins can be precipitated using

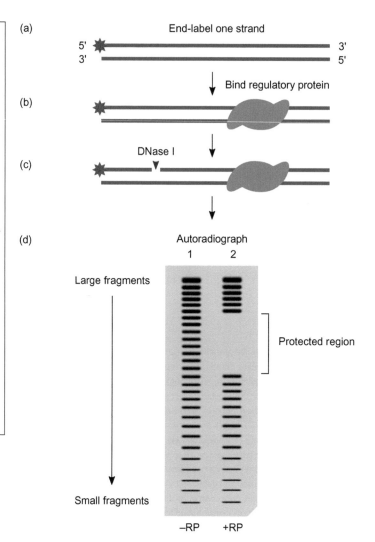

Fig. 13.2 DNA footprinting. (a) One strand of the target DNA molecule is labelled at one end (radioactive or fluorescent labels can be used). (b) The suspected regulatory protein is added and allowed to bind to its site. A control reaction without protein is also set up. (c) DNase I is used to cleave the DNA strand. Conditions are chosen, so that on average only one nick will be introduced per molecule. The region protected by the bound protein will not be digested. Given the large number of molecules involved, a set of nested fragments will be produced. (d) The reactions are run on a sequencing gel and visualised by autoradiography or a fluorophore detection method. In this example, an autoradiograph is shown. When compared to the control reaction without the regulatory protein (−RP, lane 1), the test reaction (+RP, lane 2) indicates the position of the protein on the DNA by its 'footprint'.

The 'bait and prey' concept can be used to isolate complexes formed by proteins and nucleic acids.

antibodies, and the DNA sequence determined. The bait and prey concept can also be used in a **pull-down assay** to selectively purify target complexes (these can be DNA~protein, RNA~protein or protein~protein) from mixtures such as cell lysates.

Another powerful and widely used series of methods for investigating the interaction of nucleic acids and proteins is the **yeast hybrid system**. This uses a reporter gene (either an **auxotrophic marker** such as the yeast *HIS 3* gene or a colour-based system such as *lac Z*) and a **GAL 4 promoter**. Activation of the promoter indicates interaction of the components being tested. The system has been developed for use in detecting DNA~protein binding (the **one-hybrid system** or **Y1H**), protein~protein binding (the **two-hybrid system** or **Y2H**) and RNA~protein binding (the **three-hybrid system** or **Y3H**). A significant advantage of this approach is that it is an *in vivo* technique that can be adapted for high-throughput automated screening

(a)

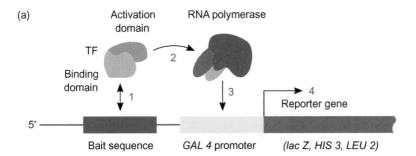

(b)

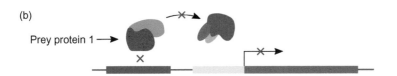

(c)

Fig. 13.3 The yeast one-hybrid system. The yeast hybrid assays are carried out *in vivo* by inserting plasmid constructs carrying the test components into yeast cells. (a) The basis of the system (for all variants) is the 'bait and prey' concept. A reporter gene (often the bacterial *lac Z* gene or a yeast auxotrophic marker such as *HIS 3* or *LEU 2*) is placed under the control of the yeast *GAL 4* promoter. In the Y1H system, a DNA sequence upstream from the promoter acts as the bait sequence; this might be kept constant to test a range of proteins or could be varied to test different sequences binding to the same protein. The transcription factor (TF) has two domains. The DNA binding domain binds to the bait sequence if it is recognised (1). The activation domain interacts with the RNA polymerase complex (2) to enhance binding to the promoter (3) and stimulate transcription of the reporter gene (4). The binding domain can be substituted by different prey proteins for testing. (b) Prey protein 1 does not interact with the bait sequence, and thus the reporter gene is not expressed. (c) Prey protein 2 does interact with the bait sequence, and thus the reporter is expressed. Generating different bait and prey libraries enables high-throughput screening of DNA~protein interaction.

of large numbers of samples. The hybrid assay technique has now also been extended into other cell types as well as the original yeast-based system. The basis of the Y1H technique is shown in Fig. 13.3.

It is often necessary to locate the start site of transcription for a particular gene, and this may not be apparent from the gene sequence data. Two methods can be used to locate the T_C start site, these being **primer extension** and **S1 mapping**. In primer extension, a cDNA is synthesised from a primer that hybridises near the 5′ end of the mRNA. By sizing the fragment that is produced, the 5′ terminus of the mRNA can be identified. If a parallel sequencing reaction is run using the genomic clone and the same primer, the T_C start site can be

Transcription start sites can be identified by primer extension and S1 mapping techniques.

located on the gene sequence. In S1 mapping, the genomic fragment that includes the T_C start site is labelled and used as a probe. The fragment is hybridised to the mRNA, and the hybrid is then digested with single-strand specific S1 nuclease. The length of the protected fragment will indicate the location of the T_C start site relative to the end of the genomic restriction fragment.

13.1.3   Investigating Gene Expression

Recombinant DNA technology has helped the study of gene expression in two main ways. Firstly, genes that have been isolated and characterised can be modified, and the effects of the modification studied. Secondly, probes that have been obtained from cloned sequences can be used to determine the level of mRNA for a particular protein under various conditions. These two approaches, and their sophisticated extensions generated as the technology developed, have provided much useful information about how gene expression is regulated in a wide variety of cell types.

> Deletion analysis can be used to determine which parts of upstream control regions are important in regulation of gene expression.

One method of determining which regions are important in controlling gene expression is **deletion analysis**. Sequences lying upstream from the T_C start site are progressively deleted using a nuclease such as exonuclease III or BAL 31 (see Section 5.2.1), to generate a series of deletions (Fig. 13.4). The effects of the various deletions can be studied by monitoring the level of expression of the

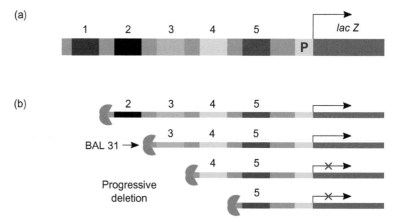

Fig. 13.4   Deletion analysis in the study of gene expression. (a) In this example, five regions (1 to 5) lying upstream from the promoter (P) are being studied to find those that are important in controlling gene expression. Often the *lac Z* gene is used as a reporter gene for detection of gene expression using the X-gal system. (b) Deletions are created using a nuclease such as BAL 31. In this example, four deletion constructs have been made, with progressively more upstream sequence removed in each construct. The effects of these deletions can be monitored by the detection of β-galactosidase activity, and thus the positions of upstream controlling elements can be determined. In this example, activity is not detected when region 3 is deleted, so this may be a binding site for a regulatory protein and can be investigated further. As an alternative to using BAL 31, restriction fragments can be removed from the controlling region.

gene itself, or of a reporter gene such as *lac Z*. In this way, regions that increase or decrease transcription can be located, although the complete picture may be difficult to decipher if multiple control sequences are involved in the regulation of transcription.

Measurement of mRNA levels is an important aspect of studying gene expression, and is often done using cDNA probes that have been cloned and characterised. The mRNA samples for probing may be from different tissue types or from cells under different physiological conditions, or may represent a time-course if induction of a particular protein is being examined. If the samples have been subjected to electrophoresis, a Northern blot can be prepared, which gives information about the size of transcripts as well as their relative abundance. Alternatively, a dot-blot can be prepared and used to provide quantitative information about transcript levels by determining the amount of radioactivity in each 'dot'. This provides an estimate of the amount of specific mRNA in each sample.

> The amount of a particular mRNA can be measured using cDNA probes and Northern or dot-blot techniques.

Early studies on gene expression tended to concentrate on one or two genes rather than on several or many genes in a particular cell or tissue type – perhaps a cDNA probe was available to facilitate the measurement of transcript levels, or an inducible system enabled differential gene expression to be examined. In such applications, the protein and gene have usually been characterised, at least to some extent, and thus the investigator can be sure of what is being measured. However, only a few genes can be investigated in this way in any reasonable scale of experiment. Thus, extending gene expression studies to try to analyse gene expression at the *genome* level (rather than at the single-gene level) required something of a technological breakthrough, which came with the development of the DNA microarray.

> In recent years, the emphasis has shifted towards analysing genomes rather than genes, using powerful techniques based on DNA microarrays and DNA chips to investigate gene expression in a whole-genome context.

The story behind this is really one of technological development to exploit the base-pairing features of complementary nucleic acid sequences. In essence, the technique requires immobilisation of a large number of different sequences on a support medium, such as a glass microscope slide (Fig. 13.5). This can be achieved in different

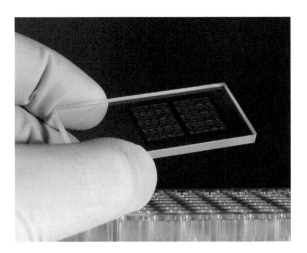

Fig. 13.5 DNA microarray on a glass slide. A spot on the slide with a DNA sequence is called a field. In this example, two panels are shown, each containing 756 fields as a 27 × 28 array. *Source:* Photograph by dra_schwartz / E+ / Getty Images.

Fig. 13.6 DNA microarray technology. (a) This 97 200 field array has 48 panels. (b) Each panel is made up of a grid of 45 × 45 fields, with each field having many copies of a single DNA or oligonucleotide sequence. Samples to be analysed (often cDNA prepared from mRNA and labelled with fluorescent dyes) are pumped onto the array and allowed to hybridise. The excess is washed off, and the fluorescence pattern on the array is read using a laser-activated scanning system. (c) The pattern of hybridisation is presented on a computer display. Different levels of signal in each field indicate the level of gene expression, and are correlated with the genes on the microarray using computer analysis.

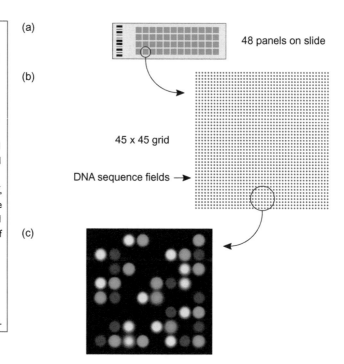

(a)

48 panels on slide

(b)

45 x 45 grid

DNA sequence fields ➜

(c)

ways, such as spotting DNA or cDNA sequences directly onto the slide using a robotic device to enable precise location of the microscopic spots of nucleic acid. An alternative is to synthesise oligonucleotide sequences directly on the slide, using technology based on computer chip manufacturing methods. When the array is ready, the pattern of sequences can be used to hybridise with a sample (often cDNA labelled with fluorescent tags). Binding of complementary sequences can be analysed, and information about the pattern of gene expression can be deduced by the level of signal detected using a laser imaging system (Fig. 13.6). There are many different ways that DNA microarrays can be used in the analysis of gene expression and in screening for mutations or looking for polymorphisms. We will look at how microarrays can be used to analyse the transcripts produced by cells in Section 13.3.1.

13.2 | Understanding Genomes

The field of **genomics** has shifted the emphasis away from studying individual genes in isolation. We are now much more likely to assess the function of a gene, or set of genes, within the context of the genome as a whole. This has required the development of a set of techniques to look at genomes in new ways, complementing early methods of characterising genome complexity. The wealth of information provided by classical genetic analysis also provided a sound basis from which more detailed studies would emerge in line with technological progress.

13.2.1 Analysing and Mapping Genomes

The comparative genome sizes presented in Table 4.3 show that eukaryotic genomes may be several orders of magnitude larger than bacterial genomes, which makes the analysis of a genome much more difficult than the analysis of a gene. Structural features such as repetitive sequence elements and intervening sequences can present difficulties of interpretation, even with the sophisticated bioinformatics resources that are now available to assemble and annotate genome sequences.

Some of the earliest indications of genome complexity were obtained by using the technique of **renaturation kinetics**. In this technique, a sample of DNA is heated to denature the double-stranded DNA. The strands are then allowed to re-associate as the mixture cools, and the uv-absorbance (A_{260}) is monitored. Single-stranded DNA has a higher absorbance than double-stranded DNA (this is called the **hyperchromic effect**), so the degree of renaturation can be assessed easily. Using this type of analysis, eukaryotic DNA was shown to be composed of several different **abundance classes**. In the case of human DNA, early estimates reported about 40 per cent of the total was either highly or moderately **repetitive sequence DNA**, which can often cause problems in the cloning and analysis of genes. Of the remaining 60 per cent, representing unique sequence and low copy number sequence elements, only around 3 per cent was thought to be actual coding sequence. As genome sequencing has progressed, these figures have been refined and the current figures are 1–2 per cent coding and 98 per cent non-coding, with repetitive sequences 50–75 per cent (depending on how these are categorised).

> Renaturation kinetics provided some early indication of the complexity of eukaryotic genomes and the presence of repetitive sequences.

Although these days many genome sequencing projects can be carried out using the whole genome shotgun method, the availability of a **genetic map** and **physical map** can be helpful. If an ordered sequencing strategy is considered, the physical map gives an overall framework against which the sequence data can be assembled, and genetic maps can help confirm the location of gene sequences responsible for diseases. An analogy may help to put the whole process in context.

Genome mapping is a bit like using a road map (either paper-based or web-based satellite versions work for this analogy). Say you wish to travel by car from one city to another – perhaps Glasgow to London, or New York to Denver. You could (theoretically) get hold of all the street maps for towns that exist along a line between your start and finish points and follow these. However, there would be gaps, and you might go in the wrong direction from time to time. A much more sensible strategy would be to look at a map that showed the whole of the journey, and pick out a route that follows major landmarks like cities or towns along the way. At this initial stage, you would not be too concerned about accurate distances, but would like to know which

> Genetic mapping is similar to geographical mapping in many respects, with scale and level of detail being two important considerations. For a map to be useful, the level of detail has to be appropriate for the map to function.

roads to take. You would then split the journey into stages where local detail becomes very important, such as how far it is to the next petrol station. For a stopover, the address of a hotel would guide you to the precise location. In genome analysis, the techniques of genetic and physical mapping provide the equivalent of large- and small-scale maps to enable progress to be made. Determination of the sequence itself is the level of detail that enables precise location of genome 'landmarks' such as genes and control regions.

Genetic mapping has provided a lot of information about the relative positions of genes on the chromosomes of organisms that can be used to set up experimental genetic crosses. The technique is based on the analysis of **recombination frequency** during meiosis, and is often called **linkage mapping**. This approach relies on having **genetic markers** that are detectable. The marker alleles must be heterozygous so that meiotic recombination can be detected. If two genes are on different chromosomes, they are unlinked and will assort independently during meiosis. However, genes on the same chromosome are physically linked together, and a crossover between them during prophase I of meiosis can generate non-parental genotypes. The chance of this happening depends on how far apart they are – if very close together, it is unlikely that a crossover will occur between them. If far apart, they may behave as though they are essentially unlinked. By working out the recombination frequency, it is therefore possible to produce a map of the relative locations of the marker genes.

> The classic method of mapping genes using recombination frequency analysis is still useful as a low-resolution method of gene localisation.

Physical mapping of genomes builds on genetic mapping and adds a further level of detail. As with genetic maps, construction of a physical map requires markers that can be mapped to a specific location on the DNA sequence. The aim of a physical map is to cover the genome with these identifiable **physical markers** that are spaced appropriately. If the markers are too far apart, the map will not provide sufficient additional information to be useful. If markers are not spread across the genome, there may be sections that have too few markers, whilst others have more than might be required for that particular stage of the investigation.

> The development of reliable and efficient physical mapping techniques was an important part of the ordered strategy for sequencing the human genome.

Physical mapping of genomes is not a trivial task, and until the early-1980s, it was thought that a physical map of the human genome was unlikely to be achieved. However, this proved to be incorrect, and techniques for physical mapping of genomes were developed relatively quickly. Physical maps of the genome can be constructed in a number of ways, all of which have the aim of generating a map in which the distances between markers are known with reasonable accuracy. The various methods that can be used for physical mapping are shown in Table 13.1. **Sequence-tagged site (STS)** mapping has become the most useful method, as it can be applied to any area of sequence that is unique in the genome and for which DNA sequence information is available. One advantage of STS mapping is that it can tie together physical map information with that generated by other methods. The technique essentially uses the sequence itself as the marker, identifying it

Table 13.1 | Some methods for physical mapping of genomes

Technique name	Application
Clone mapping	Defines the order of cloned DNA fragments by matching up overlapping areas in different clones. This generates a set of contiguous clones known as contigs. Clone mapping may be used with large or small cloned fragments as appropriate.
Radiation hybrid mapping	Fragments the genome into large pieces and locates markers on the same piece of DNA. The technique requires rodent cell lines to construct hybrid genomes as part of the process.
Fluorescent *in situ* hybridisation (FISH)	Locates DNA fragments by hybridisation, plotting the chromosomal position of the sequence by analysing the fluorescent markers used.
Long-range restriction mapping	The method is similar to any other restriction mapping procedure, but enzymes that cut infrequently in the DNA are used to enable long-range maps to be constructed.
Sequence-tagged site (STS) mapping	The most powerful technique, which can complement the other techniques of genetic and physical mapping. Can be applied to any part of the DNA sequence, as long as some sequence information is available. An STS is simply a unique identifiable region in the genome.
Expressed sequence tag (EST) mapping	A variant of STS mapping in which expressed sequences (*i.e.* gene sequences) are mapped. More limited than STS mapping, but useful in that genes are located by this method.
Variable number tandem repeat (VNTR) mapping	VNTRs (minisatellites) can be used in genotyping and mapping, and in forensic analysis. The original DNA fingerprinting technique was based on VNTR analysis.
Short tandem repeat (STR) mapping	STRs are also known as microsatellites or simple sequence repeats (SSRs). As with any physical marker, they can be useful in mapping studies. They are now used widely with standard protocols for DNA profiling.
Single nucleotide polymorphism (SNP) mapping	The smallest polymorphism and the most common genetic variant in humans. NGS techniques and bioinformatics have enabled SNPs to be used in a variety of ways in mapping studies. Their relatively even spread and high density across the genome is an advantage.

either by hybridisation techniques or, more easily, by amplifying the sequence using PCR. With regard to the human genome, STS mapping enabled the construction of a useful genetic and physical map of the human genome by the late-1990s.

The final level of detail in genome mapping is usually provided by some sort of restriction map. Long-range restriction mapping using enzymes that cut infrequently is a useful physical mapping technique, but more detailed restriction maps are required when cloned fragments are being analysed prior to sequencing. The overall link between genetic, physical and clone maps is shown in Fig. 13.7.

Restriction mapping provides further detail at the physical level. Genetic, physical and restriction maps are used in concert to provide a detailed picture of where genes are located.

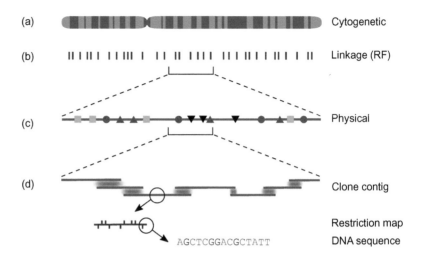

(a) — Cytogenetic

(b) — Linkage (RF)

(c) — Physical

(d) — Clone contig

Restriction map

DNA sequence

AGCTCGGACGCTATT

Fig. 13.7 Levels of genome mapping. (a) A cytogenetic map of the chromosome showing banding patterns. (b) Linkage mapping positions genetic markers on the chromosome, based on the recombination frequency (RF) between them during meiotic segregation. This requires suitable phenotypic markers and is a relative form of mapping. (c) A section of the chromosome is shown as a physical map, with different types of physical marker represented as different shapes. Various methods may be used to assign physical markers to their chromosomal locations (Table 13.1). (d) The clone map of a section of the physical map is shown, with overlapping DNA fragments known as contigs. The clone map can be used to enable restriction mapping and ultimately DNA sequencing. This progressive refinement of detail is the basis for ordered DNA sequencing strategies.

13.2.2 An Audacious Idea

The roots of genome sequencing go back to the early-1980s, with the publication of the sequence of bacteriophage lambda (λ) in 1982. Although the genome of phage ΦX174 had been sequenced in 1977, the 48 502 base-pair λ genome was the first 'large' sequence to be completed with some level of efficiency – still a significant undertaking, but impressive progress. At some point, unknown and undocumented, someone somewhere will have mused 'why can't we sequence the human genome?' and thought no more about it; biologists at that time simply did not have any way of visualising a project on the scale required.

The idea of sequencing the genome first gained momentum in the mid-1980s, when **Robert Sinsheimer**, **Renato Dulbecco** and **Charles DeLisi** independently, with different motives, began to think that it might be feasible. The annual Cold Spring Harbor Symposium of 1986 had a session to discuss the possibilities, with **Paul Berg** and **Walter Gilbert** proposing the audacious idea that it could be done, and Gilbert estimating that it would cost $3.5 billion. The initial reaction was somewhat mixed, but by 1988, the project had become more than a fanciful idea, with the US Department of Energy and the National Institutes of Health involved, and the establishment of the

The scale and cost of the Human Genome Project was unprecedented in biology, and required some audacious 'what if?' and 'why not?' thinking to become a reality. It remains a staggering achievement when considered in the context of the technology available at the time.

Human Genome Organisation (HUGO) to coordinate the major international collaborative effort required to achieve the aims of the project.

13.2.3 The Human Genome Project

The **Human Genome Project (HGP)** was officially launched in October 1990, and presented a task of almost unimaginable complexity and scale. Molecular biology has traditionally involved small groups of workers in individual laboratories, and most of the key discoveries in the pre-genomic era were made in this way. Sequencing the 3×10^9 base-pairs of the human genome was in another league altogether, and for the first time brought two novel approaches to a biological project. Firstly, in a pragmatic sense, this was certainly a 'big science' approach; the levels of funding, infrastructure, technical challenge, personnel, collaboration and just the immense scale of the task were all completely alien to biologists at the time. Secondly, and perhaps more importantly from an intellectual perspective, this was the first example of large-scale hypothesis-free science in biology; very little information about the genome and the genes it contained was available. Many of the objections to the genome project were in fact not so much about the information that would be useful, but about the generation of lots of information that would *not* be useful and the 'waste' of resources that this would entail.

As the technical, ethical and societal aspects of the HGP have been documented extensively and can be accessed easily (just search 'human genome project'), only a brief summary of the key points will be presented here. The project, as originally planned, was projected to last for 15 years (1990–2005) and cost three billion USD. In fact, it was 'completed' in 2003 and came in slightly under the projected budget at around 2.7 billion USD. The first draft sequence was announced on 26 June 2000, and the 'finished' sequence on 14 April 2003 (note that in reality, the terms *completed* and *finished* are not accurate; the initial sequences had many gaps that could not be sequenced easily). A number of genome releases since 2003 have reduced the number of gaps, but it was not until 2022 that the **Telomere-to-Telomere (T2T) consortium** published what is accepted as a complete version of the genome.

In many ways, the critical phase of the genome project was not the actual sequencing, but the genetic and physical mapping required to enable the sequence to be compiled against a reference map. Experimental crosses cannot be set up in humans, and therefore, mapping must be a retrospective activity. In many cases, tracing the inheritance pattern of genetic markers associated with disease can provide much useful information. The Centre d'Étude du Polymorphisme Humain (CEPH) in Paris maintains a set of reference cell lines extending over three generations (families of four grandparents, two parents and at least six children). The CEPH played a critical role in the HGP by coordinating the production of the genetic map

Although the HGP ended formally in 2003, the sequence was not fully complete, with many gaps and anomalies. It was not until 2022 that the T2T consortium generated what is generally accepted as the complete sequence.

that underpinned the assembly of the genome sequence, which was a significant achievement in itself.

Many diseases of genetic origin have been identified in which the defect is traceable as a **monogenic** (single gene) disorder. Currently over 26 000 entries (of various types) are listed in **Online Mendelian Inheritance in Man (OMIM)**, which is the central database site for this information (search 'OMIM' to find the site). Many of these diseases have already been studied extensively using the retrospective technique of **pedigree analysis**. However, to generate useful genetic map data, it is not always necessary to be able to trace the actual gene; if a polymorphic marker can be identified that almost always segregates with the target gene, this can be just as useful. These are called **neutral molecular polymorphisms**. In the early-1980s, a marker of this group called a **restriction fragment length polymorphism (RFLP)** began to be used to map genes. RFLPs (known as 'rifflips') are differences in the lengths of specific restriction fragments generated when DNA is digested with a particular enzyme (Fig. 13.8). They occur when there is a variation in DNA that alters either the sequence or the location of a restriction enzyme recognition site. Thus, a point mutation might abolish a particular restriction site (or create a new one), whereas an insertion or a deletion would alter the relative positions of restriction sites. If the RFLP lies within (or close to) the locus of a gene that causes a particular disease, it is often possible to trace the defective gene by looking for the RFLP, using the Southern blotting technique in conjunction with a probe that hybridises to the region of interest. This approach is extremely powerful, and enabled many genes to be mapped to their chromosomal locations before high-resolution genetic and physical maps became available. Examples include the genes for **Huntington disease (HD)** (chromosome 4), **cystic fibrosis (CF)** (chromosome 7), **sickle-cell disease** (chromosome 11), **retinoblastoma** (chromosome 13) and **Alzheimer disease** (chromosome 21).

Using RFLPs as markers, a genetic map of the human genome was available by 1987. However, this approach was limited in terms of the degree of polymorphism (a restriction site can only be present or absent) and the level of resolution. The use of **minisatellites** and **microsatellites** enabled more detailed genetic maps to be constructed. Minisatellites are made up of tandem repeats of short (10–100 bp) sequences. The number of elements in a minisatellite region can vary, and thus these are also known as **variable number tandem repeats (VNTRs)** (Fig. 13.9). These can be used in mapping studies, and also formed the basis of **genetic fingerprinting** (now more commonly referred to as **DNA profiling**, discussed in Chapter 15). One drawback with VNTRs is that they are not evenly distributed in the genome, tending to be located at the ends of chromosomes. Microsatellites (also known as **short tandem repeats** or **STRs**) have been used to overcome this difficulty. These are much shorter repeats, and the two base-pair CA repeat has been used as the standard microsatellite in mapping studies. Regions of CA repeats can be amplified using PCR, with primers that flank the repeated elements. As the primers are derived

Mendelian Inheritance in Man (MIM) was started by Victor McKusick, who is often called 'the father of medical genetics'. MIM is a major resource for collating and distributing information about genetically based diseases. It is of course now available online as OMIM.

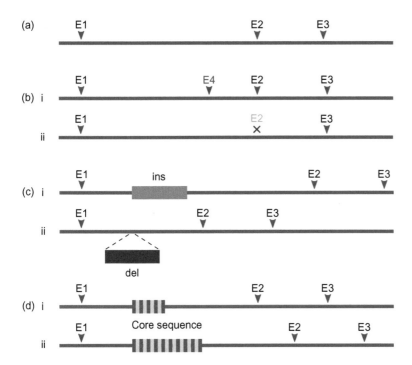

Fig. 13.8 Some ways of generating restriction fragment length polymorphisms (RFLPs). (a) Consider a DNA fragment with three *Eco* RI sites (E1, E2 and E3). On digestion with *Eco* RI, one of the fragments produced is $E_1 \rightarrow E_2$. RFLPs can be generated if the relative positions of these two sites are altered in any way. (b) The effect of point mutations. If a point mutation creates a new *Eco* RI site ((b)i, marked E4), fragment E1 $\rightarrow$ E2 is replaced with two shorter fragments: E1 $\rightarrow$ E4 and E4 $\rightarrow$ E2. If a point mutation removes the *Eco* RI site at E2, as in (b)ii, the fragment becomes E1 $\rightarrow$ E3, which is longer than the original fragment. (c) The effect of insertions or deletions. If additional DNA is inserted between E1 and E2 (ins in (c)i), fragment E1 $\rightarrow$ E2 becomes larger. Insertions might also carry additional *Eco* RI sites, which would affect fragment lengths. If DNA is deleted (del in (c)ii), the fragment is shortened. (d) The effect of variable numbers of repetitive core sequence motifs. This variation is a type of RFLP and can be used as the basis of genetic profiling. (d)i has nine copies of the repeated core sequence element (shown as alternating grey and yellow for clarity); thus, fragment E1 $\rightarrow$ E2 is smaller than that shown in (d)ii, which has 19 copies of the core sequence.

from unique-sequence regions, this essentially means that microsatellites amplified in this way are a type of STS (Table 13.1), and can therefore be used to link genetic and physical maps.

The outcome from all this activity in the 1990s was that by the end of the decade, a physical map of the genome with over 30 000 markers had been produced. Along with this increasing level of coverage and resolution, various types of software were developed to enable the map data to be correlated and viewed sensibly, and to be integrated with the emerging sequence data.

Sequencing the genome was carried out using a number of different approaches based on the automated Sanger method we looked at in Chapter 6 (see Section 6.7.4 and Fig. 6.11). In very broad terms,

Genetic mapping of the human genome was an essential part of the preparative work that eventually resulted in the completion of the genome sequencing project.

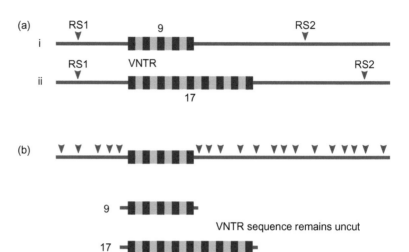

Fig. 13.9 Generation of variable number tandem repeats (VNTRs). (a) VNTRs can generate what are essentially RFLPs, in a similar way to that shown in Fig. 13.8(d). Here two restriction sites (RS1 and RS2) are separated by nine and 17 copies of a repeat element for illustration (shown as alternating light and dark red for clarity). If fragments are produced by cutting at the recognition sites for RS1 and RS2, different fragment lengths will be produced. An alternative method of analysing VNTRs is shown in (b). If a restriction enzyme is used that cuts the flanking DNA frequently (red arrowheads), but does not cut within the VNTR, the VNTR sequence is effectively isolated and trimmed, and left as a marker that can be used in DNA profiling techniques (fingerprinting).

there are two approaches that can be used to sequence a genome. In theory, it can be done using a straightforward shotgun approach. However, in the late-1990s, one limiting step in this method was the IT-based task of putting the sequences together. This is very difficult for a genome as large as the human genome, in which there are many repetitive sequences. A hierarchical shotgun approach (sometimes called directed shotgun), in which the sequenced fragments are collated against the map data, is better but is still likely to generate some anomalies. These two approaches are outlined in Fig. 13.10.

The public genome consortium used the hierarchical shotgun strategy to enable a clone contig approach to be used when assembling the sequence data from BAC clone libraries of the genome. However, this was not the only approach to sequencing the human genome. Enter Celera Genomics, a private company established by **Craig Venter** in 1998 with Perkin-Elmer Corporation. Venter, with a far-sighted vision, claimed that he could sequence the genome much more efficiently than the public consortium, by using a whole genome shotgun procedure. There was a good deal of what euphemistically might be called 'debate' when Venter put forward his ideas, particularly as he was going to use the freely available public consortium sequence data to assist his own efforts, which would result in Celera Genomics releasing its own data on a commercial basis. The

The 'race' between the public and private sequencing groups provided an interesting personal and political dimension to the genome project, which had until then been mainly seen as a technical and scientific challenge.

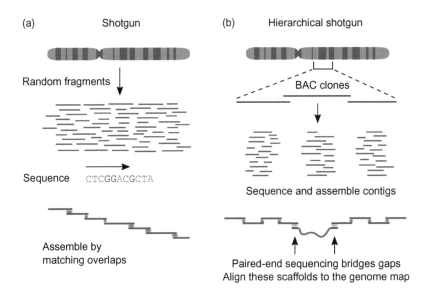

(a) Shotgun

Random fragments

Sequence   CTCGGACGCTA

Assemble by
matching overlaps

(b) Hierarchical shotgun

BAC clones

Sequence and assemble contigs

Paired-end sequencing bridges gaps
Align these scaffolds to the genome map

Fig. 13.10 Strategies for sequencing genomes. A simplified outline of shotgun and hierarchical shotgun strategies. (a) The shotgun approach breaks the genome into random fragments that are then cloned and sequenced. The sequencing technology being used determines the fragment size. The sequence is assembled into contigs by matching the overlapping sequences of the fragments. (b) A hierarchical shotgun approach generates a set of large cloned fragments (e.g. BAC clones) that are ordered into a tiling path, thus keeping track of the clones in relation to the genome map. Each BAC clone is fragmented and sequenced using a shotgun approach, and the contigs assembled. Where there are gaps between BAC clone sequences, paired-end sequencing can be used to link the contigs together into a scaffold. This is then assembled against the physical map of the genome.

whole matter became rather public and somewhat heated at times. However, it is undoubtedly true that Venter's appearance on the scene added to the urgency of the project (hence the so-called 'race' to finish the sequence). The fact that Venter, along with **Francis Collins** of the public consortium, shared in the US announcement of the first draft of the sequence in June 2000 shows that Celera Genomics had made a significant input and that their contribution was at least recognised, if not fully applauded or appreciated.

13.2.4 Other Genome Projects

The HGP has generated most widespread interest from the public – this is entirely understandable, given the potential benefits that have, and will, come from the project and its successors. However, there are many other genome projects that have delivered complete genome sequences before, during and after the HGP. Some of these helped to establish that the HGP was in fact an achievable project, acting in effect as proof of concept pilot projects, as well as being of great importance in their own right. Others have emerged because NGS technologies have made sequencing much faster and cheaper than

The first genome sequencing projects were based on so-called 'model organisms' that were already well characterised; genome sequencing was the next logical stage in investigation of these organisms.

was the case when the human genome was being sequenced for the first time.

The simple question 'how many different genomes have been sequenced' is in fact not trivial. As with any field that has expanded so quickly, the question is almost redundant; there will be no end-point for genome sequencing any time soon, where we could perhaps calculate a number that was reasonably accurate and that made some sort of sense. So although we know that many thousands of organisms have had their genomes sequenced, and that estimates can determine that as of December 2021 some 3 278 unique animal species had been sequenced, we have almost gone past the point where these are useful pieces of information. I have not therefore included a table of 'genome sequences completed', but would suggest that you have a look again at Table 4.3 (comparative genome sizes) and investigate which of these has been sequenced, and when. A useful resource to help with this can be found by searching for 'milestones in genomic sequencing', a timeline published by the journal *Nature*, which gives some very helpful insights into the scope and importance of various genome projects.

Presentation interfaces and annotation schemes turn raw genome sequence data into a resource of almost unlimited scope.

13.3 | 'Otheromes'

Over the past few years, there has been a significant expansion in 'omes' and 'omics' – originating from the term *biome*, the suffix is now used to describe a range of concepts and applications.

The molecular biology era has generated a whole raft of new disciplines under the 'omics' umbrella. However, we should remember that the ecological term **biome** was devised in 1916, and *genome* was first used by **Hans Winkler** in 1920. It was inevitable that once the genome (and later, genomics) had become established as useful concepts, the terminology would be adopted for other purposes. A list of some of the many additional 'omes' that are active research areas at the forefront of cell and molecular biology is shown in Table 13.2. We will consider some of these in the final sections of this chapter.

13.3.1 The Transcriptome

The genome represents the entire collection of genes, regulatory regions, repetitive sequences and intergenic regions. However, in any particular cell, only a subset of genes in the genome will be actively expressed at any one time. In Section 4.5.3, we saw that some genes have general 'housekeeping' functions and are not highly regulated, whilst others can be adaptively or developmentally regulated, depending on the type of cell and the environment in which it finds itself. The set of transcripts produced from genes active in a cell is called the **transcriptome**, and obviously the analysis of the transcriptome has a key part to play in the investigation of gene expression. One way of achieving this is to use DNA microarray technology.

There are several different ways in which microarray technology can be used for transcriptome analysis. One typical application is to examine the pattern of transcripts produced by two different cell samples, as shown in Fig. 13.11. The samples could be cells grown

DNA microarray technology revolutionised the way in which we carry out analysis of the transcriptome. It is now complemented, and in some cases replaced, by NGS techniques such as RNA-Seq transcriptome sequencing.

Table 13.2 Examples of 'omics' in molecular biology

Discipline	Sub-disciplines	Comment
Genomics	Comparative genomics Epigenomics Functional genomics Metagenomics Pangenomics Personal genomics Pharmacogenomics	Epigenomics is related to the control of the genome by modifications to DNA and regulatory proteins. Metagenomics analyses genomes from organisms in a bulk sample. Pangenomics studies all the genomes across a species within a clade. Personal genomics – individuals' genomes likely to be used for personalising medical treatments, such as targeted therapies using pharmacogenomic analysis
Proteomics	Structural genomics (DNA) Structural genomics (proteins)	Structural genomics can refer to the study of the genome itself or (more common usage) as part of proteomics, to determine the 3-D structure of all the proteins produced from a genome
	Proteogenomics	Developing area integrating genomics and proteomics
	Analysis of protein~protein interactions (PPIs)	
Transcriptomics		Involves analysis of gene expression
Interactomics	Analysis of PPIs Pathway analysis (*Reactome* database)	Interactions of many different sorts are critical in determining how the information in the genome is expressed
Metabolomics		Study of the distribution and role of metabolites in the cell
Metallomics		Study of the distribution and role of metal ions in biological systems, e.g. metalloproteins

Note: The term '-ome' is the suffix for the noun describing the *entity* being studied (*e.g.* genome), and '-omics' is the suffix used to describe the *discipline* or field of study (*e.g.* genomics). Many other examples of the use of the term(s) can be found that are not directly related to molecular biology, such as *biome, microbiome, foodomics, cellomics* and *ethomics*. More extensive lists and definitions can be found at the following URLs: [https://en.wikipedia.org/wiki/Omics], [http://omics.org] and [www.genomicglossaries.com]. Alternatively, use the search term 'omics' to find resources.

under different conditions, or cells from different tissues. A microarray is constructed with either cDNAs or oligonucleotides, and used to investigate the pattern of transcript binding in each of the two cell types. In this way, it is possible to identify gene sequences that are expressed in one sample, but not in the other.

A second approach to transcript analysis that is often used to analyse normal and disease states (such as in cancer cells), is shown in Fig. 13.12. In this case, the microarray or **DNA chip** is challenged with a mixture of cDNAs produced from samples of mRNA taken from normal and disease cells. The mRNAs from each cell are used to produce fluorescent-tagged cDNAs, with the normal and disease samples being tagged with different coloured dyes. The samples are mixed and incubated with the microarray. The cDNAs will bind to their complementary sequences on the microarray, and the pattern

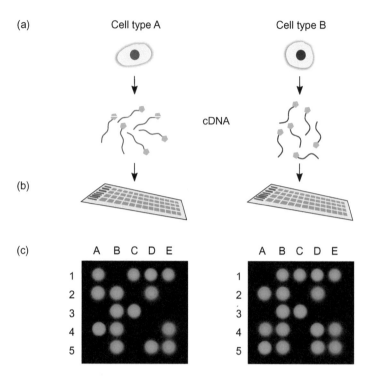

Fig. 13.11 Transcriptome analysis in two cell types using a microarray. An array is generated with the sequences from the genome of interest attached to the support matrix. (a) Samples are prepared from the two cell types A and B. Often this is done by synthesising cDNA, tagged with a fluorescent label (green in this example). (b) The cDNA samples are hybridised separately to the microarray, and the signals analysed. Results are compared as shown in (c). No signal in a field indicates that the gene was not expressed in the cell at the time of sampling. In this example, a positive result is shown by the green circles. Comparison of the results shows where genes are differentially expressed. Thus, the gene in position A1 is expressed in cell A, but not in cell B, whilst B1 is the reverse of this. Some genes (e.g. C2) are not expressed in either of the cell types, and some (e.g. E5) are expressed in both.

produced can be used to identify genes that are expressed differentially in normal and cancer cells.

Two other terms can be used to describe variations of the transcriptome concept. The **exome** is a subset of sequences within the transcriptome. It is defined as all the expressed **exons** within the protein-coding genes in a genome, and is therefore a useful target for investigation if protein synthesis and its regulation are of interest. Another term that is sometimes used is the **expressome**, which is the entire complement of products of gene expression, including transcripts, proteins and other components of cells. This term is also sometimes applied to tissues, organisms and even species. To add yet further confusion, it is also used to describe the supramolecular

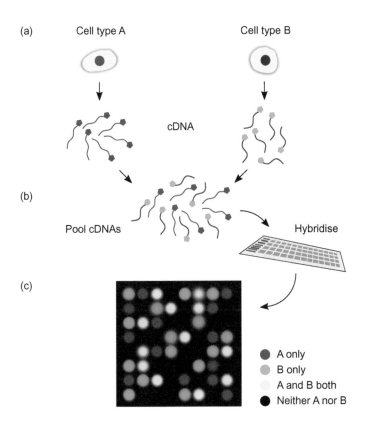

(a) Cell type A   Cell type B

cDNA

(b) Pool cDNAs   Hybridise

(c)

● A only
○ B only
○ A and B both
● Neither A nor B

Fig. 13.12 Transcriptome analysis using pooled cDNAs. A single array can be used to investigate differential gene expression. (a) In this example, mRNA from the two cell types is used to prepare cDNA that is labelled with two different fluorescent dyes (red and green). (b) The cDNAs are pooled and hybridised to the array. The pattern of signal detected is shown in (c). The various colours produced indicate the levels of transcript in each of the cells. No signal indicates no expression in either cell at the time of sampling. Single colours show relative levels of expression in the respective cells – red for A and green for B. Where both cells produce transcripts, a yellow spot is generated. Analysing the range of intensities produced gives an indication of differential gene expression using a single microarray.

complex of RNA polymerase and ribosomes involved in coupled transcription and translation in prokaryotes.

13.3.2 The Proteome

The concept of the cell as a molecular machine is one that goes some way to describing the emphasis of modern molecular biology, in which genes are just part of the overall picture. The term proteome is used to describe the set of proteins encoded by the genetic information in a cell, and great advances have been made in recent years in this area of research. The biochemical interactions between the various proteins of the cell are arguably even more important than the information in the genes themselves, as the information is silent if not expressed. The ultimate aim of cell and molecular biology is to understand cells fully at the molecular level, and in many ways, this is the biologist's holy grail, similar in scope to the search in physics for the unifying theory that will link the various branches of the discipline together.

Some indication of the complexity of the proteome in a cell can be obtained by using basic techniques such as polyacrylamide gel electrophoresis (PAGE). Although the standard one-dimensional

Several different techniques are required to get towards a full understanding of the structure and function of the many proteins that make up the proteome.

PAGE technique is useful, it is limited in terms of the number of proteins that can be resolved. **Two-dimensional electrophoresis** (2-D PAGE) gives much greater resolution. In this, proteins are separated firstly in one direction. The proteins are then separated in a second direction, often using a different gel system. One common technique uses isoelectric focusing in the first dimension, followed by SDS-PAGE in the second. In this way, several hundred proteins can usually be identified, although the technique can be limited by the sensitivity of the detection methods used. Additional methods such as mass spectrometry, X-ray crystallography and nuclear magnetic resonance can add valuable information about protein structure.

The advent of DNA microarrays enabled researchers to produce arrays using different types of molecules, which has facilitated studies of the proteome using this technology. One way of achieving this is to fix antibodies to the support matrix and use this to monitor the binding of proteins to their respective antibodies. This technology can also be used for other applications in proteomics research, which is now well established as a discipline and has given rise to the Human Proteome Organization (HUPO). This was founded in 2001 to provide a focus for the study of the proteome, and to characterise all the proteins in the human genome. There is also focus on proteomics in most major molecular biology laboratories and institutes, such as the EMBL Proteomics Core Facility. More information can be found on the websites – find them by using appropriate search terms.

13.3.3   Metabolomes, Interactomes and More

Extension of the 'ome' concept to other contents of the cell can be useful. If the aim of modern cell biology is to attempt to understand the way in which a cell works, then all the reactions and interactions need to be considered. Two additional terms that can be used to describe facets of cell function are the metabolome and the interactome. The metabolome describes the small molecules that are found in the cell, and includes the various metabolites that are essential for cell function. Thus, with the genome, transcriptome, proteome and metabolome, we have essentially a way of describing the components of cells. The next step towards a full understanding of cell function is to see how these components interact with each other. This gave rise to the concept of what became known as the interactome. Analysis of the interactome can be carried out by constructing a map of **protein~protein interactions** (PPIs), which provides some insight into how a cell functions. PPIs are most often investigated using the yeast two-hybrid system (Y2H) that we noted in Section 13.1.2. The basis of the Y2H method is shown in Fig. 13.13.

As with research into the transcriptome and the proteome, analysis of cellular interactions is supported by a wide range of online

Assembling data for all the various interactions that occur in cells is opening up a new level of understanding of how cells function.

(a)

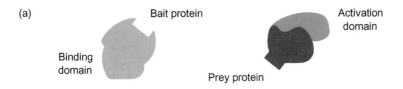

(b)

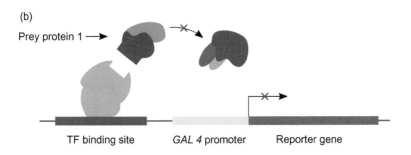

(c)

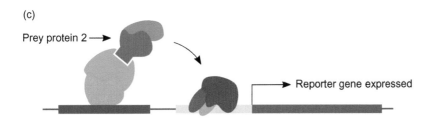

Fig. 13.13 The yeast two-hybrid system. Components of the Y2H assay system are as described in Fig. 13.3(a). The two-hybrid system investigates protein~protein interactions (PPIs); thus, the bait and prey are both proteins. (a) The bait protein is fused with the DNA binding domain of the transcription factor, and the prey protein with the activation domain. (b) The bait protein complex binds to the transcription factor binding site. In this case, prey protein 1 does not interact with the bait protein; thus, the reporter gene is not expressed. (c) Prey protein 2 does interact with the bait protein; the activation domain enhances RNA polymerase binding to the promoter, and the reporter gene is expressed. As with the one-hybrid system for DNA~protein interactions, large numbers of potential PPIs can be tested using libraries to generate different combinations of bait and prey proteins.

resources, including HuRI (see Section 11.2.5; www.interactome-atlas .org/) and Reactome (https://reactome.org/), which extends the concept into pathway mapping. Another key database is STRING (Search Tool for the Retrieval of Interacting Genes/Proteins; https://string-db.org/). These tools are sophisticated and powerful, and use data sets from a variety of sources. An example of mapping PPIs to investigate proteome dynamics is shown in Fig. 13.14.

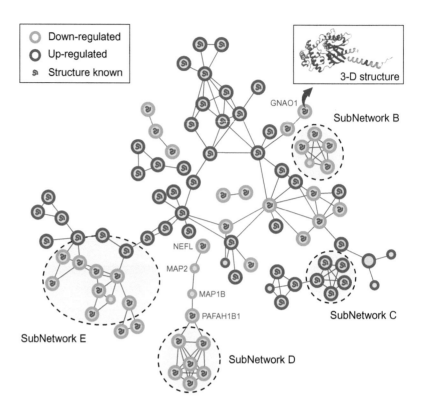

Fig. 13.14 Mapping interactions between proteins. One version of a map of protein~protein interactions (PPIs). Proteins are shown in circles (nodes) and are linked by lines (edges). This example is from a deceased patient with Alzheimer disease, and shows PPIs in the olfactory bulb area of the brain at the intermediate stage of the disease. The circles show 'up-' (red circles) and 'down-' (green circles) regulation status. The edges (blue lines) can be coloured differently to reflect the information known about the interaction (predicted, verified experimentally or based on existing data). Protein identifiers are associated with the nodes in the live version (only a few are shown here in blue text). Four sub-networks are identified in this example (dashed circles). Where a protein structure is available, this is indicated in the node. Clicking on the node opens up a 3-D structure version of the protein (panel at top right, for protein GNAO1). *Source:* Modified from Zelaya, M. V. et al. (2015). *Oncotarget*, 6, 39 437–39 456. DOI [https://doi.org/10.18632/oncotarget .6254]. Used under Licence CC-BY-2.5 [https://creativecommons.org/licenses/by/2.5/].

13.4 | Life in the Post-genomic Era

Over the past 15 years or so, NGS technology (and interpretation of the genome sequences it has generated) has driven major developments in all the 'omics' areas discussed in Section 13.3. As we saw in Section 13.2, emphasis has shifted away from looking at genes in isolation to considering how they function within a genome and contribute to the holistic expression pattern that defines cells, tissues and organisms. NGS is now *relatively* accessible and cost-effective, and the development of sophisticated web-based IT resources for collation, annotation and analysis of sequences has moved genome biology firmly into the large-scale bioinformatics domain.

Sequencing also reflects the various levels of genome organisation and expression. **Whole genome sequencing (WGS)** is no longer the daunting prospect it once was, and more targeted approaches can be used in **exome sequencing** (sometimes called **whole exome sequencing** or **WES**) or **transcriptome sequencing (RNA-Seq)**. Sequencing the transcriptome is an alternative way to monitor gene expression (compared to the use of microarray technology) and is beginning to have a significant impact in the field.

Variations on genome sequencing can be used to adapt the technology to the aims of the procedure; sequencing exomes and transcriptomes is another way to investigate gene expression at the genome level.

In a practical sense, it is not possible to provide a detailed picture of modern genomics in a book of this sort, so we won't try! Rather, we will have a brief look at some of the key elements of three aspects of the topic; although usually considered as separate sub-disciplines of genomics, in a practical sense, there are significant areas of overlap.

13.4.1 Structural Genomics and Proteomics

Structural genomics is concerned with characterisation of the physical structure of the genome and the proteins that it encodes. Genome sequencing is one obvious aspect of this approach, but determining the structure of proteins is perhaps not so obvious. In particular, the aim of structural genomics is to try to determine the 3-D structure of every protein in a given cell, and how protein folding generates these structures. Structural genomics therefore has considerable overlap with proteomics, extending the technical analyses more deeply into the areas we looked at in Section 13.3.2.

13.4.2 Functional Genomics

Functional genomics is concerned with gene expression, rather than the structural features of genes, genomes and proteins. However, one key difference between functional genomics and more traditional methods of investigating how a gene is expressed is that the aim is to see how a gene works in a particular context or against a particular background. This requires consideration of the environment in which the gene functions – perhaps a particular stage of development of the organism, or as a contributing factor to the presentation of a disease at the cellular or organismal level. One question that underpins this approach is 'why do some cells do X rather than Y in response to a particular set of conditions?'. It therefore bridges into areas of **comparative genomics** (see Section 13.4.3 below) and **epigenomics**.

13.4.3 Comparative Genomics

In addition to deciphering the genetics of any particular organism, genome sequencing also opens up the field of **comparative genome analysis**, which can help to understand how genomes evolve and how many genes are similar in all organisms. Humans have genomes that

Comparing genomes, between and within species, has been made possible by increased use of NGS technology and sophisticated IT applications.

are approximately 0.1 per cent different between individuals, and approximately 1–2 per cent different from the chimpanzee, and contain many of the same genes as bacteria. Comparative genomics involves the study of differences between species (*interspecific* analysis) and between individuals of the same species (*intraspecific* analysis).

Intraspecific analysis of disease-causing alleles will have a particularly important role in the future as **personalised medicine** becomes more established. One of the outcomes of sequencing genomes was that additional information about genetic variation was obtained. Single nucleotide polymorphisms (SNPs, pronounced 'snips') are, as the name suggests, regions in which a single base is different between individuals, and they represent the major source of distributed sequence variation. Thus, one sequence might be AGTTCGATGCG, and in another person AGTTAGATGCG, with the C at position 5 changed to an A. SNPs are (by definition) so small that they have not been subjected to the usual pressures of evolution, as most of these changes will not affect the reproductive fitness of an individual. Thus, they have become scattered evenly across the genome, on average about every thousand base-pairs. Whilst most will not be directly attributable as the cause of a particular disease, they can be used as markers to bring an increased level of subtlety to genome-based diagnosis. Detection of SNPs requires sequencing DNA from many individuals, usually many times, to ensure the accuracy of the detection. This has been made much easier with the availability of NGS methods.

Around the mid-2000s, a feature of the genome was discovered that has profound implications for the analysis of disease. A number of studies found that, compared to small-scale variations such as SNPs, larger-scale variations were more common than expected. The striking finding was that there are many more differences between individual genomes than was first thought to be the case, and that these are caused by what are known as copy number variants (CNVs). These are usually defined as being between 1 kbp and 5 Mbp, caused by rearrangements such as insertions, deletions, duplications and inversions. They are detectable by techniques such as fluorescent *in situ* hybridisation (FISH) (Table 13.1). One particularly interesting aspect of the work was the correlation between some of the CNVs found and the occurrence of disease traits as listed in the OMIM database. The term CNV is also used to define short-length variations such as the **trinucleotide repeat** that is repeated in the gene for Huntington disease.

One method of looking for genomic variation that has developed since the first study in 2005 is the use of a genome-wide association study (GWAS). This is a method to screen many thousands of genomes from individuals who are healthy (controls) or present with a disease (cases). As the name suggests, a GWAS looks for *associations* without any pre-conceived expectation or supposition; it is therefore another example of hypothesis-free investigation. Association is measured using a statistical analysis to link the prevalence of a disease with genetic markers such as SNPs. Whilst it is a very powerful technique, translation

An extension to the analysis of SNPs and DNA sequences is the discovery of copy number variants (CNVs) in the genome; this is likely to be of major importance in helping us understand the puzzle of differential susceptibility to various diseases.

of GWAS data into clinically useful applications is proving to be a challenge. Integration of GWAS and functional genomics data is likely to be one way of realising the tremendous potential of this approach.

Comparative genomics is not a small-scale undertaking. Between 2008 and 2015, the **1000 Genomes Project** (self-explanatory really!) produced the first major assembly of data on genomic variation. The resource is now managed by EMBL-EBI as the International Genome Sample Resource (IGSR). An even more ambitious task was set in motion in 2012 by the then UK Prime Minister **David Cameron**, who set up Genomics England to manage the **100 000 Genomes Project**, using samples from around 85 000 NHS (UK National Health Service) patients with cancers and rare genetically based diseases. In an astonishing feat of coordinated effort, the project was completed in 2018 and is a major part of the development of a genomics approach to healthcare. We will look more closely at how comparative genomics can be used to detect disease alleles in Chapter 15.

> Projects to sequence 1 000 and 100 000 genomes have been completed; the progress made in the 20 or so years since the human genome was sequenced is astonishing. This work, and extensions of it, will continue to provide a resource for medical science for decades to come.

13.5 | Conclusion: The Central Role of the Genome

In looking back at the topics covered in this chapter, a central theme of *transition from gene to genome* is evident, in both *structural* and *functional* contexts. As I write this, in 2022, there is a neat symmetry that reflects the eras we looked at in Fig. 1.4. In the 20 years between 1972 and 1992, the focus was largely on investigating the structure and expression of individual genes. The genomics decade (in this case, let's call this 1992–2002) saw a transition to a *genome-wide* focus. Although this period was defined by the *Human Genome Project*, the analysis and sequencing of the genomes of other model organisms also made a major contribution to our knowledge of genomes across a wide taxonomic spectrum.

The 20 years since 2002 have seen an astonishing expansion of genomics, with NGS technology making genome sequencing widely accessible and cost-effective. As a result of the increase in the amount of sequence data generated (and the IT capacity to analyse the large data sets), new applications in bioinformatics and genomic analysis were developed, and we now have established disciplines in a range of '-*omics*' that simply did not exist 30 years ago.

The use of *microarrays* and *DNA chips* has already had a profound effect on the analysis of genome structure and expression, and development of new applications in sequencing such as *WGS* (whole genome), *WES* (exome) and *RNA-Seq* (transcriptome) will continue to provide insights into how genes work within the context of the genome as a whole.

Genome analysis has required revision of the conceptual framework in some areas. Early estimates of the number of genes in the human genome put the figure as high as 150 000, although 100 000 was a more generally accepted figure. We now know that we have fewer than 20 000 protein-coding genes, and that we share the majority of these with other organisms. This should not make us feel genetically under-resourced, but should strengthen the idea that it is the *expression* of the components of the genome that is the critical factor in defining how cells, tissues, organs and organisms function.

Undoubtedly a major focus will continue to be on the medical advances that are made as a result of genome research. The wealth of information that already exists for genetically based diseases will be supplemented with more widespread and detailed analysis of individual patients' genomes at the molecular level, with a *comparative genomics* approach setting the information against an atlas of genetic variation generated by techniques such as mapping *SNPs* and carrying out *GWAS*. This will facilitate both *personalised medicine* and the development of new treatments in areas such as drug design, gene therapy and potential gene editing targets.

Further Reading

Al-Amrani, S. *et al.* (2021). Proteomics: concepts and applications in human medicine. *World J. Biol. Chem.*, 12 (5), 57–69. URL [www.wjgnet.com/1949-8454/full/v12/i5/57.htm]. DOI [https://doi.org/10.4331/wjbc.v12.i5.57].

Ghadie, M. and Xia, Y. (2019). Estimating dispensable content in the human interactome. *Nat. Commun.*, 10, 3205. URL [www.nature.com/articles/s41467-019-11180-2#citeas]. DOI [https://doi.org/10.1038/s41467-019-11180-2].

Krassowski, M. *et al.* (2020). State of the field in multi-omics research: from computational needs to data mining and sharing. *Front. Genetics*, 11, 610798. URL [www.frontiersin.org/article/10.3389/fgene.2020.610798]. DOI [https://doi.org/10.3389/fgene.2020.610798].

Oak Ridge National Laboratory. Human Genome Project Information Archive 1990–2003. URL [https://web.ornl.gov/sci/techresources/Human_Genome/contact.shtml].

Rodriguez-Esteban, R. and Jiang, X. (2017). Differential gene expression in disease: a comparison between high-throughput studies and the literature. *BMC Med. Genomics*, 10, 59. URL [https://bmcmedgenomics.biomedcentral.com/articles/10.1186/s12920-017-0293-y#citeas]. DOI [https://doi.org/10.1186/s12920-017-0293-y].

Websearch

People

Two of the many prominent scientists involved with the HGP were *Sir John Sulston* and *Craig Venter*. See what you can find out about their careers and the differing views they had about the project. What are the differences and similarities between their story and the Watson and Crick dynamic of 50 years earlier?

Places

The *Centre d'Étude du Polymorphisme Humain (CEPH)* in Paris played a key role in mapping the human genome. Have a look at the contribution the CEPH made; this will give you a little diversion from the molecular biology of DNA.

Processes

The yeast hybrid system has been used extensively as an *in vivo* system to examine molecular recognition interactions. In the chapter, we looked at the Y1H and Y2H systems for investigating DNA~protein and protein~protein interactions. Complete the trio by finding out how the *yeast three-hybrid (Y3H) system* works.

Reflections

What topics in this chapter have you found most challenging? Look for resources that help to illustrate the key points.

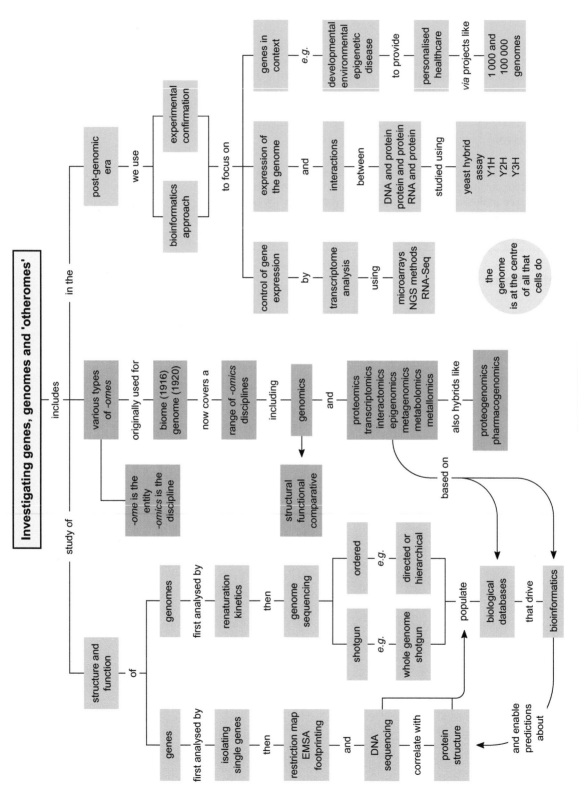

Investigating genes, genomes and 'otheromes'

includes

in the

study of

post-genomic era — *we use* — **bioinformatics approach** / **experimental confirmation** — *to focus on*

control of gene expression — *by* — **transcriptome analysis** — *using* — microarrays NGS methods RNA-Seq

expression of the genome *and* **interactions** — *between* — DNA and protein protein and protein RNA and protein — *studied using* — yeast hybrid assay Y1H Y2H Y3H

genes in context — *e.g.* — developmental environmental epigenetic disease — *to provide* — **personalised healthcare** — *via* projects like — 1 000 and 100 000 genomes

the genome is at the centre of all that cells do

various types of -omes — *originally used for* — **biome (1916) genome (1920)** — *now covers a* — **range of -omics disciplines** — *including* — **genomics** / **proteomics transcriptomics interactomics epigenomics metagenomics metabolomics metallomics** — *also hybrids like* — **proteogenomics pharmacogenomics**

-ome is the entity *-omics* is the discipline

structural functional comparative

based on

structure and function

of

genomes — *first analysed by* — **renaturation kinetics** — *then* — **genome sequencing** — shotgun / ordered — *e.g.* — whole genome shotgun / directed or hierarchical — *populate* — **biological databases** — *that drive* — **bioinformatics**

genes — *first analysed by* — **isolating single genes** — *then* — **restriction map EMSA footprinting** — *and* — **DNA sequencing** — *correlate with* — **protein structure**

and enable predictions about

Concept Map 13

Chapter 14 Summary

Learning Objectives

When you have completed this chapter, you will be able to:

- Define the range and scope of biotechnology
- Evaluate the impact of rDNA technology on biotechnological applications
- Describe how recombinant DNA is used to produce proteins
- Discuss the aims of and techniques used for protein engineering
- Outline the issues surrounding scaling up to commercial production capacity
- Describe a range of rDNA-based biotechnological applications

Key Words

Biotechnology, post-translational modification (PTM), process engineering, promoter, terminators, ribosome binding sites, Shine–Dalgarno sequences, *Saccharomyces cerevisiae*, expression vector system, cassette, native proteins, fusion proteins, novel catalytic combinations, tags, green fluorescent protein (GFP), affinity tags, reading frame, alcohol oxidase, AOX1, baculoviruses, polyhedra, polyhedrin, transfer vector, downstream processing (DSP), protein engineering, protein~protein interactions (PPIs), DNA~protein interactions, muteins, rational design, directed evolution, mutagenesis *in vitro*, oligonucleotide-directed mutagenesis, site-directed mutagenesis, error-prone PCR (epPCR), DNA shuffling, small to medium enterprise (SME), seedcorn funding, venture capital (VC), proof of concept, process economics, capital costs, operational costs, separation, concentration, purification, formulation, filtration, centrifugation, sedimentation, precipitation, evaporation, adsorption, chromatography, crystallisation, freeze-drying, rennet, chymosin, proteases, lipases, Lipolase, recombinant bovine somatotropin (rBST), technology transfer, market acceptance, bovine somatotropin (BST), bovine growth hormone, Posilac, insulin-like growth factor (IGF-1), medical diagnostics, replacement, supplementation, specific disease therapy, recombinant vaccines, diabetes mellitus (DM), insulin, type I DM, insulin-dependent DM (IDDM), type II DM, non-insulin-dependent DM (NIDDM), A-chain, B-chain, C-chain, proinsulin, Frederick Banting, Charles Best, Humulin, Lilly, tissue plasminogen activator (TPA), plasminogen, plasmin, fibrin, Genentech, Activase, Exondys 51, antisense oligonucleotide (ASO), Duchenne muscular dystrophy (DMD), small interfering RNA (siRNA), Onpattro, hereditary transthyretin amyloidosis polyneuropathy, hepatitis B, HBsAg, quadrivalent recombinant influenza vaccines, transgenic plants, bioreactor, lateral flow testing, compound annual growth rate (CAGR).

Genetic Engineering and Biotechnology

Biotechnology is one of those terms that can mean different things to different people. In its broadest context, it is essentially the use of a biological system or process to improve the lot of humankind. More specifically, this can involve using a cell or organism (maybe a mammalian cell or a microorganism) or a biologically derived substance (usually an enzyme) in a production or conversion process. Brewing and wine-making, and the production of bread, cheese and yoghurt, are processes that have been used for thousands of years. More highly developed food processing and manufacturing techniques came on the scene much later, and biotechnology began to reach full scientific maturity when the biological and biochemical aspects of the topic began to be understood more fully. This led to the diverse range of modern applications, which (in addition to continued food and beverage production) include the production of specialist chemicals and biochemicals, pharmaceutical and therapeutic products, and environmental applications such as treatment of sewage and pollutants. The range and scope of biotechnology is outlined in Fig 14.1.

In many biotechnological applications, the organism or enzyme is used in its natural form and is not modified, apart from perhaps having been subjected to selection methods to enable the best strain or type of enzyme to be used for a particular application. Thus, whilst there is no *requirement* for genetic engineering to be associated with the discipline, modern biotechnology is often linked with the use of genetically modified systems. In this chapter, we will consider the impact that gene manipulation technology has had on some areas of biotechnology. The impact of gene manipulation in medicine and transgenics will be considered in more detail in Chapters 15 and 16.

> Biotechnology has a very long history and is not dependent on gene manipulation technology. However, the advent of rDNA techniques has helped the development of biotechnology in a way that would not have been possible otherwise.

14.1 | Making Proteins

The synthesis and purification of proteins from cloned genes is one of the most important aspects of genetic manipulation, particularly where high-value therapeutic proteins are concerned. Many such proteins have already been produced by recombinant DNA (rDNA)

> The production of proteins using rDNA technology was one of the earliest applications of gene manipulation in the biotechnology industry.

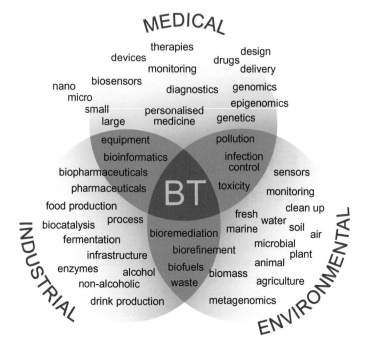

Fig. 14.1 The range and scope of biotechnology. The term is very broad, and there are many ways to categorise the various disciplines and sub-disciplines. In this example, three main areas are shown (medical, environmental and industrial). In reality, there is significant overlap between these. Thus, a manufacturing firm with expertise in small-scale fabrication might make sensors that are used in many different areas. Alternatively, a highly specialist company might make only one type of instrument, or provide a specialist service such as genome sequencing. Companies range from small and medium enterprises (SMEs) to enormous multinationals that may encompass a wide range of areas of biotechnology, often at different sites and/or in different countries.

techniques, and are already in widespread use. We will consider some examples later in this chapter. In many cases, a bacterial host cell can be used for the expression of cloned genes, but often a eukaryotic host is required for particular purposes. Eukaryotic proteins are often subjected to **post-translational modification (PTM)** *in vivo*, and it is important that any modifications are achieved in an expression system if a functional protein is to be produced.

In protein production, there are two aspects which require optimisation: (1) the biology of the system, and (2) the production process itself. Careful design of both these aspects is required if the overall process is to be commercially viable, which is necessary if large-scale production and marketing of the protein are the aim. Thus, biotechnological applications require both a biological input and a **process engineering** input if success is to be achieved, and one of the key challenges in establishing a particular application is the scale-up from laboratory to production plant. We will consider this aspect more fully in Section 14.3.

In biotechnology, the biological aspects of the application have to be complemented by the development of suitable engineering processes if commercial success is to be realised.

14.1.1 Native and Fusion Proteins

For efficient expression of cloned DNA, the gene must be inserted into a vector that has a suitable promoter. Ideally the promoter should be one that can be regulated by environmental conditions (see Table 8.2 for examples of inducible promoters), and that can be introduced into an appropriate host such as *Escherichia coli*. Although this organism is not ideal for expressing eukaryotic genes, many of the problems of using *E. coli* can be overcome by constructing the recombinant so that the expression signals are recognised by the host cell. Key signals include promoters and terminators for transcription, and ribosome binding sites (Shine–Dalgarno sequences) for translation. Alternatively, a eukaryotic host such as the yeast *Saccharomyces cerevisiae*, or mammalian cells in tissue culture, may be more suitable for certain proteins.

For eukaryotic proteins, the coding sequence is usually derived from a cDNA clone of the mRNA. This is particularly important if the gene contains introns, as these will not be processed out of the primary transcript in a prokaryotic host. When the cDNA has been obtained, a suitable expression vector system or cassette must be chosen from the wide variety available. Proteins synthesised from vector constructs can be either native proteins or fusion proteins (Fig. 14.2). Native proteins are synthesised directly from the N-terminus of the cDNA, whereas fusion proteins contain short N-terminal amino acid sequences encoded by the vector. In some cases, these may be important for protein stability or secretion, and are thus not necessarily a problem. However, such sequences can be removed if the recombinant is constructed so that the fusion protein contains a methionine residue at the point of fusion. The chemical cyanogen bromide (CNBr) can be used to cleave the protein at the methionine residue, thus releasing the desired peptide. A problem with this approach occurs if the protein contains one or more internal methionine residues, as this will result in unwanted cleavage by CNBr.

> Two main types of proteins that can be generated using rDNA techniques are native and fusion proteins.

The term fusion protein is also used to describe engineered proteins that are derived by combining different genes or gene fragments to produce a particular combination. The constructs used to generate the bait and prey protein fusions in the yeast two-hybrid system is one example (Fig 13.13). Other applications of fusion proteins are to generate novel catalytic combinations, add tags to proteins to enable detection (*e.g.* fluorescent tags such as the green fluorescent protein or GFP) or facilitate post-synthesis chromatographic purification (affinity tags). A spacer region may also need to be included in the construct to ensure correct folding of each of the functional domains in the fusion.

When constructing a recombinant for the synthesis of a fusion protein, it is important that the cDNA sequence is inserted into the vector in a position that maintains the correct reading frame. The addition or deletion of one or two base-pairs at the vector/insert junction may be necessary to ensure this, although there are vectors that have been constructed so that all three potential reading frames are represented for a particular vector/insert combination. Thus, by

> When constructing a recombinant to express a protein, it is vital that the correct reading frame is maintained; otherwise the protein may not be synthesised correctly.

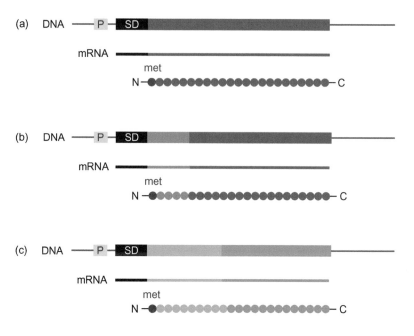

Fig. 14.2 Native and fusion proteins. (a) The control elements shown are the promoter (P) and the ribosome binding site (Shine–Dalgarno sequence, SD). The coding sequence for the cloned gene (shaded red) is not preceded by the bacterial coding sequence; thus, the mRNA encodes only insert-specified amino acid residues. This produces a native protein, synthesised from its own N-terminal methionine residue. (b) The gene fusion contains bacterial codons (shaded purple); therefore, the protein contains part of the bacterial protein. In this example, the methionine and the first four N-terminal amino acid residues are of bacterial origin. (c) A fusion protein derived from two separate coding regions to generate a hybrid protein. In this case, a 'tag' is added using a green fluorescent protein (GFP) gene (shaded green) to enable identification of the gene cloned downstream from the GFP fusion site (shaded blue).

using the three variants of the vector, the correct in-frame fusion can be obtained. The importance of the reading frame is shown in Fig. 14.3.

14.1.2 Yeast Expression Systems

As discussed in Chapter 7, the yeast *S. cerevisiae* has been the favoured microbial eukaryote in the development of rDNA technology. Whilst *S. cerevisiae* is still useful for gene expression studies and protein production, other yeasts may offer advantages in terms of growth characteristics, yield of heterologous protein and large-scale fermentation characteristics. Species that are used include *Schizosaccharomyces pombe*, *Pichia pastoris*, *Hansela polymorpha*, *Kluyveromyces lactis* and Yarrowia lipolytica. These demonstrate many of the characteristics of bacteria with respect to ease of use – they grow rapidly on relatively inexpensive media, and a range of different mutant strains and vectors is available for various applications. In some cases, scale-up fermentations present some difficulties, compared to bacteria, but

There are many different yeasts that can be used in biotechnological processes, each of which has its own particular characteristics, which may suit a range of applications.

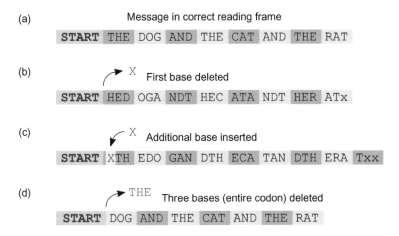

(a) Message in correct reading frame

START THE DOG AND THE CAT AND THE RAT

(b) X First base deleted

START HED OGA NDT HEC ATA NDT HER ATx

(c) X Additional base inserted

START XTH EDO GAN DTH ECA TAN DTH ERA Txx

(d) THE Three bases (entire codon) deleted

START DOG AND THE CAT AND THE RAT

Fig. 14.3 The importance of the reading frame. A simple sentence is used for illustration. In (a), the message has been 'cloned' downstream from the start site and is readable, as it is in the correct reading frame. (b) A deletion of one base at the start is enough to knock out the sense completely. Addition of an extra base also causes problems, as shown in (c). Given the triplet nature of the code, insertion or deletion of two bases (sequentially, *i.e.* as a dinucleotide) causes the same issue. (d) If an entire triplet is deleted, some sense is lost, but the message is still readable. This also applies to a triplet insertion. To avoid problems during translation, sequences at the vector/insert insertion site must be in-frame. This also applies at spacer region junctions or dual coding sequences in a hybrid protein.

these can usually be overcome by careful design and monitoring of the process. Yields of heterologous proteins of around $12\,\text{g}\,\text{L}^{-1}$ (10–100 times more than in *S. cerevisiae*) have been obtained using *P. pastoris*, which can be grown on methanol as sole carbon source. In this situation growth is regulated by the enzyme **alcohol oxidase**, which has a low specific activity and is consequently overproduced in these cells, making up around 30 per cent of total soluble protein. By placing heterologous genes downstream from the alcohol oxidase promoter (**AOX1**), high levels of expression are achieved.

One of the advantages of using yeast, as opposed to bacterial hosts, is that proteins are subjected to post-translational modifications such as glycosylation. In addition, there is usually a higher degree of 'authenticity' with respect to 3-D conformation and the immunogenic properties of the protein. Thus, in a situation where the biological properties of the protein are critical, yeasts may provide a better product than prokaryotic hosts.

14.1.3 The Baculovirus Expression System

Baculoviruses infect insects, and as such can be used as biocontrol agents for management of pests. In a protein synthesis context, they are also useful for the production of rDNA-derived proteins where bacterial or yeast hosts may not be suitable. Baculoviruses do not appear to have any negative effects on human cells. During normal

The baculovirus expression system, based on insect-infecting viruses, offers an alternative to microbial systems. Despite being more complex to use, this system can be useful for expressing certain types of eukaryotic protein.

infection of insect cells, virus particles are packaged within **polyhedra**, which are nuclear inclusion bodies composed mostly of the protein **polyhedrin**. This is synthesised late in the virus infection cycle, and can represent as much as 50 per cent of infected cell protein when fully expressed. Whilst polyhedra are required for infection of insects themselves, they are not required to maintain infection of cultured cells. Thus, the polyhedrin gene is an obvious candidate for construction of an expression vector, as it encodes a late-expressed dispensable protein that is synthesised in large amounts.

The baculovirus genome is a circular double-stranded DNA molecule. Genome size is from 88 to 200 kbp, depending on the particular virus, and the genome is therefore too large to be manipulated easily and routinely. Thus, kit-based approaches have been developed to make the use of baculovirus expression systems more accessible – one such is the *Bac-to-Bac* (!) system from ThermoFisher/Gibco. Insertion of foreign DNA into a baculovirus system is usually accomplished by using an intermediate known as a **transfer vector**. These are based on *E. coli* plasmids, and carry the promoter for the polyhedrin gene (or for another viral gene) and any other essential expression signals. The cloned gene for expression is inserted into the transfer vector, and the recombinant is used to co-transfect insect cells with non-recombinant viral DNA. Homologous recombination between the viral DNA and the transfer vector results in the generation of recombinant viral genomes, which can be selected for, and used to produce, the protein of interest. Systems based on baculoviruses demonstrate transient gene expression, in which the protein of interest is synthesised as part of the infection cycle of the viral-based vector system. Stable insect cell expression systems have also been developed, in which the cells can be used for continuous expression of protein.

One disadvantage of using insect cells, as opposed to bacteria or yeast, is that they require more complex growth media for maintenance and production. The cells are also less robust than the microbial cells, and thus require careful handling if success is to be achieved. However, there are advantages such as a greater degree of fidelity of expression and post-translational modifications that are more likely to reflect the situation *in vivo*.

14.1.4 Mammalian Cell Lines

Where the expression of recombinant human proteins is concerned, it might seem obvious that a mammalian host cell would be a better system than bacteria, eukaryotic microbes or insect cells. However, the use of such cell lines in protein production presents some problems. As with insect cell lines, the media required to sustain growth of mammalian cells are more complex and expensive than for *E. coli* and yeast, and the cells are relatively fragile when compared with microbial cells, particularly where large-scale fermentation is involved. There may also be challenges in processing the products (**downstream processing (DSP)** is used to describe post-synthesis operations).

Insect and mammalian cells require more complex media and are generally less robust than microbial cells such as bacteria and yeasts.

Despite these difficulties, many vectors are now available for protein expression in mammalian cells. They exhibit characteristics that will by now be familiar – often based on a viral system, vectors utilise selectable markers (often drug resistance markers) and have promoters that enable expression of the cloned gene sequence. Common promoters are based on simian virus (SV40) or cytomegalovirus (CMV). Some examples were presented in Table 7.3.

14.2 | Protein Engineering

One of the most exciting applications of gene manipulation lies in the field of **protein engineering**. This involves altering the structure of proteins *via* alterations to the gene sequence, and has become possible due to the availability of a range of techniques, as well as a deeper understanding of the structural and functional characteristics of proteins. The great advances made through investigating **protein~protein interactions** (PPIs) and **DNA~protein interactions** have enabled workers to pinpoint the essential amino acid residues in a protein sequence, and thus alterations can be carried out at these positions and their effects studied. The desired effect might be alteration of the catalytic activity of an enzyme by modification of the residues around the active site, an improvement in the nutritional status of a storage protein or an improvement in the stability of a protein used in industry or medicine. Proteins that have been engineered by the incorporation of mutational changes have become known as muteins.

In broad terms, there are two approaches that can be used to engineer proteins. These are sometimes called rational design and directed evolution. We will consider these separately, although recent developments in bioinformatics have begun to bring together the two approaches in novel procedures that make best use of the characteristics of both.

14.2.1 Rational Design

The use of a rational design protocol depends on some detailed information about the protein being available. Typically the target protein may have been characterised biochemically, and its gene cloned and sequenced. Thus, the mRNA coding sequence, predicted or actual amino acid sequence, 3-D structure, folding characteristics, *etc.* may be available. This information is used to predict what the effect of changing part of the protein might be; the change can then be made, and the altered protein tested to see if the desired changes have been incorporated. A procedure known as mutagenesis *in vitro* enables specific mutations to be introduced into a gene sequence. One original method for this is oligonucleotide-directed or site-directed mutagenesis, and is elegantly simple in concept (Fig. 14.4). The requirements are a single-stranded template containing the gene to be altered, and an oligonucleotide (usually 15–30 nucleotides in

Altering the characteristics of proteins by modifying gene sequences is a powerful and widely used technique that can greatly speed up the identification of novel protein variants.

Mutagenesis *in vitro* uses a range of elegant techniques to introduce defined mutations into a cloned DNA sequence, to alter the amino acid sequence of the protein that it encodes and to evaluate the effect of the changes made.

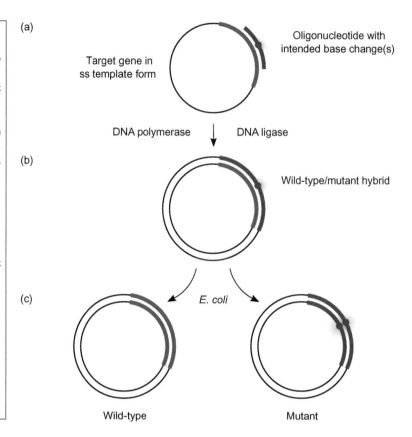

Fig. 14.4 Oligonucleotide-directed mutagenesis. (a) The requirement for mutagenesis *in vitro* is a single-stranded (ss) DNA template containing a cloned target gene. An oligonucleotide is synthesised that is complementary to the part of the gene that is to be mutated and which incorporates the desired mutation. In this case, a base change (substitution) is being introduced. The oligonucleotide is annealed to the template. (b) The molecule is made double-stranded in a reaction using DNA polymerase and ligase, which produces a hybrid wild-type/mutant DNA molecule with a mismatch in the mutated region. (c) On introduction into *E. coli*, the molecule is replicated, thus producing double-stranded copies of the wild-type and mutant forms. The mutant carries the original mutation and its complementary base or sequence (highlighted green).

length) that is complementary to the region of interest. The oligonucleotide is synthesised with the desired mutation as part of the sequence. The single-stranded template is often produced using the M13 cloning system, which produces single-stranded DNA. The template and oligonucleotide are annealed (the mutation site will mismatch, but the flanking sequences will confer stability), and the template is then copied using DNA polymerase. This gives rise to a double-stranded DNA. When this is replicated, it will generate two daughter molecules, one of which will contain the desired mutation.

Other methods for generating mutations in gene sequences have been developed. Cassette-based systems and PCR methods are now used widely, often with assembly cloning methods for the construction of expression vectors to test new variants. There is now an extensive suite of protocols to generate many different types of alterations to gene sequences, to drive both rational design and directed evolution approaches. Cassette methods are outlined in Fig. 14.5, and the technique of error-prone PCR (epPCR) in Fig. 14.6. Other PCR-based methods are shown in Fig. 14.7.

Having altered a gene by mutagenesis, the protein is produced using an expression system. Often a vector incorporating the *lac* promoter is used, so that transcription can be controlled by the

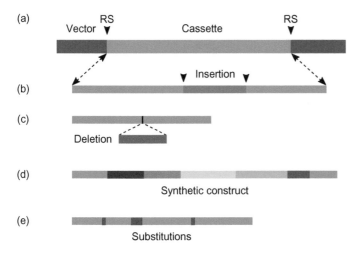

(a)

(b)

(c)

(d)

(e)

Fig. 14.5 Cassette-based mutagenesis. At its most basic, a cassette is simply a fragment that can be isolated from a vector and replaced with a modified sequence, although there can be other regulatory or cloning sites present. The target sequence can be modified in a number of ways using *in vitro* methods. (a) In this example, a fragment is flanked by two restriction sites (if these are for different enzymes, then directional insertion of the fragment after modification can be achieved). (b) A fragment is inserted to create a modified sequence; this might introduce new amino acid codons into the sequence or could introduce a termination codon. (c) A deletion can remove codons or suspected regulatory sites. (d) A range of techniques can be used to generate a synthetic construct. The sizes of the fragments joined can vary from small oligonucleotides (to introduce one or two codon changes) to parts of coding sequences from different genes, to gene sequences synthesised *in vitro* (this approach is not so much mutagenesis, but more synthetic biology). (e) One or more base substitutions can be generated and introduced; sometimes a series of random changes is screened to see what effects the changes have on protein structure.

addition of IPTG. Alternatively, the λP_L promoter can be used with a temperature-sensitive λcI repressor, so that expression of the mutant gene is repressed at 30°C but is permitted at 42°C. Analysis of the mutant protein is carried out by comparison with the wild-type protein. In this way, proteins can be 'engineered' by incorporating subtle structural changes that alter their functional characteristics. An overview of the rational design concept is shown in Fig. 14.8.

14.2.2 Directed Evolution

One of the main disadvantages of the rational design approach to protein engineering is that a very large range of potential structures can be derived by modification of a protein sequence in various ways. Thus, it is very difficult to be sure that the modification that is being incorporated will have the desired effect – and the process is labour-intensive. Directed evolution is a different approach that has increased the range and scope of producing new protein variants. As the name suggests, the technique is more like an evolutionary

The technique of directed evolution removes the need for predictive alterations to DNA and protein sequences, and is potentially more powerful than site-directed methods.

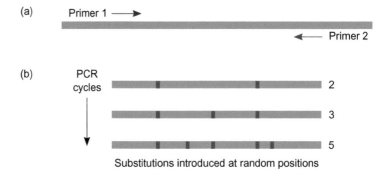

(a) Primer 1 ⟶

⟵ Primer 2

(b) PCR cycles

Substitutions introduced at random positions

Fig. 14.6 Error-prone PCR mutagenesis. This technique exploits the relatively low fidelity of *Taq* polymerase. This can be reduced further by altering the Mg^{2+}, Mn^{2+} and specific dNTP concentrations in the reaction. (a) A target fragment flanked by forward (primer 1, red) and reverse (primer 2, blue) PCR primers. (b) A PCR is set up to favour error-prone synthesis. As the reaction proceeds, base substitutions are generated, essentially at random sites. The extent of mutation can be controlled to some extent by the number of cycles to produce multiply-mutant fragments for testing.

Fig. 14.7 Mutagenesis by PCR. Using the PCR to generate mutations in target gene sequences is a powerful tool that can be used in a number of ways. Kits (such as the New England Biolabs Q5® Site-Directed Mutagenesis Kit) can make things relatively straightforward. The technique is based on inverse PCR amplification of entire plasmids containing the target sequence. In this example, four methods are shown. (a) A substitution can be introduced by inclusion of the new base in one of the primers, resulting in a final product with the base change incorporated. (b) Deletions are generated by designing primers to flank the area to be deleted. (c) Small insertions (six nucleotides or fewer) can be incorporated by adding the target insertion to the 5′ end of the primer. (d) Longer insertions use a dual approach, with both primers having target insert sequences added to their 5′ ends. *Source:* Based on an illustration from [www.neb.com] (2022). Reproduced with permission from New England Biolabs, Inc.

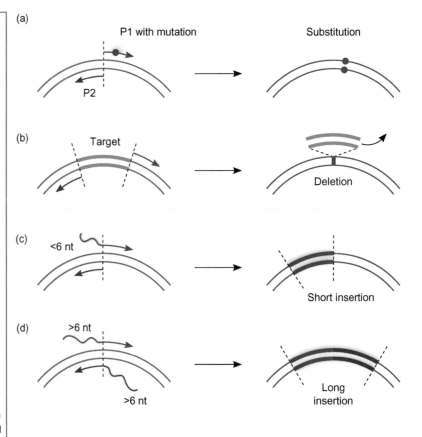

(a) P1 with mutation — Substitution

(b) Target — Deletion

(c) <6 nt — Short insertion

(d) >6 nt — >6 nt — Long insertion

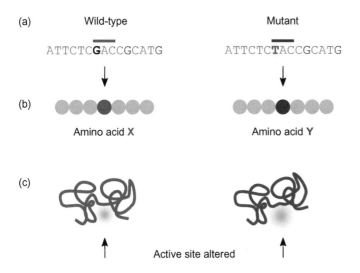

(a) Wild-type

ATTCTCGACCGCATG

Mutant

ATTCTCTACCGCATG

(b)

Amino acid X

Amino acid Y

(c)

Active site altered

Fig. 14.8 Protein engineering by the rational design method. Bioinformatics databases and IT-tools enable an *in silico* approach to predict the effect that the proposed change to the gene sequence will have on the 3-D structure of the protein. (a) A change is introduced into the gene, with the wild-type codon GAC altered to TAC by changing the first base in the codon (highlighted yellow). (b) This causes a change in the amino acid sequence, shown as X to Y (in this case, from aspartic acid to tyrosine). (c) The altered amino acid sequence causes a change in the way that the protein folds. In this example, the active site (shown by the arrows, highlighted green) becomes larger in the altered molecule.

process, rather than the incorporation of a specific alteration in a defined part of the protein. The increase in potential benefit comes not from the fact that any *particular* structural alteration is created, but that a large number of different alterations are generated and the desired variant is selected by a process that mimics natural selection. Thus, the need for predictive structural alteration is removed, as the system itself enables large numbers of changes to be generated and screened efficiently. Thus, directed evolution lends itself to high-throughput methods and automated screening.

The process generates a library of recombinants that encode the protein of interest, with random mutagenesis (often using techniques such as error-prone PCR) applied to generate the variants. Thus, a large number of different sequences can be generated, some of which will produce the desired effect in the protein. These can be selected by expressing the gene and analysing the protein using a suitable assay system, to select the desired variants. Additional rounds of mutagenesis and selection can be applied if necessary. An extension to this technique called **DNA shuffling** can be used to mix pieces of DNA from variants that show desired characteristics. This mimics the effect of recombination that would occur *in vivo*, and can be an effective way of 'fast-tracking' the directed evolution technique.

14.3 | From Laboratory to Production Plant

The range of outputs generated
from biotechnology applications
is extensive, and covers a wide
spectrum of different types of
product for use in areas as
diverse as food processing,
agriculture, healthcare and
scientific research.

We saw in the introduction to this chapter that biotechnology covers a range of disciplines and is not easily defined. In this section, we will take a wide-ranging view, and consider that the field covers any biological system, product or process that is developed at a technological level and exploited on a commercial basis. Some examples could be the production of amino acids or enzymes by traditional fermentation technology methods, drug discovery and production, products for use in other industries, food additives and healthcare products. On this basis, we would exclude any application that is not significantly high-tech; thus, bulk gathering and processing of materials would not be classed as a biotechnological process, even though it may have a significant economic impact (*e.g.* food production and processing, salmon farming and other agricultural applications). So what is needed to make the transition from laboratory to production plant?

14.3.1 Thinking Big – The Biotechnology Industry

Despite the range of applications and disciplines that make up the biotechnology sector, there are certain similarities across all areas. At the heart of any biotechnology process, there has to be first-class science, and many applications are developed from fundamental research carried out in universities and research institutes. Often 'spin-off' companies are associated with the academic institution that employs the scientists who have developed the idea, and science parks help to ensure that appropriate location, infrastructure and access to expertise are available to companies that locate to the area.

Although biotechnology
companies may be very different
in terms of what they do, there
are certain key elements (inputs
and outputs) that define their
basic operational characteristics.

The anatomy of a biotechnology company can be summarised as a set of basic requirements or inputs, and a potential set of outputs. This is shown in Fig. 14.9. At the outset, a company will usually have a particular output in mind – it may be a high-value product, the development of a cloning technology or perhaps some other service provision. With the availability of next-generation sequencing technology, the number of genomics-related companies has expanded significantly in the past few years, offering a range of services to public and private sectors and to the public generally.

A critical aspect of converting an idea into a marketable product or service is financing of the business. This is particularly crucial in the early stages of development, as many potentially sound business ideas have failed to get further than the initial stages due to a drying up of available funds. It is at this stage that a 'new start' small company (often known as an SME or **small to medium enterprise**) is at its most vulnerable. Companies that are extensions to the business of large multinational parent companies can often avoid these threats in the early stages.

Raising finance for a new biotechnology venture is not an easy process, and it can involve several stages. Usually there will have

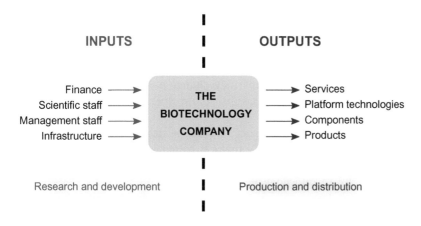

INPUTS | OUTPUTS

Finance → | THE BIOTECHNOLOGY COMPANY | → Services
Scientific staff → | | → Platform technologies
Management staff → | | → Components
Infrastructure → | | → Products

Research and development | Production and distribution

Fig. 14.9 The anatomy of a biotechnology company. The requirements are shown as inputs on the left-hand side of the diagram, and include finance, staff and infrastructure (premises, utilities, services, logistics and supply chains). The outputs are shown on the right-hand side. The output may be a single product or service, or might be a range of related products. Broadly speaking, the balance of effort on the input side is in research and development (R&D). When a product is ready for market, the emphasis shifts to production and distribution (P&D). In practical terms, many companies of medium size and upwards have a pipeline of products; thus, R&D and P&D are continuous and overlapping. The critical aspect that will determine how successful a new company will be of course what happens inside the box – a good product, with good management of staff and cost-effective production processes, may lead to sustainable production and ultimately profit.

already been a significant investment in an idea before it ever gets to the biotechnological application stage – this is the investment in the basic science that led to the proposal for a biotech company. It is impossible to quantify this level of funding in most cases, as it can involve several years of effort with many research staff in universities or research institutes. Once an idea has been developed to the point where it can be exploited, finance to develop the core idea is required. This is sometimes known as seedcorn funding, and may be supplied by government agencies (such as local enterprise companies), corporate lenders or private investors. Seedcorn funding provides the cash-flow that is essential to get a company up and running, but is usually not sufficient to enable more than establishing the company and initial development of the idea. The second phase of funding is often provided by venture capital (VC) sources. This stage of funding often marks the transition from 'good idea' to a viable product, and involves the proof of concept stage and development of production capability. This is usually the most exposed stage for investment input, as it carries a high level of risk (and potential high reward) that characterises this stage of development. Investors who manage VC funds will not necessarily be expecting a short-term return on investment, but they will want to investigate the company, its business plan and its staff before providing support. It is at this stage that the quality of the

Although a sound and commercially viable process or product is obviously essential, the most important factor in enabling a company to achieve success is the provision of realistic levels of funding for each of the stages of its development.

As a company grows and develops, the roles of staff will change as responsibilities become more specific and demanding.

staff is most important, as there may be only a few people involved at this stage and their commitment and tenacity may be what marks the difference between success and failure. In the early stages, the scientists who are involved with the project are often the major part of the management team, but this may change as the company begins to grow and there is a need for more specialised and formalised management structures.

Having survived the proof of concept stage, there is still no certainty that the product will make it to market and be profitable. At this stage, there is often the need for another large injection of funds, perhaps by second-stage venture capital funding or by developing links with a larger corporate partner. This period can often be one of the trickiest in a company, as it may involve clinical trials or steering the product through regulatory procedures in the countries in which it will be marketed. It is only when all aspects of the product development and approval are in place that commercial production and distribution can begin. An illustration of funding levels and timescales for a biotechnology company in the early stages of development is shown in Fig. 14.10.

14.3.2 Production Systems

In addition to securing financial backing for a new biotechnology company, there are of course many aspects of the *science* to be considered if success is to be achieved. The fact that an idea has reached the stage of potential commercial exploitation usually means that the basic science has been tested and perhaps a patent obtained. During early development, the emphasis is usually on getting the processes up and running, with little thought for efficiency or cost of materials. However, when considering the development to production scale, these aspects become critical, as controlling costs will have a major impact on the selling price, and thus on profitability and potential market share.

In addition to securing appropriate financing for a new company, the start-up staff must ensure that the product or process is developed from initial concept to commercial production capability.

One of the first considerations is the system to be used for production. Usually this is fairly clear-cut and is determined by the process itself; this will have a number of specific requirements, which in turn will suggest a range of operational possibilities. Production of a recombinant protein by fermentation of a microbial culture will require a different type of system than production of monoclonal antibodies. Even established and reliable systems may require fine tuning for a particular application, with initial testing and development often carried out at laboratory scale levels in the first instance.

14.3.3 Scale-Up Considerations

There are several major aspects that have to be considered when moving from pilot testing to production scale. Firstly, the physical requirements of the production plant need to be established. This will

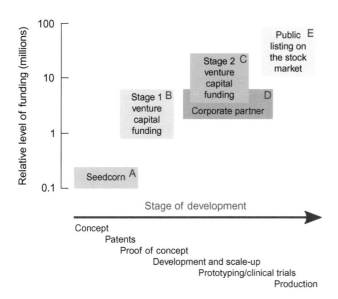

Concept
   Patents
      Proof of concept
         Development and scale-up
            Prototyping/clinical trials
               Production

Fig. 14.10 Funding for the initial phases of a start-up biotechnology company. The Y axis shows the order-of-magnitude level of funding (millions of £/$/€). In reality, the amount will vary, depending on a range of factors. The X axis is a timeline of the various stages of development; again, this is variable, but it is usually several years from inception to full public offering. At the beginning, start-up or seedcorn funding is required (levels and duration indicated by box A). This is usually followed by the first major input of investment (box B). When the company is in the developmental stage, perhaps with a patented product that is going onto the trial stage, additional funds are usually essential and are provided by either second-stage venture capital funding or perhaps by a corporate partner (C and D). The final stage is full operational production, and may be accompanied by a flotation on the appropriate stock market (E). *Source:* Modified from Bains and Evans. (2001), The business of biotechnology. In: Ratledge and Kristiansen, Editors. *Basic Biotechnology*, 2nd Edition. Cambridge University Press, Cambridge. Reproduced with permission.

include bioreactor design, control and monitoring systems, infrastructure and services and the disposal of waste products. In parallel with this, the biological aspects have to be considered – will the process work most efficiently at 100, 1 000 or 10 000 litres? Is there a contamination issue that is best resolved by using a relatively small-scale system that is easy to sterilise, even if this does not provide the most efficient production? Finally, there are aspects of **process economics** that have to be considered, with careful analysis of input costs and management of the process to reduce wastage. Both **capital costs** (setup and recurrent) and **operational costs** are important in these calculations.

The scale-up stage is critical in that an efficient system is needed to produce the product in a cost-effective way and realise a profit. Given the number of things that need to be considered, evaluated, designed and controlled, this is one of the most complex challenges in the development of a biotechnology company.

14.3.4   Downstream Processing

Having reached the stage of actually making something in a well-designed production plant operating at optimum efficiency, the next problem is how to process, package, market and distribute the product. These stages are collectively called **downstream processing**, and involve separation of the product from any co-products or wastes, and bringing it to a stage where the formulation is suitable for distribution. Further packaging and distribution may be carried out on site, or the product may be transported in bulk to a separate packaging and distribution site. Often the volume and value of a product will have a bearing on its post-production fate, with high-value/low-volume products easily packaged at, or close to, the site of production.

Downstream processing encompasses a range of techniques, many of which are well established and have been developed for use in the chemical engineering industry. Often the product will be suspended or dissolved in a liquid, perhaps the output of a batch fermentation process. It may be secreted from cells into the medium, or it may remain in the cells and may need to be released from them. Cells may be relatively fragile (*e.g.* mammalian cell cultures) or may be tougher yeast or bacterial cells. The product may be heat-sensitive or susceptible to chemical denaturation (*e.g.* a protein), or it may be resistant to many forms of chemical attack. All of these aspects have to be considered when designing a downstream processing procedure, although in most cases, there are four key stages: **separation**, **concentration**, **purification** and formulation.

Separation includes processes such as filtration, centrifugation and **sedimentation**. The aim of such a procedure is usually to clarify a suspension. The product may remain in the clarified filtrate or supernatant, or may be pelleted or collected on the filter if it is an intracellular product. If in the supernatant, concentration will usually be required, using processes such as **precipitation**, **evaporation** or adsorption. By this stage, the product is reaching the final stages of preparation, and final purification and formulation can involve chromatography, crystallisation or freeze-drying.

Like any other manufacturing process, once a biotechnology product has been made, there are several additional steps required to purify, concentrate, formulate, package and distribute the material.

Downstream processing usually involves four key stages, each of which may require a number of procedures, before a product is ready for packaging and distribution.

14.4   Examples of Biotechnological Applications of rDNA Technology

In the final part of this chapter, we will consider some examples of the types of product that can be produced using rDNA technology in biotechnological processes. This is a major and rapidly developing area that is likely to become increasingly important in the future, particularly in medicine and general healthcare, with an increasingly diverse range of products being brought to market.

14.4.1 Production of Enzymes

Enzymes are used in a number of industrial applications, including brewing, food processing, textile manufacture, the leather industry, washing powders, medical applications and scientific research. Although some enzymes continue to be prepared from natural sources, rDNA methods are now well established. Broadly speaking, enzymes are either high-volume/low-cost preparations for use in industrial-scale operations, or are low-volume/high-value products for specialist applications.

There is a nice twist to the gene manipulation story in that some of the enzymes used in the procedures are now themselves produced using rDNA methods. Suppliers produce recombinant variants of the common enzymes, and many are now engineered so that their characteristics fit the criteria for a particular process better than the natural enzyme, which increases their fidelity and efficiency.

In the food industry, one area that has involved the use of recombinant enzyme is the production of cheese. In cheese manufacture, **rennet** (also known as rennin, chymase or chymosin) has been used as part of the process. Chymosin is a protease that is involved in the coagulation of milk casein following fermentation by lactic acid bacteria. It was traditionally prepared from animal (bovine or pig) or fungal sources. In the 1960s, the Food and Agriculture Organization of the United Nations predicted that a shortage of calf rennet would develop as more calves were reared to maturity to satisfy increasing demands for meat and meat products. Today there are six sources for natural chymosin – veal calves, adult cows and pigs and the fungi *Rhizomucor miehei*, *Endothia parasitica* and *Rhizomucor pusillus*. Chymosin is now also available as a recombinant-derived preparation from *E. coli*, *K. lactis* and *Aspergillus niger*. Recombinant chymosin was first developed in 1981 and approved in 1988, and is now used to prepare around 90 per cent of hard cheeses in the UK.

Although the public acceptance of what is loosely called 'GM cheese' has not presented as many problems as has been the case with other foodstuffs, there are still concerns that need to be addressed. In cheese manufacture, three possible objections can be raised by those who are concerned about GM foods: (1) milk could have been produced from cows treated with recombinant growth hormone (see Section 14.4.2 below), (2) the cows could have been fed with GM soya or maize, and (3) the use of recombinant-derived chymosin. Despite these concerns, most consumers are content that cheese is not itself genetically modified, but is the *product* of a *product* of a genetically modified organism. Such subtlety is often lost in the debates around GM food.

A final example of recombinant-derived proteins in consumer products is the use of enzymes in washing powder. Proteases and lipases are commonly used to assist cleaning by degradation of protein and lipid-based staining. A recombinant lipase was developed in

The preparation of enzymes is a central part of biotechnology, and ranges from the production of large amounts of low-cost preparations for bulk applications to highly specialised enzymes for use in diagnostics or other molecular biology techniques.

Some aspects of biotechnology, such as the use of genetically modified organisms in food preparation or modification, may lead to public concern about the potential impact on health. This is an important aspect that we all have a role in debating.

1988 by Novo Nordisk A/V (now known as Novozymes). The company is the largest supplier of enzymes for commercial use in cleaning applications. Their recombinant lipase was known as **Lipolase**, which was the first commercial enzyme developed using rDNA technology and the first lipase used in detergents.

14.4.2   The BST Story

Not all rDNA biotechnology projects have a smooth passage from inception to commercial success. The story of **recombinant bovine somatotropin (rBST)** illustrates some of the problems that may be encountered once the scientific part of the process has been achieved. In bringing a recombinant product such as rBST to market, many aspects have to be considered. The basic science has to be carried out, followed by technology transfer to get the process to a commercially viable stage. Approval by regulatory bodies may be required, and finally (and most critical from a commercial standpoint) the product has to gain **market acceptance** and establish a consumer base. We can find all of these aspects in the BST story.

Bovine somatotropin (BST) is also known as **bovine growth hormone**, and is a naturally occurring protein that acts as a growth promoter in cattle. Milk production can be increased substantially by administering BST, and thus it was an attractive target for cloning and production for use in the dairy industry. The basic science of rBST was relatively straightforward, and scientists were already working on this in the early-1980s. The BST gene was in fact one of the first mammalian genes to be cloned and expressed, using bacterial cells for production of the protein. Thus, the production of rBST at a commercial level, involving the basic science and technology transfer stages, was achieved without too much difficulty. A summary of the process is shown in Fig. 14.11.

With respect to approval of new rDNA products, each country has its own system. In the USA, the Food and Drug Administration (FDA) is the central regulatory body, and in 1994, approval was given for the commercial distribution of rBST, marketed by Monsanto under the trade name Posilac. At that time, the European Union did not approve the product, but this was partly for socio-economic reasons (increasing milk production was not necessary) rather than for any concerns about the science. Evaluation of evidence at that time suggested that milk from rBST-treated cows was identical to normal untreated milk, and it was therefore unlikely that any negative effects would be seen in consumers.

The effects of rBST must be considered in three different contexts – the effect on milk production, the effects on the animals themselves and the possible effects on the consumer. Milk production is usually increased by around 10–15 per cent in treated cows, although yield increases of much more than this have been reported. Thus, from a dairy herd management viewpoint, use of rBST would seem to be beneficial. However, as is usually the case with any new development that is aimed at 'improving' what we eat or drink, public concern

BST was one of the early successes of biotechnology, in that recombinant BST was the result of achieving the aim of producing a useful protein by expressing cloned DNA in a bacterial host.

(a)

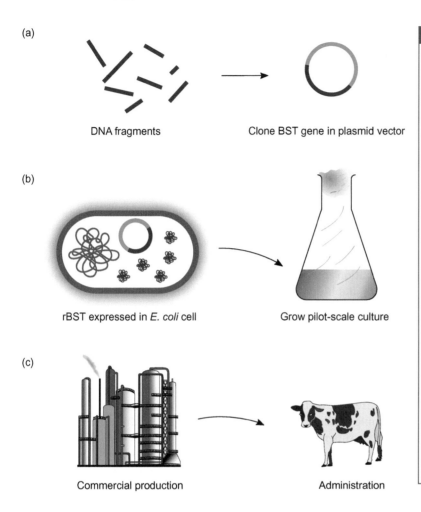

DNA fragments   Clone BST gene in plasmid vector

(b)

rBST expressed in *E. coli* cell   Grow pilot-scale culture

(c)

Commercial production   Administration

Fig. 14.11 Production of recombinant bovine growth hormone (rBST). (a) The BST gene coding sequence was cloned into a plasmid vector to generate the recombinant. (b) The plasmid construct produces rBST protein in the bacterial cell following transformation. Pilot-scale cultures will be used to verify the construct and the production of the protein. (c) Starter cultures seed large-scale cultures for production of the protein, and downstream processing procedures purify the protein from the growth medium. When regulatory approval is granted, the product is ready for sale and administration to livestock. The whole process from basic science to market usually takes several/many years from start to finish, with a large amount of investment capital required. This procedure is typical across a range of high-value proteins. Modern techniques can shorten the development time considerably, but the process of bringing a new product to market is not a trivial undertaking and may represent a significant risk to the company.

grew along with the technology. The concerns fuel a debate that is still ongoing, and is at times emotive. One area that is hotly debated is the effect of rBST on the cows themselves. Administering rBST can produce localised swelling at the site of injection, and can exacerbate problems with foot infections, mastitis and reproduction. The counter-argument is that many of these problems occur anyway, even in herds that are rBST-free. On balance, the evidence does, however, suggest that animal welfare is compromised to some extent when rBST is used.

The possible effects of rBST use on human health is another area of concern and debate. The natural hormone (and therefore the recombinant version also) affects milk production by increasing the levels of **insulin-like growth factor (IGF-1)**, which causes increased milk production. Administration of rBST generates elevated levels of IGF-1, and there is evidence that IGF-1 can stimulate the growth of cancer cells. Thus, the concern is that using rBST could pose a risk to health. The counter-argument in this case is that the levels of IGF-1 in the early stages of lactation are higher than those generated by the use of rBST in cows 100 days after lactation begins, which is often when it is administered. This arguably means that milk from early lactating cows should not be drunk at all if there are any concerns about IGF-1. Those who oppose the use of rBST point out that, unlike a therapeutic protein that would be used for a limited number of patients, milk is consumed by most people and any inherent risk, no matter how small, is therefore unacceptable. On the basis of this uncertainty, many countries, including the European Union, Canada and Australia, banned the use of rBST, citing both the animal welfare issue and the potential risk to health as reasons. There are also consumer-led pressures that impact on the use of rBST in the USA, even though the FDA guidance states that there is no risk to health from its use. However, many farmers now do not use rBST and the market has reduced substantially over the past few years. The debate remains polarised, at times supported by inaccurate or out-of-context information and interpretation (from all sides), as commercial, animal welfare and human health interests clash.

> Concerns about a biotechnology product or process are often multifaceted and can generate emotive debate; it is sometimes difficult to separate evidence from speculation.

14.4.3   Therapeutic Products for Use in Human Healthcare

Although the production of recombinant-derived proteins for use in medical applications does raise some ethical concerns, there is little serious criticism aimed at this area of biotechnology. The reason is largely that therapeutic products and strategies are designed to alleviate suffering, or to improve the quality of life for those who have a treatable medical condition. In addition, the products are used under medical supervision, and there is a perception that the corporate interests that tend to be highlighted in the food debate have less of an impact in the diagnosis and treatment of disease. In fact, there is just as much competition and investment risk associated with the medical products field, as is the case in agricultural applications; there does, however, seem to be less emotive debate in this area. It is therefore apparently much more acceptable to the public. In addition to the actual *treatment* of conditions, the area of **medical diagnostics** is a large and fast-growing sector of the biotechnology market, with rDNA and associated technologies involved in many aspects of this.

> Recombinant DNA products for medical applications are often more easily accepted by the public than is the case for genetically modified organisms used in food production.

Recombinant DNA products for use in medical therapy can be divided into three main categories. Firstly, protein products may be used for **replacement** or **supplementation** of human proteins that may be absent or ineffective in patients with a particular illness. Secondly, proteins can be used in **specific disease therapy**, to alleviate a disease state by intervention. Thirdly, the production of **recombinant vaccines** is an area that is developing rapidly and offers great promise. Some examples of therapeutic proteins produced using rDNA technology are listed in Table 14.1. We will consider examples from each of the three areas outlined above to illustrate the type of approach taken in developing a therapeutic product.

The widespread condition diabetes mellitus (DM) is usually caused either by β-cells in the islets of Langerhans in the pancreas failing to produce adequate amounts of the hormone insulin, or by

Table 14.1 | **Selected recombinant DNA-derived therapeutic products for use in humans**

Product	Type	Trade name	App.	Company	Use
Insulin	R/S	Humulin	1982	Eli Lilly	Diabetes treatment
Growth hormone	R/S	Protropin	1985	Genentech	Growth hormone deficiency in children
α-interferon	SDT	Intron A	1986	Schering-Plough	Hairy cell leukaemia
Hepatitis B vaccine	V	Recombivax HB	1986	Merck & Co.	Hepatitis B prevention
Tissue plasminogen activator	SDT	Activase	1987	Genentech	Myocardial infarction
Growth hormone	R/S	Humatrope	1987	Eli Lilly	Growth hormone deficiency in children
Hepatitis B vaccine	V	Engerix-B	1989	SmithKline Beecham	Hepatitis B prevention
Factor VIII	R/S	Recombinant anti-haemophiliac factor (rAHF)	1992	Baxter Healthcare	Treatment of haemophilia
Factor VIII	R/S	Kogenate	1993	Bayer	Treatment of haemophilia
DNase	SDT	Pulmozyme	1993	Genentech	Treatment of cystic fibrosis symptoms
Erythropoietin	R/S	Epogen	1993	Amgen	Anaemia
Imiglucerase	R/S	Cerezyme	1994	Genezyme	Type I Gaucher disease
Insulin	R/S	Humalog	1996	El Lilly	Diabetes treatment
Factor IX	R/S	BeneFix	1997	Wyeth	Treatment of haemophilia
Antisense oligonucleotide	SDT	Vitravene	1998	Isis/Novartis	Treatment of CMV retinitis in patients who are immunocompromised
Factor VIIa	R/S	NovoSeven	1999	Novo Nordisk	Treatment of haemophilia bleeding episodes
Insulin analogue	R/S	NovoLog	2000	Novo Nordisk	Diabetes treatment
Insulin analogue	R/S	Lantus	2000	Aventis	Diabetes treatment
Hepatitis A and B vaccine	V	Twinrix	2001	SmithKline Beecham	Mixed vaccine for hepatitis A and B
Laronidase	R/S	Aldurazyme	2003	BioMarin Pharmaceuticals & Genzyme	Enzyme replacement therapy for mucopolysaccharidosis I
Galsulphase	R/S	Naglazyme	2005	BioMarin Pharmaceuticals	Enzyme replacement therapy for mucopolysaccharidosis VI
Hyaluronidase	–	Hylenex	2005	Halozyme Therapeutics	Adjuvant for use with other drugs
Insulin	R/S	Exubera	2006	Pfizer	Inhalable form of insulin for diabetes treatment
Quadrivalent influenza vaccine	V	Flublok	2013	Protein Science	Influenza, 4-strain specific
Antisense oligonucleotide (ASO)	SDT	Exondys 51	2016	Sarepta Therapeutics	Duchenne muscular dystrophy
siRNA	SDT	Onpattro	2018	Alnylam Pharmaceuticals	First RNA-based therapy approved (for hATTR polyneuropathy)
Quadrivalent influenza vaccine	V	Supemtek	2020[a]	Sanofi Pasteur	Influenza, 4-strain specific

Note: Type refers to replacement/supplementation (R/S), specific disease therapy (SDT) or vaccine (V). Dates in column 4 refer to year of first approval by the FDA for use in the USA (Supemtek, 2020[a], EU-approved). Subsequent approvals for additional uses, modifications or withdrawals are not shown. Products are listed chronologically with respect to year of first approval. Trade names are registered trademarks of the companies involved; company names are as given in the approval, and may have changed due to corporate policy or merger, *etc.* Further information can be found on the FDA website at URL [www.fda.gov] and the European Medicines Agency at URL [www.ema.europa.eu].

target cells not being able to respond to the hormone. It is estimated that around 400–450 million people worldwide are affected by DM, projected to exceed 650 million by 2040. The condition is classed as either **type I DM** (formerly known as **insulin-dependent DM or IDDM**) or **type II DM** (formerly **non-insulin-dependent DM or NIDDM**). Some 10 per cent of patients have type I DM, with around 90 per cent having type II. There are also some other variants of the disease that are much less common. Type I patients obviously require the hormone, but many type II patients also use insulin for satisfactory control of their condition. Delivery of insulin is achieved by injection (traditional syringe or 'pen'-type devices), infusion using a small pump and catheter or inhalation of powdered insulin.

Insulin is composed of two amino acid chains – the **A-chain** (acidic, 21 amino acids) and **B-chain** (basic, 30 amino acids). When synthesised naturally, these chains are linked by a peptide called the **C-chain** (length varies slightly between species – 35 amino acids in humans, 30 in cows). This precursor molecule is known as **proinsulin** (Fig. 14.12). The A- and B-chains are linked together by disulphide bonds between cysteine residues, and the proinsulin is cleaved by a protease to produce the active hormone. Insulin was the first protein to be sequenced, by Frederick Sanger in the mid-1950s.

As DM is caused by a problem with a normal body constituent (insulin), therapy falls into the category of replacement or supplementation. Insulin therapy was developed in 1921 by **Frederick Banting** and **Charles Best**, and for the next 60 or so years, diabetics were dependent on natural sources of insulin, with the attendant problems of supply and quality. In the late-1970s and early-1980s, rDNA technology enabled scientists to synthesise insulin in bacteria, with the first approvals granted by 1982. Recombinant-derived insulin is now

> Recombinant insulin for use in the treatment of diabetes is one of the major success stories of rDNA-based biotechnology, in that its availability has had a major impact on the lives of millions of people.

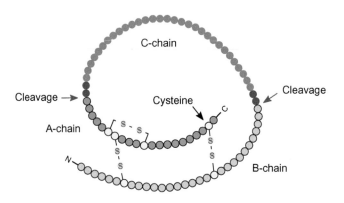

Cleavage

Cleavage

C-chain

Cysteine

A-chain

B-chain

Fig. 14.12 Proinsulin. This molecular precursor of human insulin is synthesised as an 86-amino acid polypeptide. In human proinsulin, the A-, B- and C-chains have 21, 30 and 35 amino acids, respectively. There is also a signal peptide at the amino (N-) terminus that is cleaved during initial processing in the cell. To produce the functional hormone, the C-peptide sequence is removed by a protease (P) to leave the A- and B-chains to form the final insulin molecule.

available in several forms and has a major impact on diabetes therapy. One of the most widely used forms is marketed under the name Humulin by the company **Lilly** (formerly known as Eli Lilly).

In an early method for the production of recombinant insulin, the insulin A- and B-chains were synthesised separately in two bacterial strains. The insulin A- and B-chain gene sequences were in fact not cloned genes, as the mRNA or gene sequences were not known or isolated at that time. Instead, the coding sequences were synthesised from the amino acid data and placed under the control of the *lac* promoter, so that expression of the coding sequences could be switched on by using lactose as the inducer. Following purification of the A- and B-chains, they were linked together by a chemical process to produce the final insulin molecule. The process is shown in Fig. 14.13. A development of this method involves the synthesis of the entire proinsulin polypeptide (shown in Fig. 14.12) from a single gene sequence. The product is converted to insulin enzymatically.

There are many recombinant proteins for use in specific disease therapy. One example of this type of protein is tissue plasminogen activator (TPA). This is a protease that occurs naturally and functions in breaking down blood clots. TPA acts on an inactive precursor protease called plasminogen, which is converted to the active form called plasmin. This protease attacks the clot by breaking up fibrin, the protein that is involved in clot formation. TPA is used as a treatment for heart attack victims. If administered soon after an attack, it can help reduce the damage caused by coronary thrombosis.

Recombinant TPA was produced in the early-1980s by the company **Genentech** using cDNA technology. It was licensed in the USA in 1987, under the trade name Activase, for use in treatment of acute myocardial infarction. It was the first recombinant-derived

Tissue plasminogen activator is another example of a valuable therapeutic protein that is produced by rDNA technology.

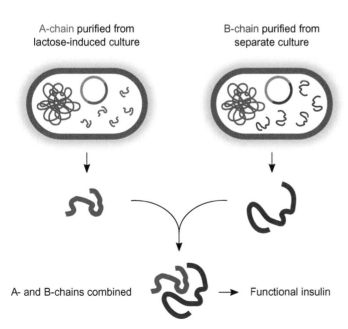

A-chain purified from lactose-induced culture

B-chain purified from separate culture

A- and B-chains combined → Functional insulin

Fig. 14.13 Production of recombinant-derived insulin A- and B-chains. The A- and B-chain gene sequences are expressed in separate fermentations. Lactose is used to induce transcription from the *lac* promoter. Following translation, the A- and B-chains are purified and combined chemically to give the final product. An alternative production method is to synthesise proinsulin as an intact polypeptide.

therapeutic protein to be produced from cultured mammalian cells, which secrete recombinant TPA when grown under appropriate conditions. The amount of recombinant TPA produced in this way was sufficient for therapeutic use; thus, a major advance in coronary care was achieved. Further uses were approved in 1990 (for acute massive pulmonary embolism) and 1996 (for acute ischaemic stroke).

In addition to protein-based therapeutics, interventions using oligonucleotide- and RNA-based systems have started to have an impact on the range of therapies available. In 2016, the FDA approved a therapy called **Exondys 51**, using antisense oligonucleotide (ASO) technology to silence gene expression in the treatment of Duchenne muscular dystrophy (DMD). The first approval for mRNA-based therapy using small interfering RNA (siRNA) was granted in 2018 for a therapeutic agent called Onpattro, for the treatment of hereditary transthyretin amyloidosis polyneuropathy.

Our final area to consider is vaccine development and recombinant technology. There are now many vaccines available for animals, and the development of human vaccines is also beginning to have an impact in healthcare programmes. The first vaccine approved for human use was for hepatitis **B**, in 1986. The yeast *S. cerevisiae* is used to express the surface antigen of the hepatitis B virus (HBsAg), under the control of the alcohol dehydrogenase promoter. The protein can then be purified from the fermentation culture and used for inoculation. This removes the possibility of contamination of the vaccine by blood-borne viruses or toxins, which is a risk if natural sources are used for vaccine production. More recently, quadrivalent recombinant influenza vaccines were approved in the USA (*Flublok*, in 2013) and in the European Union (*Supemtek*, in 2020).

A further development in vaccine technology involves using **transgenic plants** as a delivery mechanism. This area of research and development has tremendous potential, particularly for vaccine delivery in underdeveloped countries where traditional methods of vaccination may be not fully effective due to cost and distribution problems. Two approaches can be used for plant-based vaccines. Firstly, the plant can be used as a bioreactor to produce vaccines that are then processed from the plant after harvesting. This has already proved effective for the development of a COVID-19 vaccine (Table 14.2). A second aim would be to enable vaccine delivery through consuming the plant itself – the attraction of having a vaccine-containing foodstuff is clear, and development and trials are currently under way for a variety of plant vaccines.

> Vaccine production in plants, and delivery by incorporation into foods, is an elegant concept that has the potential to benefit millions of people in countries where large-scale vaccination by traditional methods is difficult.

14.4.4 Meeting the COVID-19 Challenge

From early-2020, the terrible COVID-19 pandemic did not bring much good news, but it did illustrate the importance of a strong molecular biology and biotechnology sector in modern healthcare. Biotech companies are not usually known for high levels of corporate co-operation, usually doing their best to ensure that competitive advantage is maintained. This translates into a degree of closed-mindedness, and often

Table 14.2 | COVID-19 vaccines within WHO EUL/PQ[a] process

	Manufacturer/WHO EUL holder[b]	Vaccine	Type	NRA[c]	Date of first approval[d]
1	**Pfizer-BioNTech/** BioNTec manufacturing GmbH	BNT162b2/COMIRNATY Tozinameran (INN)	Nucleoside modified mRNA	EMA	31 December 2020
2	**AstraZeneca/** AstraZeneca AB	AZD1222 Vaxzevria	Recombinant ChAdOx1 adenoviral vector encoding the Spike protein antigen of SARS-CoV-2	EMA	16 April 2021
8	**Janssen** (Johnson & Johnson) Janssen Cilag Int. NV	Ad26.COV2.S	Recombinant, replication incompetent adenovirus type 26 (Ad26) vectored vaccine encoding the (SARS-CoV-2) Spike (S) protein	EMA	12 March 2021
9	**Moderna** Moderna Biotech	mRNA-1273	mRNA-based vaccine encapsulated in lipid nanoparticle (LNP)	EMA	30 April 2021
10	**Sinopharm-BIBP/** Beijing Institute of Biological Products Co., Ltd.	SARS-CoV-2 Vaccine (Vero Cell), Inactivated (InCoV)	Inactivated, produced in Vero cells	NMPA	7 May 2021
11	**Sinovac/** Sinovac Life Sciences Co., Ltd.	COVID-19 Vaccine (Vero Cell), Inactivated/ Coronavac	Inactivated, produced in Vero cells	NMPA	1 June 2021
13	**Serum Institute of India PVT. Ltd./** Cyrus Poonawalla Group	NVX-CoV2373/Covovax	Recombinant nanoparticle prefusion Spike protein formulated with Matrix-M adjuvant	DCGI	17 December 2021
31	**Medicago** *Application withdrawn*[e]	COVIFENZ	Plant-based virus-like particle (VLP), recombinant, adjuvanted	Health Canada	N/A

Note: Numbers in the first column are those assigned to the vaccines, as listed in the WHO document cited below. [a] WHO EUL/PQ is World Health Organization Emergency Use Listing and Prequalification evaluation procedure. [b] Manufacturer shown in bold, and EUL holder in normal text. [c] NRA is the National Regulatory Authority (EMA = European Medicines Agency; NMPA = Chinese National Medical Products Administration; DCGI = Drugs Controller General of India). [d] Date of first approval, subsequent approvals or amendments/additions to approval are not listed in this table. [e] The Medicago vaccine is an interesting story, outlined further in Section 16.2.3.

Source: Information from *COVID-19 vaccines within WHO EUL/PQ evaluation process, Guidance Document,* dated 26 May 2022. [https://extranet.who.int/pqweb/sites/default/files/documents/Status_COVID_VAX_26May2022.pdf]. Accessed 2 June 2022. Reproduced by permission of the World Health Organization.

The global response to the COVID-19 pandemic from 2020 showed the importance of worldwide collaboration in developing diagnostics, therapeutics and vaccines when faced with an unprecedented challenge.

fairly swift litigation if a negative impact from some statement from a competitor is perceived. Biotech is now undoubtedly big business, with all the pros and cons that this entails – this includes a somewhat mixed public perception of companies making large profits during the COVID-19 response in areas as diverse as vaccine development and the supply of personal protective equipment, medical devices, hand sanitisers and other cleaning and disinfectant products. If we can see past these concerns, the response from the sector to COVID-19 was an astonishing example of how an international crisis can generate unprecedented levels of collaboration and bring a shared sense of purpose towards a common goal.

Two aspects of the response to COVID-19 are worth of note. The scale and reach of diagnostic testing for the SARS-CoV-2 virus, using both PCR-based technology to detect viral RNA and **lateral flow testing** for antigen detection, were required at a level previously unheard of. Even more impressive was the development of vaccines in a time frame that was compressed into months instead of years. This was only possible because of the focus and determination of the various laboratories and companies involved, and their ability to make rapid progress across a range of different methods for vaccine development. Some of the vaccines developed in response to COVID-19 are shown in Table 14.2.

14.5 | Conclusion: Industrial-Scale Biology

In this chapter, we looked at some of the areas of biotechnology in which rDNA has made a significant impact. The biotechnology industry is now a major global sector, approaching a market value of around one trillion USD. Significant expansion in countries such as China and India continues to drive growth, with a **compound annual growth rate (CAGR)** of around 16 per cent. The sector covers an enormous range of diverse applications, illustrated in Fig. 14.1.

The production of proteins has been a mainstay of the biotechnology industry since industrial-scale processes were established, well before the development of rDNA technology. However, the manipulation of protein structure has become one of the most active areas of research and development. We saw how the concepts of *rational design* and *directed evolution*, supported by methods for targeted *mutagenesis* of gene sequences, can enable proteins to be engineered for a range of purposes. Many of the enzymes used in rDNA are made by rDNA techniques, as are *therapeutic proteins* and *vaccines*.

Not all biotech ventures are wholly successful, as illustrated by the production of *recombinant bovine somatotropin (rBST)*. Public perception is a powerful force in acceptance or rejection of a new technology, and this often translates into national (or multinational, as in the European Union) responses to the science; sometimes this results in a technology (such as rBST) being banned in some countries. Debates are often polarised and divisive, and can illustrate the worst aspects of biotechnology rather than the best.

The use of rDNA techniques in the biotechnology industry continues to be a significant area of applied science. In addition to the scientific and engineering aspects of the work, the financing of biotechnology companies is an area that presents its own risks and potential rewards. The stakes are therefore high, and many fledgling companies fail to survive their first few years of operation. Even established and well-financed companies are not immune to the risk associated with the development of a new and untried product. The next few years will certainly be interesting for this sector of the applied science industry, as it continues to grow and develop as a major contributor to the global economy.

Further Reading

Brüssow, H. (2021). mRNA vaccines against COVID-19: a showcase for the importance of microbial biotechnology. *Microb. Biotechnol.*, 15 (1). URL [https://sfamjournals.onlinelibrary.wiley.com/doi/10.1111/1751-7915 .13974]. DOI [https://doi.org/10.1111/1751-7915.13974].

Engqvist, M. K. M. and Rabe, K. S. (2019). Applications of protein engineering and directed evolution in plant research. *Plant Physiol.*, 179 (3), 907–17. URL [https://academic.oup.com/plphys/article/179/3/907/6116682]. DOI [https://doi .org/10.1104/pp.18.01534].

Felberbaum, R. S. (2015). The baculovirus expression vector system: a commercial manufacturing platform for viral vaccines and gene therapy vectors. *Biotechnol. J.*, 10 (5), 702–14. URL [www.ncbi.nlm.nih.gov/pmc/ articles/PMC7159335/]. DOI [https://doi.org/10.1002/biot.201400438].

National Academies of Sciences, Engineering, and Medicine (2017). *Preparing for Future Products of Biotechnology*. The National Academies Press, Washington, DC. URL [https://nap.nationalacademies.org/catalog/24605/ preparing-for-future-products-of-biotechnology]. DOI [https://doi.org/10 .17226/24605].

Riggs, A. D. (2021). Making, cloning, and the expression of human insulin genes in bacteria: the path to Humulin. *Endocr. Rev.*, 42 (3), 374–80. URL [https://academic.oup.com/edrv/article/42/3/374/6042201]. DOI [https://doi .org/10.1210/endrev/bnaa029].

Rosano, G. L. *et al.* (2019). New tools for recombinant protein production in *Escherichia coli*: a 5-year update. *Protein Sci.*, 28 (8), 1412–22. URL [https:// onlinelibrary.wiley.com/doi/10.1002/pro.3668]. DOI [https://doi.org/10.1002/ pro.3668].

Singh, R. *et al.* (2016). Microbial enzymes: industrial progress in 21st century. *3 Biotech.*, 6 (2), 174. URL [www.ncbi.nlm.nih.gov/pmc/articles/PMC4991975/]. DOI [https://doi.org/10.1007/s13205-016-0485-8].

Websearch

People

Although not the first biotech company, Genentech is often thought of as the pioneer. It was founded by *Robert Swanson* and *Herbert Boyer* in 1976. Have a look at how these two key figures complemented each other to enable a successful company to be established.

Places

Personalise this search by looking for biotechnology companies in the *area where you live* at the moment. How far do you have to go before you find one? Is there a science park near you where incubator companies can be set up? What are the job opportunities in the companies that you find?

Processes

Have a look in a little more detail at the process of *directed evolution* and its role in investigating and developing modified proteins – the Nobel Prize in 2018 is a good start point.

Reflections

What topics in this chapter have you found most challenging? Look for resources that help to illustrate the key points.

Genetic engineering and biotechnology

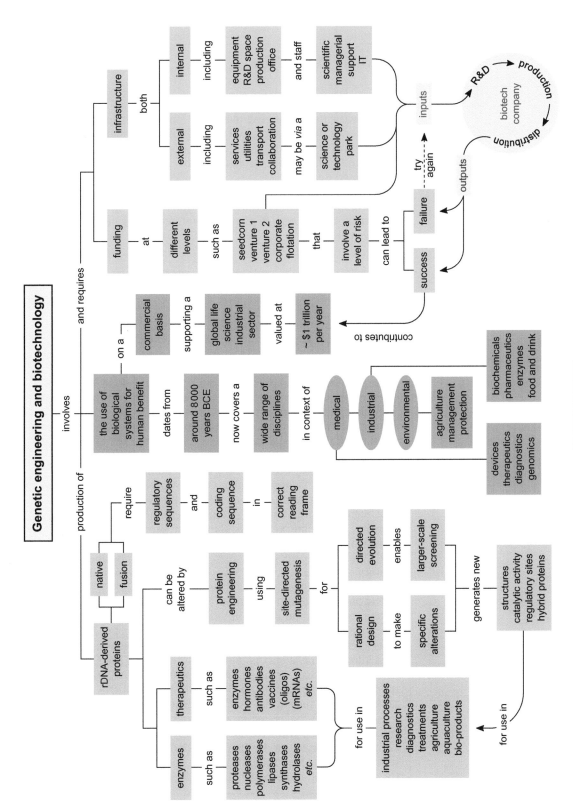

Concept Map 14

Chapter 15 Summary

Learning Objectives

When you have completed this chapter, you will be able to:

- Review the impact of gene manipulation on medicine and forensic science in, for example, the diagnosis and characterisation of medical conditions
- Explain the form and patterns of inheritance for some genetically based diseases
- Evaluate the potential of rDNA-based methods, including gene editing, in medical treatments
- Describe the use of gene manipulation technology in an investigative context

Key Words

Congenital, human immunodeficiency virus (HIV), acquired immune deficiency syndrome (AIDS), enzyme-linked immunosorbent assay (ELISA), indirect immunofluorescence assay (IFA), haploid, diploid, gametes, meiosis, zygote, polygenic, alleles, penetrance, expressivity, multifactorial, Victor McKusick, chromosomal abnormalities (aberrations), ploidy number, aneuploidy, monosomic, trisomic, Down syndrome, pancreatic exocrine deficiency, cystic fibrosis transmembrane conductance regulator (CFTR), positional cloning, ΔF508, F508del, allele-specific oligonucleotides (ASOs), Nancy Wexler, muscular dystrophy (MD), dystrophin, reference genome, monoclonal antibody (mAb), Georges Köhler, César Milstein, human anti-mouse antibody (HAMA), phage display method, George Smith, Sir Gregory Winter, tumour necrosis factor alpha (TNF-α), Humira, biosimilar, xenotransplantation, immunorejection, transgene, gene therapy (GT), gene addition therapy, antisense mRNA, liposome, lipoplexes, Ashanti DeSilva, adenosine deaminase (ADA) deficiency, severe combined immunodeficiency syndrome (SCIDS), enzyme replacement therapy (ERT), Jesse Gelsinger, ornithine transcarbamylase deficiency (OTCD), X-linked severe combined immunodeficiency syndrome (X-SCID), recessive dystrophic epidermolysis bullosa (RDEB), *COL7A1*, RNA interference (RNAi), Andrew Fire, Craig Mello, down-regulation, knockdown, gene silencing, post-transcriptional gene silencing (PTGS), dicer, slicer, small interfering RNAs (siRNAs), RNA-induced silencing complex (RISC), micro RNAs (miRNAs), antisense oligonucleotides (ASOs), sickle-cell disease (SCD), Victoria Gray, β-thalassaemia, Leber congenital amaurosis, T-cells, CAR T-cells, chimeric antigen receptor (CAR), autologous, allogenic, allorejection, He Jiankui, monozygotic, Sir Alec Jeffreys, multi-locus probe, single-locus probes, Joseph Mengele, Czar Nicholas II, molecular palaeontology, molecular ecology.

Chapter 15

Medical and Forensic Applications of Gene Manipulation

Genetic manipulation has had a significant impact on the diagnosis and treatment of human disease. As outlined in Chapter 14, many therapeutic proteins are now made by recombinant DNA (rDNA) methods, and the use of recombinant-derived products is already well established. In this chapter, we will look at how the techniques of gene manipulation impact more directly on medical diagnosis and treatment, and will also examine the use of rDNA technology in forensic science.

15.1 | Diagnosis and Treatment of Medical Conditions

Genetically based diseases (often called simply 'genetic diseases') are one of the most important classes of disease, particularly in children. A disorder present at birth is termed a congenital abnormality, and around 5 per cent of newborn babies will suffer from a serious medical problem of this type. A significant proportion of diseases presenting in later life also have a genetic cause or predisposition. From the early-1900s, medical genetics has had a major impact on the diagnosis of disease and abnormality, with the development of molecular genetics expanding the range of techniques available for diagnosis and rDNA-based therapies.

As many disease conditions have a major genetic component, gene manipulation technology has provided new tools for investigating and treating what are sometimes called 'genetic diseases'.

15.1.1 Diagnosis of Infection

In addition to genetic conditions, rDNA technology is also important in the diagnosis of certain types of infection. Normally, bacterial infection is relatively simple to diagnose with standard medical and microbiological tests. Viral infections can be more difficult, particularly where an early diagnosis is helpful in determining a treatment regime or to restrict wider transmission. One example is infection by the human immunodeficiency virus (HIV), the causative agent of acquired immune deficiency syndrome (AIDS).

In some cases, viral infections can be diagnosed by using rDNA techniques (such as PCR) to identify viral DNA before antibodies have reached detectable levels.

Immunological detection of anti-HIV antibodies, using techniques such as **enzyme-linked immunosorbent assay (ELISA)**, sometimes known as enzyme immunoassay (EIA), Western blot and indirect immunofluorescence assay (IFA), may miss early infection, as antibodies may not be detectable in an infected person until weeks after initial infection. A test where no positive result is obtained, even though the individual is infected, is a false negative. When nucleic acid probes and PCR technology became available, early detection of HIV RNA became possible, thus permitting a diagnosis before an antibody response is generated.

Other examples of the use of rDNA technology in diagnosing infections include tuberculosis (caused by the bacterium *Mycobacterium tuberculosis*), human papilloma virus (HPV) infection and Lyme disease (caused by the spirochaete *Borrelia burgdorferi*). The most recent and high-profile example of diagnosis of infection is the response to the COVID-19 pandemic, outlined in Section 14.4.4.

15.1.2 Patterns of Inheritance

Transmission genetics, the principles of which were first established by Gregor Mendel, is still an important part of modern medicine. Molecular genetics complements transmission genetics to provide a powerful range of methods for genetic analysis.

We have already seen that the **haploid** human genome is made up of some 3.2 billion base-pairs of information. The **diploid** genome has 46 chromosomes, arranged as 22 pairs of *autosomes* and one pair of *sex chromosomes*. Prior to reproduction, haploid male and female **gametes** (sperm and ovum, respectively) are formed by the reduction division of **meiosis**. On fertilisation of the ovum by the sperm, diploid status is restored, with the **zygote** receiving one chromosome of each pair from the father, and one from the mother. In males, the sex chromosomes are *X* and *Y*, and in females *XX*; thus, the father determines the sex of the child.

Single-gene disease traits are known as monogenic disorders, whilst those involving many genes are **polygenic**. Inheritance of a monogenic disease trait usually follows a basic Mendelian pattern, and can therefore often be traced in family histories by pedigree analysis. A gene may have **alleles** (different forms) that may be *dominant* (exhibited when the allele is present) or *recessive* (the effect is masked by a dominant allele). With respect to a particular gene, individuals are said to be either *homozygous* (both alleles are the same) or *heterozygous* (the alleles are different, perhaps one dominant and one recessive). Patterns of inheritance of monogenic traits can be associated with the autosomes, as either *autosomal dominant* or *autosomal recessive*, or may be sex-linked (usually with the X chromosome, thus showing *X-linked* inheritance). The Mendelian patterns and ratios for these types of inheritance are shown in Fig. 15.1. In addition to the nuclear chromosomes, mutated genes associated with the mitochondrial genome can cause disease. As the mitochondria are inherited along with the egg, these traits show *maternal patterns of inheritance*. We will consider specific examples of the patterns of inheritance outlined above in the next section.

Genetic traits can be transmitted from generation to generation in different ways. These patterns of inheritance follow set 'rules' and can be useful in the diagnosis and tracing of disease patterns in families.

The effect of a gene depends not only on its allelic form and character, but also on how it is expressed. The terms **penetrance**

(a) Autosomal dominant

♂ Genotype: Aa
    Phenotype: *affected male*

♀ Genotype: aa
    Phenotype: *normal female*

> Genotype ratio is:
> 1:1 Aa aa
> Thus 50% affected, 50% normal

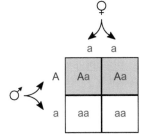

(b) Autosomal recessive

♂ Genotype: Bb
    Phenotype: *carrier male*

♀ Genotype: Bb
    Phenotype: *carrier female*

> Genotype ratio is:
> 1:2:1 BB Bb bb
> Thus 25% affected, 75% normal

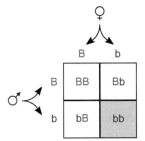

(c) X-linked

♂ Genotype: X Y
    Phenotype: *normal male*

♀ Genotype: X X^C
    Phenotype: *carrier female*

> Genotype ratio is:
> 1:1:1:1 XX XY XX^C X^{C}Y
> Thus 50% male children affected

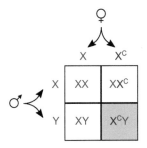

Now trace the inheritance patterns for dominant/recessive X-linked alleles H and h!

Fig. 15.1 Patterns of inheritance. (a) Autosomal dominant transmission pattern with a disease allele (**A**, red text) and the normal form (**a**, blue text). Half of the gametes from an affected individual (in this case, the male) will carry the disease allele. On mating (box diagram), the gametes can mix in the combinations shown. The result is that half the offspring will be heterozygous and therefore have the disease (shaded boxes). (b) Autosomal recessive transmission, with the disease-causing allele designated **b**, and the normal variant **B**. On mating between two carriers heterozygous for the defective allele, there is a one in four chance of having an affected child. (c) X-linked inheritance for a disease allele designated **C** (the wild-type allele is not shown on the other X chromosome, for clarity). In this case, a recessive allele is shown. Half of the male children will be affected, as there is only one X chromosome and thus no dominant allele to mask the effect. No female children are affected. However, in the case of an X-linked dominant allele, females are also affected. A useful exercise is to trace all the possible patterns of inheritance for dominant and recessive alleles (suggest use **H** and **h**).

and **expressivity** are used to describe this aspect. Penetrance is usually quoted as the percentage of individuals carrying a particular allele who demonstrate the associated phenotype. Expressivity refers to the degree to which the associated phenotype is presented (the *severity* of the phenotype is one way to think of this). Thus, alleles showing *incomplete penetrance* and/or *variable expressivity* can affect the range of phenotypes observed. Further complications arise with *multiple alleles*, or when alleles demonstrate *incomplete dominance, co-dominance* or *partial dominance*. In many cases, the route from genotype to phenotype also involves one or more environmental factors, when traits are said to be **multifactorial**.

Despite the complexities, there are many cases where a gene defect can be traced with reasonable certainty. As noted in Section 13.2.3, data for transmission of disease traits are collated in Online Mendelian Inheritance in Man (OMIM), which now runs to over 26 000 entries covering some 7 000 disease phenotypes. OMIM is the electronic version of *Mendelian Inheritance in Man*, first published by **Victor McKusick** of Johns Hopkins University in 1966. McKusick is rightly considered to be the father of medical genetics.

> Online Mendelian Inheritance in Man is another good example of how the availability of powerful desktop computers and the internet has transformed the way in which we deal with complex data sets.

15.1.3 Genetically Based Disease Conditions

Genetic problems may arise from either **chromosomal abnormalities (aberrations)** or gene mutations. An abnormal chromosome complement can involve whole chromosome sets (variation in the **ploidy number**, such as triploid, tetraploid, *etc.*) or individual chromosomes (**aneuploidy**). Any such variation usually has very serious consequences, often resulting in spontaneous abortion. Multiple chromosome sets are rare in most animals, but are often found in plants. As gamete formation involves meiotic cell division in which homologous chromosomes separate during the reduction division, even-numbered multiple sets are most commonly found in polyploid plant species that remain stable.

Aneuploidy is a much more common form of chromosomal variation in humans, but is still relatively rare in terms of live-birth presentations. A missing chromosome gives rise to a **monosomic** condition, which is usually so severe that the foetus fails to develop fully. An additional chromosome gives a **trisomic** condition, which is more likely to persist to term. Monosomy and trisomy can affect both autosomes and sex chromosomes, with several recognised syndromes such as **Down syndrome** (trisomy 21). Most cases involving changes to chromosome number are caused by non-disjunction at meiosis during gamete formation. In addition to variation in chromosome number, structural changes can affect parts of chromosomes and can cause a range of conditions. Some examples of chromosomal aberrations in humans are shown in Table 15.1.

> Chromosomal abnormalities often have very serious effects on the organism, as the disruption to normal genetic balance is usually severe.

Although chromosomal abnormalities are a very important type of genetic defect, it is in the analysis of gene mutations that molecular genetics has had most impact. Many diseases have now been characterised fully, with their mode of transmission and action defined at both the chromosomal and molecular levels. Table 15.2 lists some of

Table 15.1 | Examples of types of chromosomal aberration in humans

Condition	Chromosome designation	Syndrome	Frequency per live births
Autosomal			
Trisomy-13	47, 13+	Patau syndrome	1:12 500–1:22 000
Trisomy-18	47, 18+	Edwards syndrome	1:6 000–1:10 000
Trisomy-21	47, 21+	Down syndrome	1:800
Sex chromosome variation			
Missing Y	45, X	Turner syndrome	1:3 000 female births
Additional X	47, XXX	Triplo-X	1:1 200 female births
Additional X	47, XXY	Klinefelter syndrome	1:500 male births
Additional Y	47, XYY	Jacobs syndrome	1:1 000 male births

Structural defects	**Cause**
Deletion	Part of chromosome deleted, e.g. AB**CDE**FGH → ABFGH
Duplication	Part of chromosome duplicated, e.g. A**BCD**EFGH → A**BCDBCD**EFGH
Inversion	Part of chromosome inverted, e.g. ABC**DEF**GH → ABC**FED**GH
Translocation	Fragment moved to different chromosome, e.g. ABC**DEF**GH → PQR**DEF**STUV
Fragile-X syndrome	Region of X-chromosome susceptible to breakage; known as Martin-Bell syndrome, presenting as 1:1 250 male births and 1:2 500 female births

Note: Chromosome designation lists the total number of chromosomes, followed by the specific defect. Thus, 47, 13+ indicates an additional chromosome 13, and 47, XXY a male with an additional X chromosome. The syndrome is usually named after the person who first described it; the possessive (*Down's* syndrome) is often still used, but the modern convention is to use the non-possessive (*Down* syndrome) to reflect the fact that discoverers of the syndrome did not usually have the condition.

the more common forms of monogenic disorder that affect humans. We will consider some of these in more detail to outline how a disease can be characterised in terms of the effects of a mutated gene.

Cystic fibrosis (CF) is the most common genetically based disease found in Western Caucasians, appearing with a frequency of around 1 in 2 000–2 500 live births. It is an autosomal recessive characteristic; thus, the birth of an affected child may be the first sign that there is CF in the family. The carrier frequency for the CF defective allele is around 1 in 20–25 people. The disease presents with various symptoms, the most serious of which is the clogging of respiratory passageways with thick, sticky mucus. The pancreatic duct may also be affected by CF, resulting in **pancreatic exocrine deficiency**, which causes problems with digestion.

CF is an example of a serious disease that has been studied from the viewpoint of molecular, transmission and population genetics.

CF can be traced in European folklore, from which the following puzzling statement comes: '*Woe to that child which when kissed on the forehead tastes salty. He is bewitched and soon must die.*' The condition was first described in 1938, although characterisation of the disease at the molecular level was not achieved until the gene was isolated in 1989. The defect responsible for CF affects a membrane protein involved in chloride ion transport, resulting in epithelia having insufficient

Table 15.2 Selected monogenic traits in humans

Inheritance pattern/ disease	Frequency per live births	Features of the disease condition
Autosomal recessive		
Cystic fibrosis	1:2 000–1:2 500 in Western Caucasians	Ion transport defects; lung infection and pancreatic dysfunction result
Tay–Sachs disease	1:3 000 in Ashkenazi Jews	Neurological degeneration, blindness and paralysis
Sickle-cell disease	1:50–1:100 in African populations where malaria is endemic	Sickle-cell disease affects red blood cells; heterozygous genotype confers a level of resistance to malaria
Phenylketonuria	1:2 000–1:5 000	Mental retardation due to accumulation of phenylalanine
α_1-antitrypsin deficiency	1:5 000–1:10 000	Lung tissue damage and liver failure
Autosomal dominant		
Huntington disease	1:5 000–1:10 000	Late-onset motor defects, dementia
Familial hypercholesterolaemia	1:500	Premature susceptibility to heart disease
Breast cancer genes BRCA1 and 2	1:800 (1:100 in Ashkenazi Jews)	Susceptibility to early-onset breast and ovarian cancer
Familial retinoblastoma	1:14 000	Tumours of the retina
X-linked		
Duchenne muscular dystrophy	1:3 000–1:4 000	Muscle wastage, teenage onset
Haemophilia A/B	1:10 000	Defective blood clotting mechanism
Mitochondrial		
Leber hereditary optic neuropathy (LHON)	Mitochondrial defect, maternally inherited/late onset, thus difficult to estimate	Optic nerve damage, may lead to blindness, but complex penetrance of the defective gene due to mitochondrial pattern of inheritance

surface hydration and an increase in the salt content of sweat (hence the statement). The gene/protein responsible for CF is called the **cystic fibrosis transmembrane conductance regulator (CFTR)**. So how was the gene cloned and characterised?

The hunt for the CF gene is a good example of how the technique of **positional cloning** can be used to find a gene for which the protein product is unknown, and for which there is little cytogenetic or linkage information available. Positional cloning, as the name suggests, involves identifying a gene by virtue of its position – essentially by deciphering the molecular connection between phenotype and genotype. In 1985, a linkage marker (called *met*) was found that localised the CF gene on the long arm of chromosome 7. The search for other markers uncovered two that showed no recombination with the CF locus – thus, they were much closer to the CF gene. These markers were used as the start points for a trawl through some

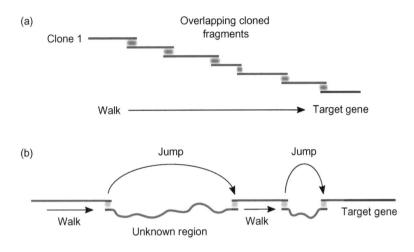

(a)

Clone 1

Overlapping cloned
fragments

Walk → Target gene

(b)

Jump   Jump

Walk   Walk   Target gene

Unknown region

Fig. 15.2 Chromosome walking and jumping. (a) Chromosome walking uses probes derived from the ends of overlapping clones to enable a 'walk' along the sequence (see also Fig. 13.10). Thus, a probe from clone 1 identifies the next clone, which then provides the probe for the next, and so on. In this way, a long contiguous sequence can be assembled. (b) In chromosome jumping, regions that are difficult to clone can be 'jumped'. The probes are prepared using a technique that enables fragments from distant sites to be isolated in a single clone by circularising a large fragment and isolating the region containing the original probe and the distant probe. This can then be used to isolate a clone containing sequences from the distant region. Alternatively, paired-end sequencing can be used to generate the probe sequences. Often a combination of walks and jumps is needed to move from a marker (such as an RFLP) to a gene sequence.

280 kbp of DNA, looking for potential CF genes. This was done using *chromosome walking* and *chromosome jumping* to search for contiguous DNA sequences from clone banks (this was before YAC/BAC vectors enabled cloning of large fragments). The basis of chromosome walking and jumping is shown in Fig. 15.2. Using these methods, four candidate genes were identified from coding sequence information, and by tracing patterns of expression of each of these in CF patients, the search uncovered the 5′ end of a large gene that was expressed in the appropriate tissues. This became known as the CFTR gene. A summary of the hunt for CFTR is shown in Fig. 15.3.

> Location and identification of the CFTR gene was a major breakthrough in medical genetics.

Having identified the CFTR gene, more detailed characterisation of its normal gene product, and the basis of the disease state, could begin. The gene is some 189 kbp in size, and encodes 27 exons that produce a protein of 1 480 amino acids. The protein is similar to the ATP-binding cassette (ABC) family of membrane transporter proteins. When the gene was being characterised, it was noted that around 70 per cent of CF cases appeared to have a similar defective region in the sequence – a 3-bp deletion in exon 10. This causes the amino acid phenylalanine to be deleted from the protein sequence. This mutation is called ΔF508 or F508del (Δ or del for deletion, F is phenylalanine and 508 is the position in the amino acid sequence). It affects the

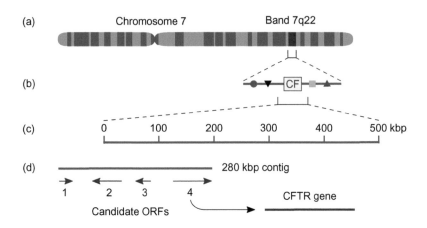

Fig. 15.3 The hunt for the cystic fibrosis gene. (a) Mapping studies placed the gene on the long arm of chromosome 7, at band position 7q22. (b) Markers associated with this region were mapped in relation to the CF gene locus. (c) A region of some 500 kbp was examined, and a contiguous sequence (clone 'contig') of 280 kbp was identified. This region contained four candidate gene sequences or open reading frames (ORFs, labelled 1–4). Further analysis of mRNA transcripts and DNA sequences eventually identified ORF 4 as the start of the 'CF gene', which was named the cystic fibrosis transmembrane conductance regulator (CFTR) gene.

folding of the CFTR protein, which means that it cannot be processed and inserted into the membrane correctly after translation. Thus, patients who carry two ΔF508 alleles do not produce any functional CFTR, with the associated disease phenotype arising as a consequence of this. The ΔF508 mutation is summarised in Fig. 15.4.

Molecular characterisation of a gene opens up the possibility of accurate diagnosis of disease alleles. Although screening for CF traditionally involved the 'sweat test', there is now a range of molecular techniques that can be used to confirm the presence of a defective CFTR allele, which can enable heterozygous carriers to be identified with certainty. Two of these early tests involve the use of PCR to amplify a fragment around the ΔF508 region to identify the 3-bp deletion, and the use of **allele-specific oligonucleotides** (ASOs) in hybridisation tests. The use of these techniques is shown in Fig. 15.5.

Although the ΔF508 mutation is the most common cause of CF, to date, around 2 000 mutations have been identified in the CFTR gene, including promoter mutations, frameshifts, amino acid replacements, defects in splicing and deletions. With more sophisticated diagnosis, patients are being diagnosed with milder presentations of CF, which may not appear as early or be as severe as the ΔF508-based disease. Thus, the CF story provides a good illustration of the scope of molecular biology in medical diagnosis, as it has enabled the common form of the disease to be characterised and has also extended our knowledge of how highly polymorphic loci can influence the range of effects that may be caused by mutation.

Many different mutations of the CFTR gene have been identified, although the most prevalent is the absence of phenylalanine at position 508 in the protein.

(a)
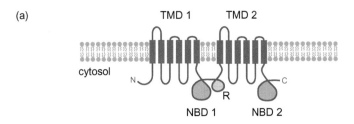

(b)
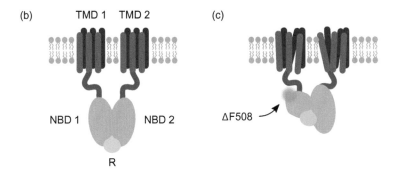

Fig. 15.4 CFTR structure and the cystic fibrosis ΔF508 mutation. (a) Representation of the structural organisation of the CFTR protein in the cell membrane. The N- and C-termini of the polypeptide are shown. The protein has five domains: two transmembrane domains (TMD1 and TMD2), two internal nucleotide binding domains (NBD1 and NBD2) and a regulatory domain (R). (b) The normal CFTR protein folds correctly and sits in the membrane with the TMDs as cylindrical arrangements of the six-domain regions for each TMD. (c) Normal CFTR has phenylalanine (F) at position 508 in the NBD1 region. In the mutant ΔF508 protein, this is deleted, causing the protein to fold incorrectly, which prevents the NBD1 domain from interacting properly with the regulatory region. The mutation also destabilises the TMDs and other interactions within the polypeptide. *Source:* Modified from Kim, S. J. and Skach, W. R. (2012), *Frontiers in Pharmacology*, 3, 1–11. DOI [https://doi.org/10.3389/fphar.2012.00201]. Used under Licence CC-BY-3.0. [https://creativecommons.org/licenses/by/3.0/].

In the area around Lake Maracaibo in Venezuela, there is a large family group of people who are descended from a woman who had migrated from Europe in the 1800s. Members of this group share a common ailment. They begin to exhibit peculiar involuntary movements, and also suffer from dementia and depression. Time of onset is usually around the age of 40–50. Their children, who were born when their parents were healthy, also develop the symptoms of this distressing condition, which is known as Huntington disease (HD; previously known as Huntington's chorea, which describes the *choreiform* movements of sufferers).

A clinical psychologist called **Nancy Wexler** has made a long-term study of thousands of HD sufferers from the Lake Maracaibo population, by carrying out an extensive pedigree analysis. This confirmed that HD follows an autosomal dominant pattern of inheritance, where the presence of a single defective allele is enough to trigger the disease state.

Pedigree analysis can be an invaluable tool for tracing the pattern of inheritance of a trait in a population.

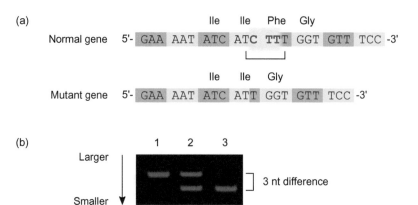

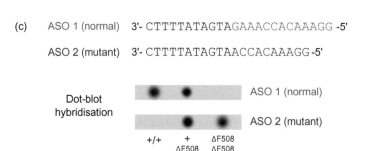

Fig. 15.5 Diagnosis of ΔF508 CF allele. (a) The normal and mutant gene sequences around position 508 (Phe) are shown. The three base-pairs that are deleted in the ΔF508 mutation are shaded yellow in the normal sequence; these are the final C of the isoleucine codon that precedes the phenylalanine, and the first two Ts of the Phe codon. The mutation causes loss of phenylalanine but maintains the reading frame and the integrity of the amino acid chain. (b) A PCR-based test for ΔF508. A 100-bp region around the deletion is amplified using PCR, and the products run on a gel that will discriminate between the normal fragment and the mutant fragment, which will be three nucleotides smaller. Lanes 1, 2 and 3 show patterns obtained for homozygous normal (+/+), heterozygous carrier (+/ΔF508) and homozygous recessive CF patient (ΔF508/ΔF508). (c) A similar pattern is seen with the use of allele-specific oligonucleotide probes (ASOs). The probe sequence is shown, derived from the gene sequences shown in (a). By amplifying DNA samples from patients using PCR and performing a dot-blot hybridisation with the radiolabelled ASOs, a simple diagnosis is possible. In this example, hybridisation with each probe separately enables the three genotypes to be determined by examining an autoradiograph.

Thus, children of an affected parent have a 50 per cent chance of inheriting the condition. As the disease presents with late onset (relative to childbearing age), many people would wish to know if they carried the defective allele, so that informed choices could be made about having a family. The search for the gene responsible for HD involved tracing a restriction fragment length polymorphism (RFLP) that is closely linked to the HD locus. The RFLP, named G8, was identified in

1983. It segregates with the HD gene in 97 per cent of cases. The HD gene itself was finally identified in 1993, located near the end of the short arm of chromosome 4. The defect involves a relatively unusual form of mutation called a trinucleotide repeat. The HD gene has multiple repeats of the sequence CAG, which codes for glutamine. In normal individuals, the gene carries up to 34 of these repeats. In HD alleles, more than 42 of the repeats indicates that the disease condition will appear. There is also a correlation between the number of repeats and the age of onset of the disease, which appears earlier in cases where larger numbers of repeats are present.

Most X-linked gene disorders are recessive. However, their pattern of inheritance means that they are *effectively* dominant in males (XY), as there is no second allele present as would be the case for females (XX), or in an autosomal diploid situation. Thus, X-linked diseases are often most serious in boys, as is the case for **muscular dystrophy (MD)**. This is a muscle wasting disease that is progressive, usually from teenage onset. The severe form of the disease is Duchenne muscular dystrophy (DMD), although there is a milder form called Becker muscular dystrophy (BMD). Both these defects map to the same location on the X-chromosome. The MD gene was isolated in 1987 using positional cloning techniques. It is extraordinarily large, covering 2.4 Mb of the X-chromosome (that's 2 400 kbp, or around 2 per cent of the total). The 79 exons in the MD gene produce a transcript of 14 kb, which encodes a protein of 3 685 amino acids called dystrophin. Its function is to link the cytoskeleton of muscle cells to the sarcolemma (membrane).

> The dystrophin gene, which is involved in muscular dystrophy, is vast – some 2.4 Mb in length.

The MD gene shows a much higher rate of mutation than is usual – some two orders of magnitude higher than other X-linked loci. This may simply be due to the extreme size of the gene, which therefore presents a 'big target' for mutation. Most of the mutations characterised so far are deletions; those that affect reading frame generally cause DMD, whilst deletions that leave reading frame intact tend to be associated with the Becker form.

15.1.4 Investigating Disease Alleles Using Comparative Genomics

The analysis of allelic variants is an important part of genomics, as it provides information that is useful in the investigation of genome variability (in a range of organisms, not only humans) and the genetic basis of disease. Coupled with other techniques such as genome-wide association studies (GWAS) and analysis of copy number variation (CNV) (see Section 13.4.3), comparison of variants across many individual genomes helps to build up a picture of the prevalence of alleles in a population. With the availability of next-generation sequencing techniques, this has become an established technique that is used widely.

> Comparative genomics enables sequence variation to be compared across many different genomes to investigate both intraspecific and interspecific variation.

Investigating allelic variants is done by comparing the genomic sequences in the area(s) of interest and aligning the sequences to compare against a reference genome. SNPs are the most common form of variation, but other gene mutations can be compared to

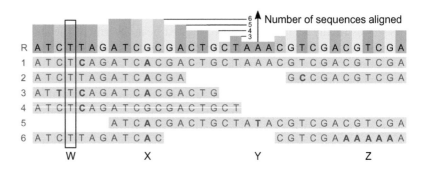

Fig. 15.6 Analysis of genome sequence variants. Variation can be analysed by aligning gene or genome sequences from different individuals (or species) against a reference sequence. This example is a short sequence from an Ensembl genome browser analysis. The reference sequence is shown at the top (R), with six aligned sequences numbered 1–6. The reference sequence here shows the base colours; the number of sequences compared at any particular base is shown by the height of the coloured column. A number of features can be identified using this approach, with changes shown in red bold text. The boxed area (W) shows a position where all the bases are the same (T). Position X shows five out of the seven sequences with A instead of the reference sequence G (this can happen because reference sequences may not show the most commonly found base at all positions). This would be a fairly convincing indication that A is likely to be the most common base at this position. This is not the case at position Y, where only two sequences are compared against the reference, and thus further comparisons would be needed to establish the frequency of base changes at this point. Finally, region Z shows multiple changes in one sequence, with a run of five adenines replacing the reference sequence of CGTCG. (Note that this is for illustration; it is not usual for a short sequence like this to show so much variation.) *Source:* Modified from [www.ebi.ac.uk/training]. Used under Licence CC-BY-4.0 [https://creativecommons.org/licenses/by/4.0/].

characterise their prevalence and effect(s) on the phenotype. An outline of SNP variant analysis is shown in Fig. 15.6.

15.1.5 Vaccine Development Using rDNA

The global response to the SARS-CoV-2 virus (see Section 14.4.4 and Table 14.2) brought vaccines into prominent public view, with progress reported regularly in the media. Arguably, given the impact of social media, this was perhaps the most widespread dissemination of vaccine-related information ever, with terms such as *mRNA vaccine* and *vaccine hesitancy* becoming part of everyday language. Despite this recent level of interest, we should remember that vaccines have been around for a long time, and the techniques of molecular genetics and rDNA have had an impact on vaccine development well before the COVID-19 era, both in terms of the characterisation of the target agent (*e.g.* a bacterium or virus) and in the development of the vaccine itself.

Vaccines are classified according to type, including *whole pathogen vaccine* (the original method of vaccination), *subunit vaccine* (including recombinant-derived), *nucleic acid vaccine* (*e.g.* mRNA vaccines) and *viral*

Although COVID-19 brought recombinant vaccine development into the public eye, the science was based on many years of work well before the SARS-CoV-2 virus appeared.

Table 15.3 Types of vaccine currently available

Type of vaccine	Examples (on UK schedule)
Whole pathogen vaccines	
Live attenuated vaccines	MMR (measles, mumps and rubella) Nasal influenza Shingles (varicella-zoster virus)
Inactivated vaccines (whole killed)	Inactivated polio vaccine (IPV) Inactivated influenza (split virion) Hepatitis A
Subunit vaccines	
Recombinant protein vaccines	Hepatitis B HPV (human papilloma virus) Men B (meningitis)
Toxoid vaccines	Diphtheria Tetanus Pertussis (whooping cough)
Conjugate vaccines	Hib (*Haemophilus influenzae* B) Men C (meningitis) PCV (children's pneumococcal vaccine) Men ACWY (meningitis)
Virus-like particles (VLPs)	Hepatitis B HPV (human papilloma virus)
Outer membrane vesicles (OMVs)	Men B (meningitis)
Nucleic acid vaccines	
RNA vaccines	COVID-19 mRNA vaccines
DNA vaccines	Under development
Viral vectored vaccines	
Replicating	Ebola
Non-replicating	Ebola COVID-19

Source: Information from The Vaccine Knowledge Project, Oxford Vaccine Group, University of Oxford [https://vk.ovg.ox.ac.uk/vk/types-of-vaccine]. Reproduced with permission.

vectored vaccine. Examples of these types of vaccine are shown in Table 15.3.

15.1.6   Therapeutic Antibodies

The development of antibody-based therapy has been one of the most active areas of therapeutics over the past few years, with a significant presence in the market. Along with vaccines, this is big business. In 2021, sales of the top 50 drugs were around $350 billion in total, with

The development of antibody therapy was transformed when monoclonal antibody technology became available.

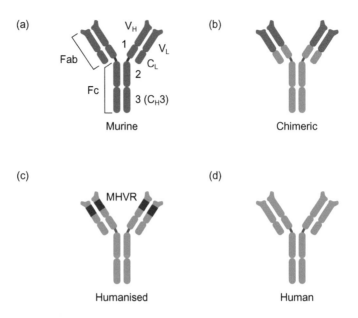

Fig. 15.7 Progressive humanising of monoclonal antibodies (mAbs). This shows the progression from fully murine (mouse) to fully human mAbs. (a) Structure of a murine (mouse) mAb showing the nomenclature of the various regions. Antibodies are composed of two heavy and two light chains (the familiar 'Y' shape diagram). Fab and Fc are produced on proteolytic cleavage. There are constant (C) regions that make up the common structural features of the antibody, shown here as C_L (light chain constant region) and 1, 2 and 3 (heavy chain constant regions, as in C_H3) products. The variable regions (V_H and V_L) determine the antigen binding specificity. (b) A chimeric antibody where the murine variable regions are joined to human constant regions. (c) A humanised antibody where the murine hypervariable region (MHVR) of V_H and V_L is the only mouse-derived component. (d) A fully human antibody with no murine components. *Source:* Modified from Lu, R.-M. *et al.* (2020). *J. Biomedical Sci.*, 27(1). DOI [https://doi.org/10.1186/s12929-019-0592-z]. Reproduced under Licence CC-BY-4.0 [https://creativecommons.org/licenses/by/4.0/].

vaccines at $97 billion (six of the top 50; note that the Pfizer/BioNTech COVID-19 vaccine sales totalled $59.12 billion alone) and antibody therapies at $98 billion (15 of the top 50). Antibody therapy was first reported in 1890, when polyclonal antisera were used to demonstrate serum-borne transfer of immunity to diphtheria in horses. The development of **monoclonal antibody (mAb)** technology by **Georges Köhler** and **César Milstein** in 1975 provided the means to develop the therapy further.

The original murine (mouse) hybridoma technology for producing mAbs caused an immune reaction in humans, known as the **human anti-mouse antibody (HAMA) response**. This led to the progressive development of methods for humanising the antibodies (Fig. 15.7) to reduce unwanted immunogenicity, and production of human antibodies in transgenic mice. A second (and very powerful) option for generating and screening antibodies is called the **phage display method**, devised by **George Smith** and **Sir Gregory Winter**

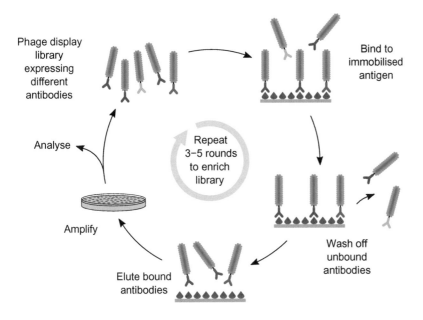

Phage display library expressing different antibodies

Bind to immobilised antigen

Repeat 3–5 rounds to enrich library

Analyse

Amplify

Elute bound antibodies

Wash off unbound antibodies

Fig. 15.8 The phage display method for mAb production. A bacteriophage such as M13 is used to express the target mAbs as fusions with a viral coat protein. This generates the phage display library. Phage are bound to immobilised target antigen; the target mAb will bind strongly (shown as red antibody), and there will be some non-specific binding (purple and brown) and some unbound mAbs (blue and green). The unbound mAbs are washed off, and the bound mAbs eluted by a more stringent wash. These are then amplified and re-screened. Usually 3–5 rounds of 'panning' are carried out to enrich for the target mAb, which is then purified and analysed. *Source:* Modified from Alfaleh, M. A. *et al.* (2020). *Frontiers in Immunology*, 11, 1986. DOI [10.3389/fimmu.2020.01986]. Used under Licence CC-BY-4.0 [https://creativecommons.org/licenses/by/4.0/].

(Fig. 15.8), for which they were awarded the Nobel Prize in 2018. Winter has been a pioneer of antibody engineering and set up Cambridge Antibody Technology, where the first fully human therapeutic mAb was developed. This was an inhibitor of **tumour necrosis factor alpha (anti-TNF-α)**, marketed under the trade name **Humira**. It has been the bestselling pharmaceutical worldwide since 2014, only knocked off the top spot in 2021 by the COVID-19 vaccine mentioned previously. Humira, and more recent **biosimilar** versions, is used to treat inflammation in a range of diseases, including rheumatoid arthritis, Crohn disease, psoriasis and ankylosing spondylitis.

15.1.7 Xenotransplantation

The shortage of donor organs for transplant surgery is a significant problem worldwide, with many potential recipients dying before a donor organ becomes available. **Xenotransplantation** is therefore an option that could alleviate much of the suffering of patients experiencing end-stage failure of major organs. Xenotransplantation is where the donor organ or tissue is from a different animal species to the recipient, compared to *autotransplantation* (tissue from the organism itself to a different site) and *allotransplantation* (tissue from a different member of the same species). The history of the field is fascinating, with early experimental surgery carried out in the first decade of the twentieth century. Results were disappointing, but detail about the cause of failure remained elusive. Successful allograft transplants of kidneys between identical twins were achieved in the early-1950s, and knowledge of the immune system was beginning to unravel some of the detail around **immunorejection** of grafts and the need for drug-mediated *immunosuppression* to address this. A number of target donor organisms (including chimpanzees, baboons and pigs) were investigated, but the discovery of a porcine (pig) retrovirus in the 1990s hindered progress. Despite this setback, pig organs remain the most likely source for human-target xenografts.

Flip forward to 7 January 2022, and a somewhat controversial procedure when a terminally ill patient, who was not suitable for allograft heart transplant, was the first person to receive a genetically modified pig heart. The controversy was not so much around the procedure (although there were, and are, critics of this, particularly from an animal rights perspective), but rather the public reaction to the recipient having been convicted for stabbing a man in 1988. This opens up a moral dilemma in some people's viewpoint, but the clinicians (rightly, given their oath as physicians) and the regulatory authorities are clear that this should have no bearing on potential treatment for any patient.

The first porcine–human heart transplant was carried out early in 2022. The patient survived for two months. In 1967, the first human–human heart transplant patient had survived for 18 days.

The pig heart came from an animal that had undergone gene editing to knock out pig genes involved in immunorejection and add human genes to increase the likelihood of acceptance. The recipient lived for two months after transplant, and died on 8 March 2022. One concerning aspect was that there were signs of infection of the heart by porcine cytomegalovirus, which was a possible contributor to the death of the patient. At the time of writing, it is not clear what the impact of this experimental surgery will be on future xenograft trials; progress in this area remains challenging.

15.2 | Treatment Using rDNA Technology – Gene Therapy

Gene therapy holds great promise that has not yet been fully realised.

Once genetic defects have been characterised, options for therapies can be considered. If the defective gene can be replaced with a functional copy (sometimes called the **transgene**, as in *transgenic*), symptoms may be reduced or prevented. This approach is known as **gene therapy (GT)**. Although it has not yet fulfilled its early expectations, it remains one of the most promising aspects of the use of gene technology in medicine. There are two possible approaches to gene therapy, with markedly different ethical implications: (1) introduction of the transgene into the somatic cells of the affected tissue, or (2) introduction into the reproductive (germ line) cells. Most scientists and clinicians consider somatic cell gene therapy an acceptable practice, no more morally troublesome than taking an aspirin. However, tinkering with the reproductive cells, with the probability of germ line transmission, is regarded as unacceptable by most people.

There are several requirements for a gene therapy protocol to be effective. Firstly, the gene defect itself will have been characterised, and the gene cloned and available in a form suitable for use in a clinical programme. Secondly, a system for getting the gene into the correct site in the patient is needed. Finally, the inserted gene must be expressed in the target cells. Ideally, the faulty gene would be replaced by a functional copy. This is known as *gene replacement therapy*, and requires recombination between the defective gene and the transgene. Due to technical difficulties in achieving this reliably, an alternative is to use *gene addition therapy*, where the transgene functions alongside the defective gene. This approach is useful only if the gene defect is not dominant, in that a dominant allele will still produce the defective protein, which may overcome any effect of the transgene. Therapy for dominant conditions could be devised using antisense mRNA, in which a reversed copy of the gene is used to produce mRNA in the antisense configuration. This can bind to the mRNA from the defective allele and effectively prevent its translation.

Although the concept of gene replacement or addition therapy is elegantly simple, it is much more difficult to achieve in reality.

15.2.1 Getting Transgenes into Patients

One challenge in gene therapy is the target cell or tissue system. In some situations, cells can be removed from a patient, altered and replaced. This approach is known as *ex vivo* gene therapy. It is mostly suitable for diseases that affect the blood system. It is not suitable for tissue-based diseases such as DMD or CF, in which the problem lies in dispersed and extensive tissue such as the lungs and pancreas (CF) or the skeletal muscles (DMD). It is difficult to see how these conditions could be treated by *ex vivo* therapy, and therefore, the technique of treating these conditions at their locations is used. This is known as *in vivo* gene therapy. Features of these two types of gene therapy are illustrated in Fig. 15.9, with both approaches having been used with some success.

Gene therapy protocols can be performed outside the body (*ex vivo*) or delivered into the target tissue in the body (*in vivo*).

As with vectors for use in cloning procedures, viruses are an attractive option for delivering genes into human cells. We can use the **term** *vector* in its cloning context, as a piece of DNA into which the transgene is inserted. The viral particle itself is often called the *vehicle* for delivery of the transgene, although some authors describe the whole system simply as a vector system. The main viral systems that have been developed for gene therapy protocols are based on retroviruses, adenoviruses, adeno-associated viruses, lentiviruses and herpes simplex viruses. The advantage of viral systems is that they provide a specific and efficient way of getting DNA into the target cells. However, care must be taken to ensure that viable virus particles are not generated during the therapy procedure.

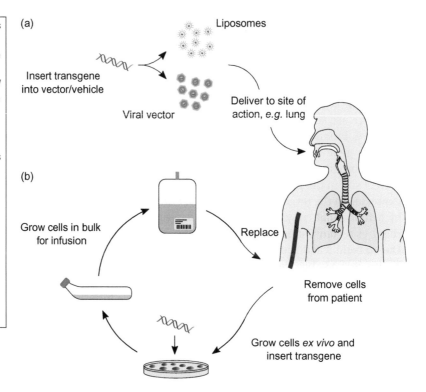

Fig. 15.9 *In vivo* and *ex vivo* routes for gene therapy. The *in vivo* approach is shown in (a). The gene is inserted into a vector such as a virus, or encapsulated in liposomes/lipoplexes, and introduced into the target tissue of the patient. In this case, the lung is the target, and an aerosol can be used to deliver the transgene. In (b), the *ex vivo* route is shown. Cells (e.g. from blood or bone marrow) are removed from the patient and grown in culture. The transgene is therefore introduced into the cells outside the body. Modified cells can be selected and amplified (as in a typical gene cloning protocol with mammalian cells) before they are infused or injected back into the patient.

In addition to viral-based systems, DNA can be delivered to target cells by non-viral methods. Naked DNA can be used directly, although this is not an efficient method. Alternatively, the DNA can be encapsulated in a lipid micelle called a **liposome**. Developments of this technique produced more complex structures that resemble viral particles, and these were given the name **lipoplexes** to distinguish them from liposomes. Some features of selected delivery systems are shown in Table 15.4.

When a delivery system is available, the patient can be exposed to the virus in a number of ways. Delivery into the lungs by aerosol inhalation is one method appropriate to *in vivo* therapy for CF, as this is the main target tissue. Injection or infusion are other methods that may be useful, particularly if an *ex vivo* protocol has been used.

15.2.2 Gene Therapy for Adenosine Deaminase Deficiency

The first human gene therapy treatment was administered in September 1990 to a four-year-old girl called **Ashanti DeSilva**, who received her own genetically altered white blood cells. Ashanti suffered from a recessive defect known as **adenosine deaminase (ADA) deficiency**, which causes the disease **severe combined immunodeficiency syndrome (SCIDS)**. Although a rare condition, this proved to be a suitable target for the first steps in gene therapy in that the gene defect was known (the 32-kbp gene for ADA is located on chromosome 20) and

Gene therapy for ADA deficiency was the first successful demonstration that the process could improve the condition.

| Table 15.4 | Vector/vehicle systems for gene therapy |

System	Features
Viral-based	
Retroviruses	RNA genome, usually used with cDNA, requires proliferating cells for incorporation of the transgene into the nuclear material. Not specific for one cell type, and can activate cellular oncogenes
Lentiviruses	RNA genome. Most lentiviral vectors (LVVs) are derived from HIV-1 genomes that have been altered significantly
Adeno-associated viruses	Single-stranded DNA genome, replication-defective, thus requires helper virus. Some benefits over adenoviral systems, may show chromosome-specific integration of transgene
Adenoviruses	Double-stranded DNA genome, viruses infect respiratory and gastrointestinal tract cells, thus effective in non- or slowly dividing cells. Generally provoke a strong immune response
Herpes simplex virus (HSV-1)	Double-stranded DNA genome. HSV vectors have been developed into three types with different properties
Non-viral	
Liposomes	System based on lipid micelles that encapsulate the DNA. Some problems with size, as micelles are generally small and may restrict the amount of DNA encapsulated. Inefficient compared to viral systems
Lipoplexes	Benefits over liposomes include increased efficiency due to charged groups present on the constituent lipids. Non-immunogenic, so benefits compared to viral systems
Naked DNA	Inefficient uptake, but may be useful in certain cases
Nanoparticles	Gold nanoparticles have lower cytotoxicity, compared to some other methods, and can be complexed with other carriers such as liposomes
Cationic polymers	Neutralise anionic DNA to form complexes (known as polyplexes) to deliver the transgene to the cell

an *ex vivo* strategy could be employed. Before gene therapy was available, patients could be treated by **enzyme replacement therapy (ERT)**. The treatment is still important as an additional supplement to gene therapy, the response to which can be variable in different patients.

For the first ADA treatments, lymphocytes were removed from the patients and exposed to recombinant retroviral vectors to deliver the functional ADA gene into the cells. The lymphocytes were then replaced in the patients. Further developments came when bone marrow cells were used for the modification. The stem cells that produce T-lymphocytes are present in bone marrow, and thus altering these progenitor cells should improve the effect of the ADA transgene, particularly with respect to the duration of the effect. The problem is that T-lymphocyte stem cells are present as only a tiny fraction of the bone marrow cells, and thus efficient delivery of the transgene is difficult. Umbilical cord blood is a more plentiful source of target cells, and this method has been used to effect ADA gene therapy in newborn infants diagnosed with the defect.

15.2.3 Gene Therapy for Cystic Fibrosis

CF is an obvious target for gene therapy, as it presents much more frequently that ADA deficiency, and is a major health problem for CF patients. Drug therapies can help to alleviate some of the symptoms of CF by digestive enzyme supplementation and the use of antibiotics to treat infection. However, as with ADA, ERT addresses the *symptoms* of the disease, rather than the *cause*. As outlined in Section 15.1.3, the CFTR gene/protein has been defined and characterised. As CF is a recessive condition, a functional copy of the CFTR gene inserted into the appropriate tissue (chiefly the lung) would restore the normal salt transport mechanism. Early indications that this could be achieved came from experiments that demonstrated that normal CFTR could be expressed in cell lines to restore defective CFTR function, thus opening up the real possibility of using this approach in patients.

Development of a suitable therapy for a disease such as CF usually involves developing an animal model for the disease, so that research can be carried out to mimic the therapy in a model system before it reaches clinical trials. In CF, the model was developed using transgenic mice that lack CFTR function. Adenovirus-based vector/vehicle systems were used, and these were shown to be functional. Thus, the system seemed to be effective, and human trials could begin. In moving into a human clinical perspective, there are several things that need to be taken into account, in addition to the science of the gene and its delivery system. For example, how can the efficacy of the technique be measured? As CF therapy involved cells deep in the lung, it is difficult to access these cells to investigate the expression of the normal CFTR transgene. Using nasal tissue can give some indications, but this is not completely reliable. Also, how effective must the transgene delivery/expression be in order to produce a clinically significant effect? Do all the affected cells in the lung have to be 'repaired', or will a certain percentage of them enable restoration of near-normal levels of ion transport? Early clinical trials using adenovirus vectors, from the first trial in 1993, showed very limited clinical benefit, with some inflammatory side-effects. Progress has not delivered the hoped-for outcomes, and whilst active research and clinical trials will continue, we are still some way from effective gene therapy for CF.

In addition to gene therapy and ERT, *CFTR modulators* can be used to alter the structure of the CFTR protein in the cell. These small molecules can alter the shape of the defective protein and restore some functionality; in terms of their place in therapy, they therefore sit between the treatment of symptoms (ERT) and cause (GT). Four groups of CFTR modulator (correctors, potentiators, stabilisers and amplifiers) are classified according to the effect on the misfolded CFTR protein.

15.2.4 What Does the Future Hold for Gene Therapy?

Gene therapy has not yet realised its potential to deliver effective clinical applications. Whilst this is disappointing, the very rapid pace of developments in modern genetics sometimes leads to the expectation of 'instant success'. At the time of writing, 23 gene/cell therapy products were listed as approved by the FDA. Given the amount of effort over the past 30 years, and more than 2 000 human clinical trials, this return might be seen as failure. However, tackling something like gene therapy presents very complex challenges, and thus we should maybe not be *too* surprised when progress is not as rapid as we would wish. At present, gene therapy remains almost entirely an experimental application delivered in a clinical trial setting.

Lack of success is of course disappointing (and often devastating) from the patient's perspective, but more distressing are cases involving the deaths of patients. Most of the difficulties tend to be associated with the use of viral vectors. In 1999, a young man called **Jesse Gelsinger** was undergoing gene therapy for **ornithine transcarbamylase deficiency (OTCD)**. Sadly, he suffered an adverse reaction to the vector and died a few days after treatment. Another setback was the development of leukaemia-like disease in patients in a French trial that had initially delivered a positive outcome for the treatment of **X-linked severe combined immunodeficiency syndrome (X-SCID)**.

Despite the challenges of establishing robust, safe and effective methods for routine clinical use of gene therapy, many dedicated scientists and clinicians worldwide remain committed to solving the problems and achieving success. One notable achievement published in March 2022 used a novel approach to treat **recessive dystrophic epidermolysis bullosa (RDEB)**, a distressing condition where the skin blisters and scars very easily. It is caused by mutations in the collagen gene *COL7A1*. Using a modified herpes simplex virus as the vector/vehicle in a gel formulation, gene function was restored by applying the gel directly to the damaged areas of skin.

> The availability of gene therapy 'medicine' in an off-the-shelf, reliable and tested form for a variety of diseases is still a long way off, despite steady progress.

> Any novel process will run into difficulties; in medical applications, the consequences are often distressing and can prove to be serious setbacks.

15.3 | RNA Interference

As we saw in Chapter 9 when we considered the PCR, from time to time, a new discovery appears that enables a step change in a discipline. The discovery of **RNA interference (RNAi)** is another example of this, for which **Andrew Fire** and **Craig Mello** were awarded the 2006 Nobel Prize. RNAi is a fascinating topic, with potentially very significant applications in both basic science and therapeutics, so it is important to at least introduce the basics of what RNAi is and what it can be used for.

> Many scientists think that RNA interference has the potential to be a major therapeutic tool in combating a wide range of diseases.

15.3.1 What Is RNAi?

RNAi was discovered in 1998 in the nematode *Caenorhabditis elegans*, although earlier work with *Petunia* pigmentation genes had raised the question of how gene addition could apparently lead to reduction in expression. RNAi is thought to have evolved as a defence against viruses and transposons, triggered by the presence of double-stranded

RNA molecules (dsRNAs). The term *interference* gives some clue as to the functioning of the RNAi system. It refers to the down-regulation (or knockdown) of gene expression, or gene silencing, which in this case is mediated *via* short RNA molecules that enable specific regulation of particular mRNAs. In plants, the process is called post-transcriptional gene silencing (PTGS). The control of gene silencing by RNAi is complex, but in essence, it works by facilitating the degradation of mRNA following a sequence-specific recognition event. Other mechanisms may prevent translation by antisense RNA binding, or may shut off transcription by methylation of bases in the promoter sequence of the gene.

> RNAi works by down-regulating or 'silencing' gene expression by acting after the mRNA has been transcribed. It is therefore sometimes called post-transcriptional gene silencing.

The mechanism of mRNA degradation involves two enzymes called dicer and slicer. In the cell, dsRNA is recognised by the dicer enzyme, which processively degrades the dsRNA into fragments around 21 nucleotides in length. These are called small interfering RNAs (siRNAs). These associate with a protein complex termed RNA-induced silencing complex (RISC), which contains a nuclease called slicer. RISC is activated when the dsRNA fragment is converted to single-strand form. The RISC contains the antisense RNA fragment, which binds to the sense strand mRNA molecule. The slicer nuclease then cuts the mRNA and the product is degraded by cellular nucleases. Thus, the expression of the gene is effectively neutralised by removal of mRNA transcripts. In some cases, RNAi is effected by the synthesis of short RNAs from control genes. These are called micro RNAs (miRNAs), and are targeted by dicer to generate the response and silence the target gene by either degradation or preventing translation of the mRNA transcript. An outline of RNAi is shown in Fig. 15.10.

> The enzymes involved in generating small interfering RNA (siRNA) have been named 'dicer' and 'slicer' – appropriate given their roles in chopping up RNA molecules.

15.3.2 Using RNAi as a Tool for Studying Gene Expression

The discovery of RNAi presented genome scientists with exactly the tool that they had been seeking for years – the ability to switch off any gene for which the DNA sequence was known. By generating dsRNAs that produce antisense RISCs for each gene, the effect of turning off the expression of the gene can be investigated. This enables investigation of gene expression at the genome level in a way that was previously impossible. Instead of knock*out*, where a gene is inactivated, RNAi knock*down* acts more like a 'dimmer switch' that gives a greater level of control over experiments on gene expression. International co-operation in RNAi research and its applications means that the future of RNAi in investigating gene expression and developing therapies seems assured for years to come.

> Genome sequencing provides the information necessary to investigate gene expression using RNAi methods.

15.3.3 RNAi as a Potential Therapy

One exciting prospect for RNAi is in the area of therapeutic applications. Some scientists are a little cautious about expecting too much too soon, bearing in mind the problems around establishing gene therapy. However, many more are convinced that RNAi technology can deliver therapies that can be adapted to almost any disease that involved expression of a gene or genes. If the gene can be identified

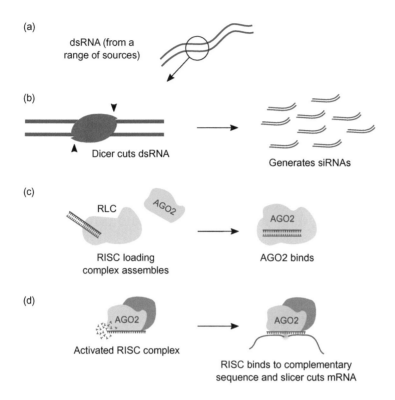

(a) dsRNA (from a range of sources)

(b) Dicer cuts dsRNA — Generates siRNAs

(c) RLC AGO2 / RISC loading complex assembles — AGO2 / AGO2 binds

(d) AGO2 / Activated RISC complex — AGO2 / RISC binds to complementary sequence and slicer cuts mRNA

Fig. 15.10 RNA interference. This is triggered by double-stranded RNA (dsRNA), as shown in (a). The RNA can arise from internal or external sources by different mechanisms. (b) The dsRNA is cut into pieces about 21 bp in length by the enzyme *dicer*. This generates small interfering RNA (siRNA). (c) The RISC loading complex (RLA) assembles the dsRNA fragment and a protein AGO2 (a member of the Argonaute protein family) which binds to the RISC. (d) The binding activates the RISC complex, and the non-complementary strand of the dsRNA is digested. The activated complex binds to its complementary sequence in an mRNA molecule. If an exact match, the *slicer* function of the AGO2 enzyme cuts the mRNA. An imperfect match can prevent translation by binding to the mRNA, but not causing the slicer activity.

and the sequence-specific dsRNA introduced into the target cells, then the RNAi system should switch on and result in the knockdown of that particular gene.

As with gene therapy, there are several stages that are critical if a therapeutic effect is to be realised. The first stage is the preparation of the nucleic acid. In gene therapy, this is usually a functional gene, which is challenging because genes are often large and complex. In RNAi, the dsRNA is much smaller and is likely to pose fewer problems. The next stage is delivering the nucleic acid to the target cells, which poses similar problems to those found in gene therapy. In some cases, direct injection of dsRNAs can be used (for accessible tissues), whilst the problems are more complex for internal tissues.

Assuming the delivery of the dsRNA can be achieved, the final step is effecting the therapeutic action. This is where RNAi has great

Despite some technical problems, typical of any developing technology, RNAi is elegantly simple in that the cell machinery takes over if the nucleic acid can be delivered to the correct cells.

potential, in that the system should trigger the effect when the dsRNA enters the cell. Thus, there is no need for recombination or expression of a gene sequence, as is the case in gene therapy. A major disadvantage is that the effects of RNAi in response to a single delivery of dsRNA are transient, and thus repeated or continuous delivery of the therapy is likely to be needed if a long-term benefit is to be achieved.

Possible targets for RNAi therapies include age-related macular degeneration (AMD), conditions such as diabetic retinopathy, cystic fibrosis, HIV infection, hepatitis C, respiratory infections, Huntington disease and many others. A number of potential therapies have reached Phase III clinical trials but have yet to be approved for clinical use. The first RNAi therapy was approved by the FDA in 2018 (Table 14.1), with two others following in 2019 and 2020.

15.3.4 Antisense Oligonucleotides

A different approach that involves binding to mRNA to down-regulate gene expression is the use of antisense oligonucleotides (ASOs) (note this is the same acronym as for allele-specific oligonucleotides; Fig. 15.5). The first approval for use was for an ASO called Vitravene, in 1998 (FDA; Table 14.1); thus, ASO therapy predates RNAi by 20 years. The ASO is a short oligonucleotide sequence, complementary to part of the mRNA, that essentially silences gene expression by knockdown of the translatable mRNA. With RNAi, ASO therapy remains promising despite relatively few approved products for either technology to date.

One complication with both RNAi and ASO approaches is that off-target effects can be significant, due to binding to similar but non-target sequences, and efforts to minimise these effects are an important part of the design of the dsRNA or ASO. It is also difficult to devise effective interference-based therapies for polygenic traits. Despite these difficulties, optimism remains high for the development of RNAi and ASO methods as perhaps the most important gene-based therapies in the years ahead.

15.4 | Medical Applications of Genome Editing

Genome editing has quickly become established and used to generate precise changes in genomes.

In Chapter 12, we looked at the methods that can be used for genome editing, and saw how the CRISPR-Cas9 system enables precise alterations to the genome. As with gene therapy and RNAi, there are some fairly clear criteria when considering a target, both in terms of the genetics of the disease and the challenges of accessing the target tissue(s), with both *in vivo* and *ex vivo* approaches being feasible. A relationship timeline for the various forms of gene-based therapy is shown in Fig. 15.11.

15.4.1 Disease Targets for Genome Editing

Although, in theory, any disease that has been characterised at the molecular level could be a target for genome editing, monogenic traits

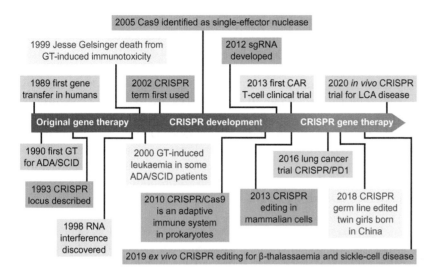

Fig. 15.11 Milestones in gene therapy. Some of the key events in the short history of gene therapy. The two setbacks in 1999 and 2000, and the CRISPR babies controversy in 2018, are shown in red text. Although not *strictly* gene therapy in the original sense, RNA interference and CAR T-cell therapy are shown in green highlight, as they represent important techniques that are considered by many to be a form of gene therapy. *Source:* Modified from Uddin, F. *et al.* (2020). *Frontiers in Oncology* 10, 1387. [DOI: 10.3389/fonc.2020.01387]. Used under Licence CC-BY-4.0 [https://creativecommons.org/licenses/by/4.0/].

present the simplest (but still significant) challenges. Many of the diseases listed in Table 15.2 are being targeted with gene editing approaches. Addressing diseases with polygenic, multifactorial and environmental/lifestyle components will prove to be more difficult, but identifying single-gene targets in diseases such as cancers and heart disease will undoubtedly impact on the options for treatment, at least to some extent.

15.4.2 Sickle-Cell Success

Sickle-cell disease (SCD) is a blood disorder where the erythrocytes (red blood cells) are not the normal flattened disc shape, but are elongated and shaped like a sickle (a curved agricultural cutting tool) or a banana. This causes restriction in blood flow as the blood is 'sticky', which leads to a number of problems, the most prominent being episodes of extreme pain caused by a *sickle-cell crisis*. The disease shows autosomal recessive inheritance and is most common in people of African descent. Paradoxically, there is a benefit in that people who are heterozygous for the disease allele (classed as having *sickle-cell trait*) have a level of protection against the serious effects of malaria. There has therefore been a continuing selection pressure that has ensured that the sickle-cell allele has remained in the gene pool. The molecular basis of SCD is a gene mutation that causes an amino acid change, and as a consequence a structural change, to haemoglobin.

In an uplifting story that parallels that of Ashanti DeSilva and ADA in 1990, **Victoria Gray** was the first patient to undergo gene editing to treat SCD. In summer of 2019, she received her own bone marrow cells that had been removed and edited *ex vivo* to restore the synthesis of foetal haemoglobin, which is normally switched off after birth during the transition to adult haemoglobin. One year after treatment, all signs were positive, with the foetal protein offsetting the problems caused by the defective adult version. At the start of 2022, Victoria continued to thrive; whilst too soon to pronounce this success as a cure for SCD, other patients have benefitted and the treatment looks very promising.

Progress is being made in other areas as well as SCD. Gene editing trials for the treatment of β-thalassaemia, carried out at the same time as the SCD work, have shown an encouraging level of success. In 2020, the first trial of gene editing delivered *in vivo* was carried out for Leber congenital amaurosis, a leading cause of blindness in childhood. Antisense technology has also been trialled for this condition, and is showing some promise. Towards the end of 2020, lipid nanoparticles were used to deliver gene editing to treat hereditary transthyretin amyloidosis. These two approaches, using *in vivo* delivery of the editing components, will most likely be key 'step-change' moments, making treatments much more easily accessible. A major drawback of the *ex vivo* approach used for the SCD and β-thalassaemia trials is that severe chemotherapy is required to kill the patient's remaining bone marrow cells to enable the edited cells to re-populate the bone marrow; this is not a pleasant experience for the patient. If the *in vivo* approach can be developed and extended, this will be a major breakthrough.

> Sickle-cell disease and β-thalassaemia were the first two diseases to be treated using gene editing.

Keeping up to date with developments in CRISPR gene editing is not easy, given the pace of progress. A very useful resource, founded by Jennifer Doudna (one of the Nobel Laureates who discovered the CRISPR system), is the Innovative Genomics Institute (https://innovativegenomics.org), where you can find much useful information, including an annual review of clinical trials of treatments based on gene editing.

15.4.3 CRISPR-Cas9 – CAR T-Cell Therapies in Cancer Treatment

The first two immunotherapies based on engineered T-cells were approved by the FDA in late 2017. They use CAR T-cells, which are T-lymphocytes that have been modified *ex vivo* to express a chimeric antigen receptor (CAR) that recognises cancer cells. The cells are grown in a bioreactor and replaced in the patient (Fig. 15.12). Most therapies to date have used autologous T-cells (from the patient). CAR T-cell therapy is a relatively new and exciting prospect for treating cancers such as leukaemia and lymphoma, and has potential for other

> CAR T-cell therapy is already having a major impact on cancer treatment.

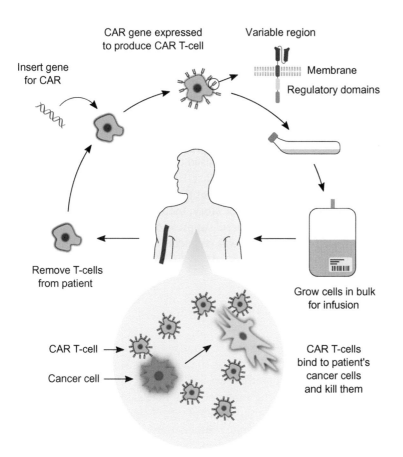

Insert gene for CAR

CAR gene expressed to produce CAR T-cell

Variable region

Membrane

Regulatory domains

Remove T-cells from patient

Grow cells in bulk for infusion

CAR T-cell →

Cancer cell →

CAR T-cells bind to patient's cancer cells and kill them

Fig. 15.12 CAR T-cell therapy. T-cells are removed from the patient and engineered *ex vivo* by inserting genes for chimeric antibody receptors (CARs) to produce CAR T-cells. These cells express antibody fragments that recognise the cancer cell antigens. The CAR region is expanded top right. The antibody variable regions are linked to a transmembrane domain (blue) and intracellular regulatory domains. Successive improvements to the structural arrangements of the CAR have been made (fifth-generation CARs are now being developed) to increase efficacy and reduce off-target toxicity. The CAR T-cells are grown in bulk and transfused into the patient to attack the cancer cells. *Source:* Modified from an open-access graphic, courtesy of the National Cancer Institute [www.cancer.gov].

tumour-based cancers if the challenges around tumour penetration can be overcome.

With the development and widespread uptake of CRISPR-Cas9 as an editing tool, engineering the CAR T-cells more precisely has become feasible, not only for generating the CAR elements, but also for improving the persistence of the T-cells and enabling allogenic T-cells (not from the patient) to be used by engineering the cells to reduce the possibility of allorejection. This should enable the development of the technology as an 'off-the-shelf' therapy, and is a good illustration of how a cell-based therapy can be coupled with a gene editing approach. Many significant advances are expected in this area over the next few years.

15.4.4 The *CCR5* Controversy

Whilst there are lots of positive developments in modern biology, some of which we have just considered, we have also seen that occasionally things do not go as expected. This is normal; the unfortunate case of Jesse Gelsinger is an example of an unforeseen tragedy, part of the accepted risk of working at the frontiers of knowledge. What is quite different is when a scientist goes against all the prevailing scientific, societal and ethical norms. Such a case became known as 'China's CRISPR babies'. We won't cover this in detail here; you can find the story easily on the web.

The chief protagonist was a Chinese scientist called **He Jiankui**, who, with others, edited germ line cells and implanted the embryos. Subsequently twins were born in 2018, followed by a third baby in 2019. When details emerged, there was outrage in the scientific community, and He was subsequently judged to have acted unlawfully and unethically, with flawed science based on a dubious aim (inactivation of the *CCR5* gene to protect against HIV infection). As well as the ethical and legal issues, almost all of the worst elements of the scientific process can be found in the story, including misrepresentation of data, false claims and censored suppression of information on the story. In December 2019, He was convicted and jailed for three years.

15.5 | DNA Profiling

As each genome is unique (apart from **monozygotic** twins), genome analysis can be used to identify an individual if a suitable technique is available. The original method was called *DNA fingerprinting*, but with improved technology, the range of tests that can be carried out has increased, and today the more general term *DNA profiling* is preferred. The technique has found many applications in both criminal cases and in disputes over whether people are related or not (paternity disputes and immigration cases are the most common). The basis of all the techniques is that a sample of DNA from a suspect (or a person in a paternity or immigration dispute) can be matched with that of the reference sample (from the victim of a crime, or a relative in a civil case). Modern techniques use the PCR; in scene-of-crime cases, this will amplify and detect minute samples of DNA from bloodstains, body fluids, skin fragments or hair roots.

> DNA profiling techniques exploit the fact that each genome is unique (apart from the genomes of monozygotic twins).

15.5.1 The History of 'Genetic Fingerprinting'

The original DNA fingerprinting technique was devised by **Alec Jeffreys** (now Sir Alec) in 1985, who realised that the work he was doing on sequences within the myoglobin gene could have wider implications. The method is based on the fact that there are highly variable regions of the genome that are specific to each individual. These are minisatellite regions, which have a variable number of short repeated sequence elements known as variable number tandem repeats (VNTRs; see Section 13.2.3 and Fig. 13.9). Within the VNTR, there are core sequence motifs that can be identified in other polymorphic VNTR loci, and also sequences that are restricted to the particular VNTR. The arrangement of the VNTR sequences and the choice of a suitable probe sequence are the key elements that enable a unique 'genetic fingerprint' to be produced.

(a)

Digest with *Hin*fl

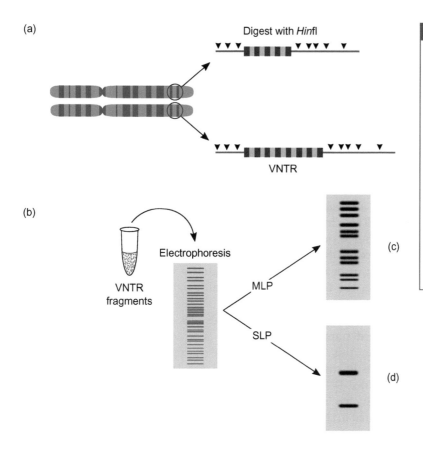

VNTR

(b)

Electrophoresis

VNTR
fragments

MLP

SLP

(c)

(d)

Fig. 15.13 Genetic fingerprinting of minisatellite DNA sequences. This is the original fingerprinting method. (a) A chromosome pair, with one minisatellite (VNTR) locus highlighted. In this case, the locus is heterozygous for VNTR length. Cutting with *Hin* fl effectively isolates the VNTRs. (b) The VNTR fragments produced (from many loci) are separated by electrophoresis and blotted. Challenging with a multi-locus probe (MLP) produces the 'barcode' pattern shown in (c). If a single-locus probe (SLP) is used, the two alleles of the specific VNTR are identified, as shown in (d).

By using a probe that hybridises to the core sequence, and carrying out the hybridisation under low stringency, polymorphic loci that bind the probe can be identified. The probe in this case is known as a multi-locus probe, as it binds to multiple sites. This generates a pattern of bands that is unique – the 'genetic fingerprint'. If probes with sequences that are specific for a particular VNTR are used (single-locus probes), a more restricted fingerprint is produced, as there will be two alleles of the sequence in each individual, one maternally derived and one paternally derived. An overview of the basis of the original method is shown in Fig. 15.13.

In forensic analysis, the original DNA profiling technique has now been replaced by a PCR-based method that amplifies parts of the DNA known as short tandem repeats (STRs; also known as **microsatellites**). These are repeats of two, three, four or five base-pairs. A major advantage over minisatellite (VNTR) repeats is that STR repeats are found throughout the genome; thus, better coverage is achieved than with minisatellites. The PCR overcomes any problems associated with the tiny amounts of sample that are often found at the crime scene, and by using fluorescent labels and automated detection equipment, a DNA profile can be generated quickly and accurately. As well as the traditional PCR, in 2018, the 'lab on a chip' concept was developed for

The original method of generating a DNA profile (sometimes called DNA fingerprinting) produces results in the now familiar 'barcode' format. More recently, STR profiling has been adopted.

PCR/STR analysis using micro-PCR (μPCR). A comparison of the outcomes of the three profiling methods (single- and multi-locus fingerprinting and STR amplification) is shown in Fig. 15.14.

15.5.2 DNA Profiling and the Law

The use of DNA profiling is now accepted as an important way of generating evidence in legal cases and there are several factors that must be considered if the evidence is to be sound. The techniques must be reliable, and accessible to trained technical staff, who must be aware of the potential problems with the use of DNA profiling. To ensure that results from DNA profile analysis are admissible as evidence in legal cases, rigorous quality control measures must be in place. These include accurate recording of samples, and careful cross-checking of the procedures. Great care must be taken to ensure that no trace of DNA contamination is present, so strict operating procedures must be observed, and laboratories inspected and authorised to conduct the tests. This is essential if public confidence in the technique is to be maintained.

> To be useful in a legal context, DNA profiling must be managed and regulated within an agreed framework so that the public can have confidence in the procedure.

The use of STR-based methods has resulted in the setting of accepted standard protocols and gene loci or target sequences. In Europe, this is overseen by the European Network of Forensic Science Institutes (ENFSI), and in the USA by the American FBI's Combined DNA Index System (CODIS). The number of loci used for profiling has been extended to offset the huge increase in information deposited in the various forensic databases (currently 17 loci in the UK and 20 in the USA). As case law has developed in forensic applications of DNA analysis, acceptance has increased and the methods are now an integral part of the criminal justice system. In addition to dealing with new cases, DNA evidence is also being used to revisit old cases and, in many of these, to either confirm or overturn original convictions.

15.5.3 Mysteries of the Past Revealed by Genetic Detectives

Identifying individuals using DNA profiling can go much further back in time than the immediate past, where perhaps a recently deceased individual is identified by DNA analysis. In the 1990s, DNA analysis enabled the identification of notable people such as **Joseph Mengele** (of Auschwitz notoriety) and **Czar Nicholas II** of Russia (search 'Romanov bones' for more information), thus solving long-running debates about their deaths. DNA has also been extracted from an Egyptian mummy (2 400 years old), and a human bone that is 5 500 years old. Thus, it is possible to use rDNA technology to study ancient DNA, recovered from museum specimens or newly discovered archaeological material. Topics such as the migration of ancient populations, the degree of relatedness between different groups of animals and the evolution of species can be addressed if there is access to DNA samples that are not too degraded. This area of work is sometimes called molecular palaeontology. With DNA having been extracted from fossils as old as 65 million years, the 'genetics of the past' looks

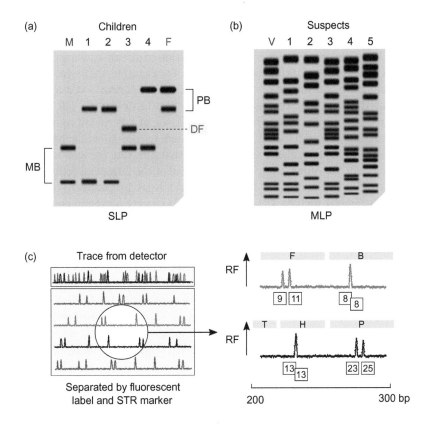

Fig. 15.14 Results from DNA profiling methods. Outcomes from three different techniques are shown. The original DNA fingerprinting technique is shown in (a) and (b), and the STR DNA profiling method in (c). In (a), a single-locus probe (SLP) has been used in a paternity test. Samples from the mother (M), four children (1–4) and father (F) are separated on a gel, blotted and hybridised with the probe, and an autoradiograph produced. The parental bands are maternal (MB) and paternal (PB). Each child inherits one allele from each parent. In lane 3, the paternal allele indicates a different father (DF). (b) Forensic analysis of suspects' DNA. Samples from the victim (V) and five suspects (1–5) are processed and hybridised with a multi-locus probe (MLP). The 'barcode' pattern indicates the multiple alleles detected in each sample. Comparing the patterns enables the perpetrator to be identified (which one?). (c) DNA profiling using PCR-amplified STR regions. Different fluorescent primers are used to amplify the STRs (the yellow dye is shown as a black trace for clarity). Up to 20 or more STR amplifications can be multiplexed. The fragments are separated by capillary electrophoresis and monitored by laser. The output measures each dye in turn, producing a collated trace (left panel, top). This is separated by label and STR for analysis (left panel, bottom). The circled area is shown in the right panel. The STR loci are identified (here by single letters) and the relative fluorescence (RF) indicates the amplified alleles. The traces show homozygous alleles as a single peak, and heterozygous alleles as a double peak. The number of repeats of the STR is often indicated (boxes). The amplified fragments are usually in the range of 100–500 bp; the region shown is 200–300 bp. Each STR will generate fragments of a particular size range, depending on the primer locations and number of repeats.

like providing evolutionary biologists, taxonomists and palaeontologists with much useful information in the future.

Use of human remains in the identification of disease-causing organisms can also be a fruitful area of research. For example, there was some debate as to the source of tuberculosis in the Americas – did it exist before the early explorers reached the New World, or was it a 'gift' from them? By analysing DNA from the lung tissue of a Peruvian mummy, researchers found DNA that corresponded to the tubercle bacillus *Mycobacterium tuberculosis*, thus proving that the disease was in fact endemic in the Americas prior to the arrival of the European settlers.

In addition to its use in forensic and legal procedures, and in tracing genetic history, DNA profiling is also a very powerful research tool that can be applied in many different contexts. Techniques such as RAPD analysis and genetic profiling are being used with many other organisms, such as cats, dogs, birds and plants. Application of the technique in an ecological context enables problems that were previously studied by classical ecological methods to be investigated at the molecular level. This use of molecular ecology is likely to have a major impact on the study of organisms in their natural environments, and (like molecular palaeontology) is a good example of the coming together of branches of science that were traditionally treated as separate disciplines.

> In addition to medical and forensic applications, the use of DNA profiling and identification techniques is proving to be of great value in many other areas of science.

15.6 | Conclusion: rDNA in Diagnosis, Analysis and Treatment

As might be expected when considering the impact of rDNA on health, disease and forensic analysis, we have covered a lot of ground in this chapter. You will have your own areas of particular interest, so some additional study will be needed to fill in the details in some cases. In some ways, we followed a fairly obvious timeline in medical genetics, moving from the early work in medical genetics to establish *patterns of inheritance* and define the cause(s) of *genetically based diseases*, through to the initial excitement around *gene therapy* and later *RNAi* and *gene editing*.

We considered how rDNA techniques have contributed to the *diagnosis* of infection and disease, and to the development of *therapeutics* such as vaccines and antibodies. In gene therapy, the initial promise has still not been realised fully, and some think that techniques like RNAi, *antisense oligonucleotides* and genome editing may supplant gene therapy in its original form. This remains to be seen; one key theme throughout the chapter has been to try to illustrate that a problem can be addressed by different techniques.

Forensic analysis is another area where rDNA techniques have changed the way in which the criminal justice system works, with *DNA profiling* (formerly DNA *fingerprinting*) now established as a key element of the evidence used to support case work. Applications of profiling are also used in many other areas, including family relationship determination, historical forensic analysis and personal genome assessment.

In addition to the very positive aspects of using rDNA applications in a medical context, we touched on a darker side when we briefly outlined the *gene-edited babies* scandal. Whilst we always need to test the boundaries between what is possible and what is permissible, there was an almost universal negative reaction across the scientific community. It remains to be seen how this impacts on the subject of human germ line editing in the years ahead.

Further Reading

Arnaud, C. H. (2017). Thirty years of DNA forensics: how DNA has revolutionized criminal investigations. *Chem. Eng. News*, 95 (37). URL [https://cen.acs.org/analytical-chemistry/Thirty-years-DNA-forensics-DNA/95/i37].

Arnold, C. (2021). Record number of gene-therapy trials, despite setbacks. *Nat. Med.*, 27, 1312–15. URL [www.nature.com/articles/s41591-021-01467-7]. DOI [https://doi.org/10.1038/s41591-021-01467-7].

Black, R. (2020). Possible dinosaur DNA has been found. *Scientific American*, 17 April 2020. URL [www.scientificamerican.com/article/possible-dinosaur-dna-has-been-found/].

Cyranoski, D. (2019). The CRISPR-baby scandal: what's next for human gene-editing? *Nature*, 566, 440–2. URL [www.nature.com/articles/d41586-019-00673-1]. DOI [https://doi.org/10.1038/d41586-019-00673-1].

Dance, A. (2017). Human artificial chromosomes offer insights, therapeutic possibilities, and challenges. *Proc. Natl. Acad. Sci. U. S. A.*, 114 (37), 9752–4. URL [www.pnas.org/doi/10.1073/pnas.1713319114]. DOI [https://doi.org/10.1073/pnas.1713319114].

Dimitri, A. *et al.* (2022). Engineering the next-generation of CAR T-cells with CRISPR-Cas9 gene editing. *Mol. Cancer*, 21, 78. URL [https://molecular-cancer.biomedcentral.com/articles/10.1186/s12943-022-01559-z#citeas]. DOI [https://doi.org/10.1186/s12943-022-01559-z].

Frangoul, H. *et al.* (2021). CRISPR-Cas9 gene editing for sickle cell disease and β-thalassemia. *N. Engl. J. Med.*, 384, 252–60. URL [www.nejm.org/doi/full/10.1056/NEJMoa2031054]. DOI [https://doi.org/10.1056/NEJMoa2031054].

Kwon, D. (2020). RNA interference comes of age. *The Scientist*, 9 December 2020, News and Opinion. URL [www.the-scientist.com/news-opinion/rna-interference-comes-of-age-68251].

Mallapaty, S. (2022). How to protect the first 'CRISPR babies' prompts ethical debate. *Nature*, 603, 213–14. URL [www.nature.com/articles/d41586-022-00512-w]. DOI [https://doi.org/10.1038/d41586-022-00512-w].

Pelc, C. and Smiley, J. D. (2022). First pig-to-human heart transplant may offer new options for patients. *Medical News Today*, 11 May 2022. URL [www.medicalnewstoday.com/articles/first-successful-pig-to-human-heart-transplant-may-offer-new-options-for-patients].

Tamura, R. and Toda, M. (2020). Historic overview of genetic engineering technologies for human gene therapy. *Neurol. Med. Chir.*, 60 (10), 483–91. URL [www.ncbi.nlm.nih.gov/pmc/articles/PMC7555159/]. DOI [https://doi.org/10.2176/nmc.ra.2020-0049].

Websearch

People

This is a big topic, so a number of people to have a look at for this area. Firstly, investigate the inspirational stories of *Ashanti DeSilva* and *Victoria Gray* who were the first people to be treated by gene therapy for ADA/SCIDS and gene editing for SCD. Secondly, find out about the career of *Sir Alec Jeffreys* who developed the original technique of DNA fingerprinting. What are the similarities and differences when you compare the discovery of DNA fingerprinting with the discovery of the PCR, which we looked at in the websearch for Chapter 9? Finally, on a less positive note, examine the controversial work of *He Jiankui* and the CRISPR babies scandal.

Places

Lake Maracaibo sits at the Northern tip of South America, in Venezuela. Have a look at the extraordinary story of how this region helped in deciphering the genetics behind Huntington disease.

Processes

A critical part of developing a new therapy involves performing *clinical trials* to test the safety and efficacy of the potential treatment. Find out how clinical trials are designed, and look in particular for any clinical trials for gene therapy that have been carried out in your country. A useful source of information that lists databases for gene therapy clinical trials in a range of locations can be found at URL (www.genetherapynet.com/clinical-trials.html).

Reflections

What topics in this chapter have you found most challenging? Look for resources that help to illustrate the key points.

Medical and forensic applications of gene manipulation

involves — analysis of a range of DNA samples — **in** — research / forensics

using:

DNA fingerprinting — **of** — minisatellites (VNTRs) — **using:**
- single-locus probe (SLP) — **for** — relationship testing — **e.g.** — paternity test
- multi-locus probe (MLP) — **produces** — 'barcode' pattern — can be used as — evidence in criminal trials

DNA profiling — **by** — PCR amplification — **of** — microsatellite DNA (STRs)
- also used for — ecology / palaeontology / geneaology / tracing GMOs

diagnosis of:

genetically based disease — **including:**

analysis and diagnosis — **using** — comparative genomics — to analyse — allelic variants — e.g. — 1000 and 100 000 genome projects — to develop — personalised medicine

patterns of inheritance — **such as:**
- monogenic
- polygenic
- multi-factorial
- dominant
- recessive
- X-linked

simple Mendelian or complex — and include — multiple alleles / penetrance / expressivity — help inform approach to → personalised medicine

and also — bacterial and viral infection — **using** — PCR — e.g. — HIV / hep B/C / CMV / HPV / SARS-CoV-2 — before — antibodies can be detected

and — treatment of disease — **by:**

- vaccines — e.g. — viral vector/nucleic acid (mRNA)
- antibodies — monoclonal — can be — murine / chimeric / humanised / fully human
- gene therapy — in/ex vivo — + 1990 Ashanti DeSilva / − 1999 Jesse Gelsinger
- RNAi — gene silencing — to — regulate gene expression
- CAR T-cells — for — cancer therapy — mostly blood cancers to date
- CRISPR-Cas — gene/genome editing

treatment of disease — e.g. — ADA / SCIDS / CF / AIDS / cancers / eye disease

scientists and clinicians collaborating → leads to better treatments for disease and possibility of a cure

unethical or illegal procedures → 2018 CRISPR baby scandal — dubious rationale e.g. CCR5/HIV

Concept Map 15

Chapter 16 Summary

Learning Objectives

When you have completed this chapter, you will be able to:

- Define the term 'transgenic'
- Outline the range and scope of transgenic technology
- Describe the uses and potential applications of transgenic plants and animals
- Identify the scientific, commercial and ethical issues surrounding transgenic organism technology

Key Words

Transgenic, intragenic, cisgenesis, genetically modified organism (GMO), overnutrition, undernutrition, United Nations (UN), selective breeding, polygenic trait, *Agrobacterium tumefaciens*, crown gall disease, Ti plasmid, T-DNA, opines, cointegration, binary vector, intermediate vector, disarmed vectors, mini-Ti, ice-forming bacteria, *Pseudomonas syringae*, Steven Lindow, ice-minus bacteria, deliberate release experiment, *Bacillis thuringiensis*, Bt plants, western corn rootworm, glyphosate, 5-enolpyruvylshikimate-3-phosphate synthase, EPSP synthase, Monsanto, Bayer, Roundup-ready, non-Hodgkin lymphoma (NHL), Flavr Savr, Calgene, antisense technology, polygalacturonase (PG), vitamin A deficiency (VAD), β-carotene, Ingo Potrykus, Peter Beyer, Golden Rice (GR), Greenpeace, Arctic Apple, polyphenol oxidase (PPO), gene protection technology, plant-made pharmaceuticals (PMPs), Medicago, virus-like particles, Sir Richard Roberts, Michael Le Page, pharm animal, pharming, xenotransplantation, nuclear transfer, pronuclei, supermouse, mosaic, chimeric, Chinook salmon (*Oncorhynchus tshawytscha*), Atlantic salmon (*Salmo salar*), Ocean Pout (*Zoarces americanus*), AquAdvantage, AquaBounty, transposable elements, oncomouse, Philip Leder, *c-myc* oncogene, mouse mammary tumour (MMT) virus, prostate mouse, severe combined immunodeficiency syndrome (SCIDS), Alzheimer disease, knockout mouse, knockin mouse, ΔF508 mutation, tissue plasminogen activator (TPA), PPL Therapeutics, α_1-antitrypsin, blood coagulation factor IX (FIX), ATryn, non-human primate (NHP), ANDi, green fluorescent protein (GFP), Huntington disease (HD), genome editing, xenogenic, gene drive, allelic fitness, *Anopheles*, *Plasmodium*, *P. falciparum*, suppress, modify, embryonic stem cell technology, transgenic mouse models, nutritional biofortification.

Chapter 16

Transgenic Plants and Animals

The production of a transgenic organism involves altering the genome so that a permanent change is effected and inherited. The term applies to genes that are introduced into a plant or animal from a non-related source organism with which the target organism cannot breed. Thus, transgenesis differs from intragenic modification, where techniques such as gene editing are used to generate changes *within* the target genome, and **cisgenesis**, where the transferred gene comes from a source that could be crossed with the target organism and therefore introduce the change by conventional breeding. Regardless of the subtleties of definition, this area of genetic engineering has caused great public concern, and there are many complex issues surrounding the development and use of transgenic organisms, with strongly held and often entrenched views. In this chapter, we will look at some of the elements of this complexity before considering the generation and use of transgenic plants and animals.

16.1 | A Complex Landscape

The widely accepted term genetically modified organism (GMO) can generate polarised opinions more than most areas of science, in particular around GMOs in the food chain. Public perception of GMOs (or indeed any topic) is influenced by a range of factors, including: the scientific community, pressure groups, activists, advocates, corporate PR, lobbying, high-profile litigation and mainstream and social media. These are also affected by the 'spin' that both protagonists and antagonists put on things. All this creates a very messy picture of GMO technology, and trying to find a consensus view among all these competing elements can be a confusing exercise; at times it can be difficult to separate factual and evidence-based information from comment and opinion (have a look at the websites listed in Table 16.1 to illustrate this issue). In areas as important and

Table 16.1	Some organisations with roles/views on the use of GMOs	
Organisation	**URL**
Bayer	www.bayer.com/en/
Canadian Biotechnology Action Network (cban)	https://cban.ca/
Food and Agricultural Organization (FAO) of the United Nations	www.fao.org/home/en/
Genetic Literacy Project	https://geneticliteracyproject.org/
GeneWatch	www.genewatch.org/
GMWatch	www.gmwatch.org/en/
Golden Rice Project	www.goldenrice.org/
Health and Environment Alliance (HEAL)	www.env-health.org/
International Agency for Research on Cancer	www.iarc.who.int/
International Life Sciences Institute (ILSI)	https://ilsi.org/
International Service for the Acquisition of Agri-biotech Applications (ISAAA)	www.isaaa.org/
US Food and Drug Administration (FDA)	www.fda.gov/food/consumers/agricultural-biotechnology

Note: A selected range of different types of organisation, with very different views on GMO technology, listed alphabetically with URLs (or you can use search terms related to the organisation instead). A quick review of Section 1.4 would be a useful refresher of what to bear in mind as you look at these websites. Two suggested threads to follow are: (1) Golden Rice and (2) glyphosate. Piece together the picture for each, considering carefully the varying points of view in the websites. When you have finished the exercise, are you clear about the facts, issues and opinions around each topic? Following a similar theme of differing points of view, have a look at the debate around the regulation (or not) of *gene-edited* crops as opposed to *transgenic* crops.

fundamental as food supply, healthcare and nutrition, there are many different constituencies, four of which are:

- Scientific
- Societal
- Corporate
- Geopolitical

As we will see, in some cases, all four of these impact on the success or failure of a particular project.

A very high level of investment is required to develop transgenic technology, with consequentially large profits to be made by biotechnology and agrochemical companies which, understandably, want a return on investment. This has led to some instances where corporate interests have overstepped the accepted norms of business and scientific ethics, which does great damage to the legitimate interests of the majority of scientists who want to make things better, not worse, for an increasing world population already facing the impact of climate change.

In addition to challenges in what we might call the 'collective affective domain', the scientific and technical problems associated with genetic engineering in higher organisms are often difficult to overcome. This is partly due to the size and complexity of the genome, and partly due to the fact that the development of plants and animals is a process that is still not fully understood at the molecular level. Despite these difficulties, methods for the generation of transgenic plants and animals are now well established.

16.2 | Transgenic Plants

All life on earth is dependent on the photosynthetic fixation of carbon dioxide by plants. We sometimes lose sight of this fact, as most people are removed from the actual process of generating our food, and the food industry generates all sorts of processed and pre-prepared foods that further neutralise the link to photosynthesis. Overnutrition in developed countries is as significant a problem as undernutrition, with lifestyle-choice diseases like cancers, diabetes and obesity of increasing concern. The inequality of distribution means that close to one billion people are partly or wholly undernourished. The Food and Agriculture Organization (FAO) of the United Nations (UN) has confirmed that we do at present produce enough food to nourish the world population, which leads to the argument that if we could fix the distribution, all would be well; this is often used by anti-GMO campaigners as a justification for not deploying genetically modified (GM) crops. It is, however, rather too simplistic, as well as being (however well intentioned) almost impossible to achieve in a sustainable way.

> It is often easy to forget that we are dependent on the photosynthetic reaction for our foods, and plants are therefore the most important part of our food supply chain.

16.2.1 Why Transgenic Plants?

For thousands of years, humans have manipulated the genetic characteristics of plants by selective breeding. However, classical plant breeding programmes rely on being able to carry out genetic crosses between individual plants. Such plants must be sexually compatible (which usually means that they have to be closely related), and thus it has not been possible to combine genetic traits from widely differing species. The advent of genetic engineering has removed this constraint, and has given the agricultural scientist a very powerful way of incorporating defined genetic changes into plants.

> In agriculture, several aspects of plant growth are potential targets for improvement, either by traditional plant breeding methods or by manipulation.

There are two main requirements for the successful genetic manipulation of plants: (1) a method for introducing the manipulated gene into the target plant, and (2) knowledge of the molecular genetics of the system that is being manipulated. In many cases, the latter is the limiting factor, particularly where the characteristic under study involves many genes (a polygenic trait). Some of the targets for the manipulation of crop plants are listed in Table 16.2. Many of these have already been achieved to some extent, at least in scientific and technical terms.

16.2.2 Making Transgenic Plants

A number of biological, physical and chemical methods can be used to insert transgenes into plant cells (Table 16.3). Currently the most widely used plant cell vectors are from *Agrobacterium tumefaciens*,

Table 16.2 Possible targets for crop plant transgenesis

Target	Benefits

Resistance to:
  diseases
  herbicides
  insects
  plant viruses
	Improve productivity of crops and reduce losses due to biological agents

Tolerance of:
  cold
  drought
  salinity (salt concentration)
	Permit growth of crops in areas that are physically unsuitable at present

Reduction of photorespiration	Increase efficiency of energy conversion
Nitrogen fixation	Capacity to fix atmospheric nitrogen extended to a wider range of species
Nutritional value	Improve nutritional value of storage proteins by protein engineering
Storage properties	Extend shelf life of fruits and vegetables
Consumer appeal	Make fruits and vegetables more appealing with respect to colour, size, shape, etc.
Vaccine development	Production of vaccines in plants, either as bioreactors for production of vaccines as virus-like particles (VLPs) or engineered to contain the vaccine for ingestion
Spread engineered traits through a natural population by a gene drive	Potential for spreading transgene-derived or CRISPR-generated traits through a population. Some concerns around unintended consequences remain

The *Agrobacterium* Ti plasmid-based vector is the most commonly used system for the introduction of recombinant DNA into plant cells.

Ti plasmids are too large to be directly useful, and so have been manipulated to have the desired characteristics of a good cloning vehicle.

which is a soil bacterium that is responsible for **crown gall disease**. The bacterium infects the plant through a wound in the stem, and a tumour of cancerous tissue develops at the crown of the plant. The agent responsible for the formation of the crown gall tumour is not the bacterium itself, but a plasmid known as the **Ti plasmid** (Ti stands for Tumour inducing). Ti plasmids are large, mostly around 200 kbp (but can be in the megabase size range). The **T-DNA** region encodes genes for tumour formation and the synthesis of **opines** (amino acid derivatives), and is the region that is integrated into the plant genome. Ti plasmids also carry genes for virulence functions. A generalised map of a Ti plasmid is shown in Fig. 16.1.

In the development of transgenic plant methodology, two approaches using Ti-based plasmids were devised: (1) **cointegration**, and (2) the **binary vector** system. In the cointegration method (Fig. 16.2), a plasmid based on pBR322 (or similar) is used to clone the gene of interest. This **intermediate vector** is then integrated into a Ti-based vector that carries the *vir* region (which specifies virulence), and has the left and right borders of T-DNA, which are important for integration of the T-DNA region. However, most of the T-DNA has

Table 16.3 | Methods for introducing transgenes into plant cells

Method	Features
Vector-mediated gene transfer	
Agrobacterium-mediated gene transfer	Most common method, efficient in dicotyledonous plants. Less efficient for monocots but can be used. Vector technology well developed
Plant virus vectors	Efficient, can give high levels of expression. Not as well developed or widely used as the *Agrobacterium* method
Direct gene transfer	
Physical methods	
Biolistics (particle bombardment or 'gene gun' method)	Widely used method using gold or tungsten particles to penetrate the cell wall. Can be used with monocots and dicots. Risk of gene rearrangement and genome damage due to fragmentation of DNA
Microinjection	Can only inject individual plant cells, and requires a high level of skill
Electroporation	Restricted to use with protoplasts. Transient pores generated by electrical pulses enable DNA uptake
Silicon carbide fibres	Requires careful handling and regenerable cell suspensions
Chemical methods	
Polyethylene glycol (PEG)-mediated	Restricted to use with protoplasts. Regeneration of viable plants can be difficult
Liposome fusion	Restricted to protoplasts; fewer problems with plant regeneration than with the PEG method
Diethylaminoethyl (DEAE) dextran-mediated	Does not result in stable transformation
Hybrid methods	
Agrolistics	As the name suggests, a combination of *Agrobacterium*-mediated and biolistic methods. Can increase efficiency of the method when used with species recalcitrant to *Agrobacterium*-mediated transfer alone.

Source: Information modified from Low, L.-Y. *et al.* (2018). In Ö. Çelik, editor, *New Visions in Plant Science*. IntechOpen, London. [DOI: 10.5772/intechopen.72517]. Reproduced under Licence CC-BY-3.0 [https://creativecommons.org/licenses/by/3.0/].

been replaced by a pBR322 sequence, which permits incorporation of the recombinant plasmid by homologous recombination. This generates a large plasmid that can facilitate integration of the cloned DNA sequence. Removal of the T-DNA has another important consequence, as cells infected with such constructs do not produce tumours and are subsequently much easier to regenerate into plants by tissue culture techniques. Ti-based plasmids lacking tumourigenic functions are known as **disarmed vectors**.

The binary vector system uses separate plasmids to supply the disarmed T-DNA and virulence functions. The **mini-Ti** plasmid is transferred to a strain of *A. tumefaciens* (which contains a compatible plasmid with the *vir* genes) by a tri-parental cross. Genes cloned into

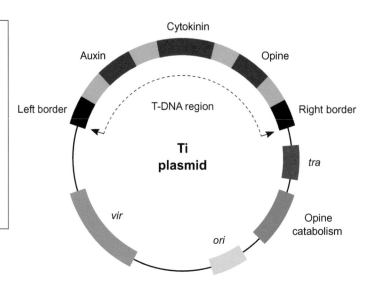

Fig. 16.1 Generalised diagram of the Ti plasmid from *Agrobacterium tumefaciens*. Regions indicated are the T-DNA, bordered by left and right border sequences (shown in black). The T-DNA region has genes for auxin, cytokinin and opine synthesis. The *tra* region (blue) is involved in conjugal transfer. The *vir* genes specify virulence. *Source:* Original by Mouagip. Modified and used under Licence CC0–1.0 [https://creativecommons.org/publicdomain/zero/1.0/deed.en].

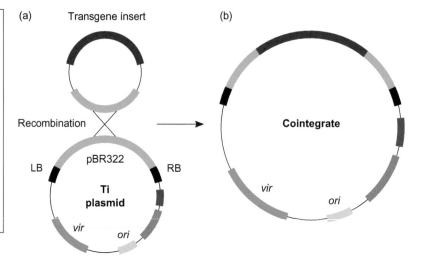

Fig. 16.2 Formation of a cointegrate Ti plasmid. (a) The transgene is cloned in a plasmid vector such as pBR322 or similar. Part of the plasmid is inserted into the T-DNA region of the Ti plasmid to enable integration by homologous recombination between the pBR regions (shown in green). (b) This generates a cointegrate vector in which the T-DNA is 'disarmed'. The cointegrate can be used to introduce the transgene into plant cells by *Agrobacterium*-mediated transfer.

The regeneration of a functional and viable organism is a critical part of generating a transgenic plant, and is often achieved using leaf discs.

mini-Ti plasmids are incorporated into the plant cell genome by *trans* complementation, where the *vir* functions are supplied by the second plasmid (Fig. 16.3).

When a suitable strain of *A. tumefaciens* has been generated, containing a disarmed recombinant Ti-derived plasmid, infection of plant tissue can be carried out. This is often done using leaf discs, from which plants can be regenerated easily, and many genes have been transferred into plants by this method. The one disadvantage of the Ti system is that it does not normally infect monocotyledonous (monocot) plants such as cereals and grasses, although it has been used successfully by additional manipulation of the system. Other methods (such as direct introduction or use of biolistics) can be used to deliver recombinant DNA to the cells of monocots. A summary of

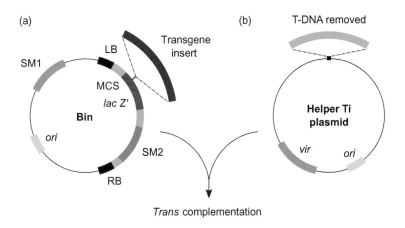

(a) (b) T-DNA removed

Fig. 16.3 The binary vector system. (a) The binary vector (Bin) has an origin of replication and one or more selectable markers (shown here as SM1 and SM2). The T-DNA region is engineered to include the gene sequence for the α-peptide of β-galactosidase (*lac Z'*, shown in red), downstream from a multiple cloning site (MCS, shown in blue), into which DNA can be cloned. The MCS/*lac Z'*/SM2 region is flanked by the T-DNA left and right border regions (LB and RB, shown in black). (b) The Bin plasmid is used in conjunction with a helper Ti plasmid, which carries the *vir* genes but has no T-DNA. The two plasmids complement each other in *trans* to enable transfer of the cloned DNA into the plant genome.

some of the methods available for generating transgenic plants is shown in Fig. 16.4.

16.2.3 Putting the Technology to Work

Transgenic plant technology has been used for a number of years, with varying degrees of success. One of the major problems we have already mentioned has been public acceptance of transgenic plants, and the consequent impact on regulatory frameworks that means some 26 countries have banned GMOs, although some permit the import of GMO-derived foods. In fact, in some cases, this resistance from the public has been more of a problem than the actual science. Towards the end of 1999, the backlash against so-called 'Frankenfoods' had reached the point where companies involved were being adversely affected, either directly by way of action against field trials, *etc.*, or indirectly in that consumers were not buying the transgenic products.

One of the first recombinant DNA experiments to be performed with plants did not in fact produce transgenic plants at all, but involved the use of GM bacteria. In nature, ice often forms at higher temperatures than normal by interacting with proteins on the surface of **ice-forming bacteria**, which are associated with many plant species. One of the most common ice-forming bacterial species is *Pseudomonas syringae*. In the late-1970s and early-1980s, **Steven Lindow** of the University of California at Berkeley removed the gene

Public confidence in, and acceptance of, transgenic plants has been a difficult area in biotechnology, with many different views and interests. A rational and balanced debate is important if the technology is ultimately to be of widespread value.

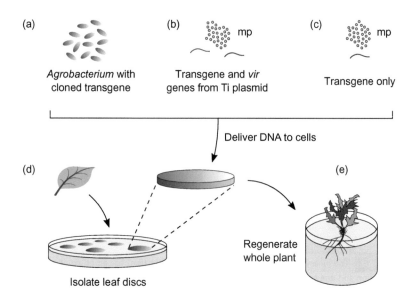

(a) *Agrobacterium* with cloned transgene

(b) mp Transgene and *vir* genes from Ti plasmid

(c) mp Transgene only

Deliver DNA to cells

(d) Isolate leaf discs

(e) Regenerate whole plant

Fig. 16.4 Transgenesis and regeneration of transgenic plants from leaf discs. Three methods for delivering a transgene into plant cells are shown. (a) The transgene is cloned into a Ti plasmid vector and introduced into *Agrobacterium* by transformation, and used to infect plant cells by a range of methods. (b) and (c) show biolistic-based methods (Fig. 7.13). The DNA, usually in plasmid form, is coated onto gold or tungsten microprojectiles (mp). In (b), the transgene is co-delivered with the *vir* genes from the Ti plasmid to enhance integration of the gene into the plant cell chromosome (this hybrid technique is sometimes called the *agrolistic* method). In (c), the standard biolistic method using the transgene DNA only is shown. (d) Introduction of the transgene into leaf disc tissue is shown. Following *Agrobacterium* infection or biolistic delivery, transformed cells are selected and grown into callus tissue and then to whole plants using plant tissue culture techniques (e).

that is responsible for synthesising the ice-forming protein, producing what became known as ice-minus bacteria. Plans to spray the ice-minus strain onto plants in field trials were ready by 1982, but approval for this first **deliberate release experiment** was delayed as the issue was debated. Finally, approval was granted in 1987, and the field trial took place despite disruption of the test site by vandals. Some success was achieved as the engineered bacteria reduced frost damage in the test treatments. However, the technology was not developed as a commercial venture, largely due to the anti-GMO culture prevalent at the time. It is interesting to note that in 2021, a renewed interest in the technology started to develop as growers face increasing costs and challenges in preventing frost damage.

The bacterium *Bacillus thuringiensis* has been used to produce transgenic plants known as **Bt plants**. The bacterium produces toxic crystals that kill caterpillar pests when they ingest the toxin. The bacterium itself has been used as an insecticide, sprayed directly onto crops. However, the gene for toxin production has been isolated and inserted into plants such as corn, cotton, soybean and potato, with the

Bt plants, in which a bacterial toxin is used to confer resistance to caterpillar pests, have been successfully established and are grown commercially in many countries.

first Bt crops planted in 1996. By 2000, over half of the soybean crop in the USA was planted with Bt-engineered plants, although there have been some problems with pests such as the **western corn rootworm** developing resistance to the Bt toxin.

One concern that has been highlighted by the planting of Bt corn is the potential risk to non-target species. In 1999, a report in *Nature* suggested that larvae of the Monarch butterfly, widely distributed in North America, could be harmed by exposure to Bt corn pollen, even though the regulatory process involved in approving the Bt corn had examined this possibility and found no significant risk. Risk is associated with both toxicity and exposure, and subsequent research demonstrated that exposure levels are too low to pose a serious threat. The debate showed all the hallmarks of the polarising nature of GMOs, with the original paper criticised on its technical merits and validity. Not surprisingly, environmental groups and biotechnology industry had different views on the issue.

Herbicide resistance is one area towards which a lot of effort has been directed. The theory is simple – if plants can be made herbicide-resistant, then weeds can be treated with a broad-spectrum herbicide without the crop plant being affected. One of the most common herbicides is glyphosate, which is available commercially in a number of formulations. Glyphosate acts by inhibiting an amino acid biosynthetic enzyme called **5-enolpyruvylshikimate-3-phosphate synthase** (**EPSP synthase** or **EPSPS**). Resistant plants have been produced by either increasing the synthesis of EPSPS by incorporating extra copies of the gene, or by using a bacterial EPSPS gene that is slightly different from the plant version and produces a protein that is resistant to the effects of glyphosate. **Monsanto** (now part of **Bayer**) produced various crop plants, such as soya, that are called **Roundup-ready**, in that they are resistant to the herbicide. Such plants are now used widely in the USA and some other countries, and herbicide resistance is the most commonly manipulated trait in GM plants.

> Herbicide resistance is an area that has been exploited by plant biotechnologists to engineer plants that are resistant to common herbicides that are used to control weeds.

Controversy is again linked with glyphosate-resistant crops, with Bayer setting aside some $11 billion in 2020 (with an additional $4.5 billion added in 2021) to settle current and future litigation based on claims that glyphosate is linked to **non-Hodgkin lymphoma** (**NHL**). There is some debate as to whether this is the case, with the International Agency for Research on Cancer (IARC), associated with the **World Health Organization** (**WHO**) classifying glyphosate as carcinogenic in 2015. This led to accusations of bias and intense lobbying by Monsanto, often by questionable means, and counter-claims of industry-sponsored trials being used to support non-carcinogenic claims. The US Environmental Protection Agency (EPA), the European Chemicals Agency (ECHA) and the European Food Safety Authority (EFSA) do not classify glyphosate as carcinogenic (as of June 2022). However, in 2019 and 2020, two meta-analyses of previous studies did find a link between glyphosate and cancers in both humans and rodents. The whole story is frankly a mess, with a cycle of claims and counter-claims by pro and anti groups. It also

> The safety of glyphosate in terms of potential carcinogenic effects has generated a polarised debate, with conflicting interpretations of the evidence base.

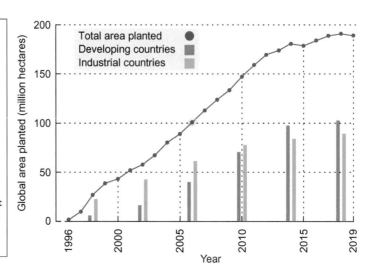

Fig. 16.5 Global area of genetically modified crops grown between 1996 and 2019. The total area of GM crops planted worldwide is shown for the period of 1996–2019. Areas are in millions of hectares (Mha). For 1998, 2002, 2006, 2010, 2014 and 2018, the balance between industrialised countries and developing countries is shown by bars. *Source:* Information from material published by ISAAA [www.isaaa.org]. Data reproduced without modification under Licence CC-BY-NC-ND [https://creativecommons.org/licenses/by-nc-nd/2.0/].

GM plants are now grown in some 29 countries, with over 190 million hectares being planted in 2019.

illustrates, not for the first time, the misuse of corporate influence by Monsanto (search for the 'Monsanto Papers' for further information). The science of evaluating the carcinogenic potential of a substance in humans is difficult enough, without adding additional confusion; clearly there is a need for continued independent evaluation of the research in this area. The concerns around glyphosate have begun to have an impact on the availability of glyphosate for home garden use, with alternatives being marketed as 'glyphosate-free'.

Since the first GM crops were planted commercially, there has been a steady growth in the area of GM plantings worldwide. Fig. 16.5 shows the total area of GM crops planted from 1996 to 2019. Two points are of interest here, in addition to the overall growth in area year-on-year. Firstly, the symbolically significant 100 million hectare (ha) barrier was broken in 2006. Secondly, developing countries are embracing GM crop plants and have now surpassed industrialised countries. This is also evident from Table 16.4, which shows data for areas planted in 2019 for each of the 19 countries growing more than 100 000 ha of GM crops commercially. The four main GM crops are soybean, maize, cotton and canola (oilseed rape). Fig. 16.6 shows areas planted with GM varieties for each of these crops in 2019.

In addition to the 'big four' GM crops, other plant species have of course been genetically modified, with one example being the tomato. One approach to delaying the ripening process illustrates how a novel idea, utilising advanced gene technology to achieve an elegant solution to a defined problem, can still fail due to other considerations – this is the story of the **Flavr Savr** (*sic*) tomato.

The biotechnology company **Calgene** developed the Flavr Savr tomato using what became known as **antisense technology**. In this approach, a gene sequence is inserted in the opposite orientation, so that on transcription, an mRNA that is complementary to the normal mRNA is produced. This antisense mRNA will therefore bind to the

Table 16.4 | Global area of GM crops planted in 2019, by country

Rank	Country	Area (10^6 ha)	Crops	Percentage of total (%)	
1	USA	71.5	Maize, soybean, cotton, alfalfa, canola (oilseed rape), sugar beet, potato, papaya, squash, apple	37.57	
2	Brazil	52.8	Soybean, maize, cotton, sugar cane	27.75	
3	Argentina	24.0	Soybean, maize, cotton, alfalfa	12.61	90.75
4	Canada	12.5	Canola (oilseed rape), soybean, maize, sugar beet, alfalfa, potato	6.57	
5	India	11.9	Cotton	6.25	
6	Paraguay	4.1	Soybean, maize, cotton	2.15	
7	China	3.2	Cotton, papaya	1.68	
8	South Africa	2.7	Maize, soybean, cotton	1.42	
9	Pakistan	2.5	Cotton	1.31	7.93
10	Bolivia	1.4	Soybean	0.74	
11	Uruguay	1.2	Soybean, maize	0.63	
12	Philippines	0.9	Maize	0.47	
13	Australia	0.6	Cotton, canola (oilseed rape), safflower (sunflower family)	0.32	
14	Myanmar	0.3	Cotton	0.16	
15	Sudan	0.2	Cotton	0.11	1.31
16	Mexico	0.2	Cotton	0.11	
17	Spain	0.1	Maize	0.05	
18	Colombia	0.1	Maize, cotton	0.05	
19	Vietnam	0.1	Maize	0.05	
	Total	**190.4**		100	100

Note: The top five countries accounted for more than 90 per cent of the total area planted. The subdivisions show countries planting more than 10, 1 and 0.1 million ha, respectively.

Source: Information from material published by ISAAA [www.isaaa.org]. Data reproduced without modification under Licence CC-BY-NC-ND [https://creativecommons.org/licenses/by-nc-nd/2.0/].

normal mRNA in the cell, inhibiting its translation and effectively shutting off expression of the gene. The principle of the method is shown in Fig. 16.7. In the Flavr Savr, the gene for the enzyme poly-galacturonase (PG) was the target. This enzyme digests pectin in the cell wall, and leads to fruit softening and the onset of rotting. The elegant theory is that inhibition of PG production should slow the decay process; thus, the fruit should be easier to handle and transport after picking. It can also be left on the vine to mature longer than is usually the case, thus improving flavour. After much development, the Flavr Savr became the first GM food to be approved for use in the USA in 1994. The level of PG was reduced to something like 1 per cent of the normal levels, and the product seemed to be set for commercial success. However, various problems with the characteristics affecting growth and picking of the crop led to the failure of the Flavr Savr in commercial terms. Calgene became part of Monsanto (itself later

Sometimes an elegant and well-designed scientific solution to a particular problem does not guarantee commercial success for the produce, with many factors impacting on the success or failure of a GM product.

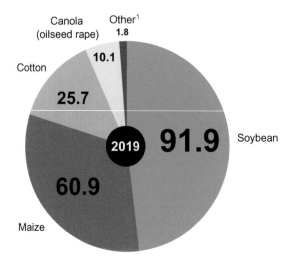

Fig. 16.6 Global areas of the main GM crops grown in 2019. The areas of the main GM crops planted worldwide in 2019 are shown. The four main crops were soybean, maize, cotton and canola (oilseed rape). Numbers are areas in millions of hectares (Mha). [1] The crops in this segment were sugar beet, potato, apple, squash, papaya and eggplant (aubergine or brinjal). *Source:* Information from material published by ISAAA [www.isaaa.org]. Data reproduced without modification under Licence CC-BY-NC-ND [https://creativecommons.org/licenses/by-nc-nd/2.0/].

Fig. 16.7 Antisense technology. (a) The gene to be targeted for antisense regulation is transcribed to produce 'sense' (normal) mRNA. (b) A copy of the gene is introduced into a separate site on the genome, but in the opposite orientation. On transcription, this produces the 'antisense' mRNA. (c) The sense and antisense mRNAs hybridise, preventing translation of the gene transcript and therefore reducing expression of the gene. Part of the sequence is shown to illustrate the complementary mRNAs.

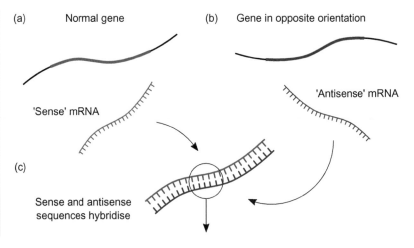

Sense 5'- GCUCAUGGCUACCAAGGUAGCUAUUUCU -3'
Antisense 3'- CGAGUACCGAUGGUUCCAUCGAUAAAGA -5'

acquired by Bayer), having been stretched too far by the development of the Flavr Savr.

Attempts to improve the nutritional quality of crops are not restricted to the commercial or health food sectors. For many millions of people around the world, access to basic nutrition is a matter of

survival rather than choice. Rice is the staple food of some 3.5 billion people, and about 10 per cent of these suffer from health problems associated with **vitamin A deficiency (VAD)**, as rice does not contain β-carotene (provitamin A). Around a third of children under 5 are affected by VAD, with blindness and premature death being the most serious consequences. The WHO estimates that between 250 000 and 500 000 children with VAD become blind each year, with half of them dying within 12 months.

Rice is of such importance as a staple food that it is an obvious target for GM technologists.

Rice has been one of the most intensively studied crop species, with a long history of development. The development of 'Miracle Rice' as a product of the green revolution of the 1960s was a significant step forward in improving yield, but widespread planting throughout South East Asia led to a rice monoculture, with increased susceptibility to disease and pests, and the increased dependence on pesticides that this brings. In 1999, **Ingo Potrykus** and **Peter Beyer** succeeded in producing 'Golden Rice' (GR) (Fig. 16.8) with β-carotene in the grain endosperm, where it is not normally found. Subsequently an improved strain was developed, called Golden Rice 2, with higher levels of β-carotene than the original GR. As β-carotene is a precursor of vitamin A, increasing the amount available by engineering rice in this way should help to alleviate some of the problems of VAD. This is obviously a positive development. However, corporate interests in patent rights to the technologies involved, and other non-scientific problems, had to be sorted out before agreement was reached that developing countries could access the technology freely. Opposition to

The development of 'Golden Rice' is an example of good science and good intent, although, as often is the case, the political agenda may have a role to play in determining how benefits are shared and distributed.

GR from anti-GMO groups, including **Greenpeace**, slowed the approval process, and despite no evidence of any adverse effects, many regulators were reluctant to approve GR. Although, in 2018, some countries had approved the rice as safe for *consumption*, it was only in mid-2021 that GR was approved for *cultivation* in the Philippines, a major step forward more than 20 years after it was first developed. Regulatory and legal wrangles continue, and in a now familiar pattern, it is once again clear that it is difficult to reconcile the competing views of the pro- and anti-GMO camps. As is always the case, careful and valid appraisal of the evidence is required to enable informed decisions to be made. Unfortunately, there is often strong and emotive language used, particularly by those antagonistic to the technology, that is not helpful. Most who have a balanced view of GR accept that it has the potential to alleviate some of the effects of VAD; many feel strongly that not using it is contributing to the avoidable deaths of children, and some even consider this as a crime against humanity, to be laid at the doors of the anti-GMO movement.

Although perhaps having less potential for a significant impact on human health than GR, the **Arctic Apple** came to market in late-2017 in the USA and Canada. The apples have been engineered to down-regulate the expression of the gene for **polyphenol oxidase (PPO)** by gene silencing using RNAi technology (see Section 15.3 for the mechanism). PPO is the enzyme that causes browning of fruits when sliced or bitten; thus, Arctic Apples can be sold as slices without preservatives and last longer than non-silenced apples. Whether this comes to be seen by the consumer as a 'must purchase' remains to be seen, but the harvest has increased each year as the area grown increases and the trees mature, so from the production point of view, this is a successful product-to-market example.

The development and implementation of protection technologies is a highly controversial and emotive area of plant genetic manipulation, with strong opinions on both sides of the argument.

A further twist in the corporate *vs.* common good debate can be seen in the so-called **gene protection technology**. This is where companies design their systems so that their use can be controlled, by some sort of manipulation that is essentially separate from the actual transgenic technology that they are designed to deliver. This caused such a level of concern among a broad spectrum of groups that the technology has not been fully commercialised. The websearch suggestion for this chapter gives you a chance to explore this area more fully.

In addition to GM crops, transgenic plants also have the potential to make a significant impact on the biotechnology of therapeutic protein production. As we saw in Chapter 14, bacteria, yeast and mammalian cells are commonly used for the production of high-value proteins by recombinant DNA technology. An attractive option is to use transgenic plants to produce such proteins. Plants are cheap to grow, compared to the high-cost requirements of microbial or mammalian cells, and this cost reduction and potentially unlimited scale-up make transgenic plants a viable alternative to other methods. The term **plant-made pharmaceuticals (PMPs)** has been coined to describe this aspect of transgenic plant technology.

The use of plants as 'factories' for the synthesis of therapeutic proteins is an area that is currently being developed to enable the production of plant-made pharmaceuticals (PMPs).

We have yet another example of conflicting views about the use of a successful technology. The Canadian company **Medicago** developed a vaccine effective against COVID-19, producing it in plant cells as **virus-like particles** (**VLPs**). It was the first plant-derived vaccine to be authorised for use by Health Canada in early 2022. However, as Medicago is partly financed by a tobacco company, which violates WHO policy, the WHO application is not being considered (as of June 2022) and was therefore withdrawn. This raises a number of interesting ethical issues that are not related to the efficacy or safety of the vaccine, but to policy, politics, corporate and financial issues.

16.2.4 Have Transgenic Plants Delivered or Disappointed?

The answer to this question is 'yes' – transgenic plants have both delivered and disappointed over the past 25 years. The science has been novel and impressive, with many notable developments. The area of GM crops now being cultivated has grown steadily since the first plantings, even if dominated by the four most prevalent crops (Figs. 16.5 and 16.6). There are also many novel modified plants that have been constructed in laboratories worldwide that could be developed into viable crop plants to address the targets outlined in Table 16.2.

The main problem with GM crops has been around their approval and deployment. Anti-GMO campaigning over a protracted period has been very successful in generating a strong public resistance to GMOs, despite the evidence from hundreds of original studies and a growing number of meta-analyses that they pose no additional risk compared with non-GM counterparts. In this regard, GM crops have disappointed – not because of bad science, but because of bad publicity. We need to fix this to enable transgenic and gene-edited plants to play a significant role in addressing food security for a growing world population in a changing climate.

The frustration at the delay in enabling GR to be grown by farmers is perhaps best summed up by quoting the words of others. In 2016, **Sir Richard Roberts** (Nobel Prize, 1993) organised a campaign to counteract the misinformation that was being spread around GMOs in general, and GR in particular. The initial letter was supported by over 100 Nobel Laureates (now listed as 159 on the website at www.supportprecisionagriculture.org/). It is worth reading. In a later article, there is one phrase that puts the opposition to GMOs into context. Roberts states: 'It is hard for me to understand how anyone could be opposed to producing a food plant with such obvious beneficial pharmaceutical effects as preventing childhood blindness.'

Opposition to Golden Rice is considered by most to be completely unjustified and not based on an objective assessment of the science.

In 2019, **Michael Le Page**, writing in *New Scientist* on 20 November (just before GR was approved as safe by regulators in the Philippines), stated: 'What shocks me is that some activists continue to misrepresent the truth about the rice. The cynic in me expects profit-driven multinationals to behave unethically, but I want to think that those voluntarily campaigning on issues they care about have higher standards.'

16.3 | Transgenic Animals

The debate around the use of GM technology becomes even more far-reaching when transgenic animals are concerned, and often raises animal welfare issues as well as the genetic modification aspects of the technology.

The generation of transgenic animals is one of the most complex aspects of genetic engineering, in terms of both technical difficulty and the ethical problems that arise. Many people, who accept that genetic manipulation of bacterial, fungal and plant species is beneficial, find difficulty in extending this acceptance when animals (particularly mammals) are involved.

16.3.1 Why Transgenic Animals?

The term transgenic is mostly used to describe whole organisms that have been modified to contain transgenes in a stable form that are inherited by transmission through the germ line.

Cell-based applications, such as the production of proteins in cultured mammalian cells and studies on the molecular genetics of diseases like cancer, are an important part of genetic engineering in animals. However, the term transgenic is usually reserved for whole organisms, and the generation of a transgenic animal is much more complex than working with cultured cells. Many of the problems have been overcome using a variety of animals, with early work involving amphibians, fish, mice, pigs and sheep.

Transgenics can be used for a variety of purposes, covering both research and biotechnological applications (Table 16.5). The study of embryological development has been extended by the ability to introduce genes into eggs or early embryos, and there is scope for the

Table 16.5 | Actual and potential applications of transgenic animal technology

Area	Application
Fundamental research	Investigation of how genes function *in vivo* Investigating biochemical pathways and physiological processes
Medicine	Transgenic models to study how diseases are caused and regulated at the molecular level Transgenics developed for testing potential therapeutic products Production of therapeutics such as monoclonal antibodies (mAbs) for cancer treatments and therapeutic proteins in the milk of sheep or goats Use of the silkworm as a bioreactor Alteration of the immunogenic properties of potential organ donor species for xenotransplantation procedures Potential use of gene drives to control insect-borne diseases such as malaria
Food and agriculture	Improvement in reproduction and fecundity Modifications to growth rates and body composition to enhance nutritional qualities AquAdvantage salmon (modified growth hormone) was the first GM animal to be approved for sale as a human food
Production of fibres	Modifications to animals such as sheep and goats to improve the qualities of their wool and hair fibres Modification of silkworms to enhance silk production using the spider silk gene in a targeted gene replacement
Household pets	GloFish

manipulation of farm animals by the incorporation of desirable traits *via* transgenesis. The use of whole organisms for the production of recombinant protein is a further possibility, and this has already been achieved in some species. The term pharm animal, or **pharming** (from *pharm*aceutical), is sometimes used when talking about the production of high-value therapeutic proteins using transgenic animal technology. Engineering pigs to avoid immunorejection following xenotransplantation (see Section 15.1.7) is another area that could potentially address the issue of lack of donor organs for transplant surgery.

16.3.2 Producing Transgenic Animals

There are several possible routes for the introduction of genes into embryos, each with its own advantages and disadvantages. Some of the methods are: (1) direct transfection or retroviral infection of embryonic stem cells followed by introduction of these cells into an embryo at the blastocyst stage of development, (2) retroviral infection of early embryos, (3) direct microinjection of DNA into oocytes, zygotes or early embryo cells, (4) sperm-mediated transfer, (5) transfer into unfertilised ova, and (6) physical techniques such as biolistics or electrofusion. In addition to these methods, the technique of nuclear transfer (used in organismal cloning, discussed in Chapter 17) is sometimes associated with a transgenesis protocol.

Early success was achieved by injecting DNA into one of the pronuclei of a fertilised egg, just prior to the fusion of the pronuclei (which produces the diploid zygote). This approach led to the production of the celebrated '**supermouse**' in the early-1980s, which represents one of the milestones of genetic engineering. Generating supermouse involved placing a copy of the rat growth hormone (GH) gene under the control of the mouse metallothionine (mMT) gene promoter. A linear fragment of the recombinant plasmid carrying the fused gene sequences (MGH) was injected into the male pronuclei of fertilised eggs (linear fragments appear to integrate into the genome more readily than circular sequences). The resulting fertilised eggs were implanted into the uteri of foster mothers, and some of the mice resulting from this expressed the GH gene. Such mice grew some 2–3 times faster than control mice, and were up to twice the size of the controls. Pronuclear microinjection is summarised in Fig. 16.9.

In generating a transgenic animal, it is desirable that all the cells in the organism receive the transgene. Presence of the transgene in the germ cells of the organism will enable the gene to be passed on to succeeding generations, and this is essential if the organism is to be useful in the long term. Thus, introduction of genes has to be carried out at a very early stage of development, ideally at the single-cell zygote stage. If this cannot be achieved, there is the possibility that a mosaic embryo will develop, in which only some of the cells carry the transgene. Another example of this type of variation is where the embryo is generated from two distinct individuals, as is the case when embryonic stem cells are used. This results in a **chimeric** organism. In

The production of 'supermouse' in the early-1980s represents one of the milestone achievements in genetic engineering.

Mosaics and chimeras, where not all the cells of the animal carry the transgene, are variations that can arise when producing transgenic animals.

Fig. 16.9 Production of 'supermouse'. (a) Fertilised eggs were removed from a female mouse. (b) DNA carrying the rat growth hormone gene/mouse metallothionein promoter construct (MGH) was injected into the male pronucleus. (c) The eggs were then implanted into a foster mother and brought to term. (d) Some of the pups expressed the MGH construct (MGH⁺), and were larger than the normal pups.

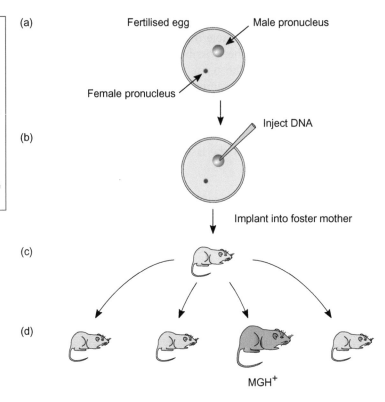

(a) Fertilised egg | Male pronucleus

Female pronucleus

(b) Inject DNA

Implant into foster mother

(c)

(d)

MGH⁺

practice, this is not necessarily a problem, as the organism can be crossed to produce offspring that are homozygous for the transgene in all cells. A chimeric organism that contains the transgene in its germ line cells will pass the gene on to its offspring, which will therefore be heterozygous for the transgene (assuming they have come from a mating with a homozygous non-transgenic). A further cross with a sibling will result in around 25 per cent of the offspring being homozygous for the transgene. This procedure is outlined for the mouse in Fig. 16.10.

16.3.3 Applications of Transgenic Animal Technology

Introduction of growth hormone genes into animal species has been carried out, notably in pigs, but in many cases, there are undesirable side-effects. Pigs with the bovine growth hormone gene show greater feed efficiency and have lower levels of subcutaneous fat than normal pigs. However, problems such as an enlarged heart, a high incidence of stomach ulcers, dermatitis, kidney disease and arthritis have demonstrated that the production of healthy transgenic farm animals is a difficult undertaking.

One concept-to-market success has been the development of transgenic salmon with an increased growth rate. The modification was made in 1989, when a growth hormone gene from a **Chinook salmon** (*Oncorhynchus tshawytscha*), a species of Pacific salmon, was inserted

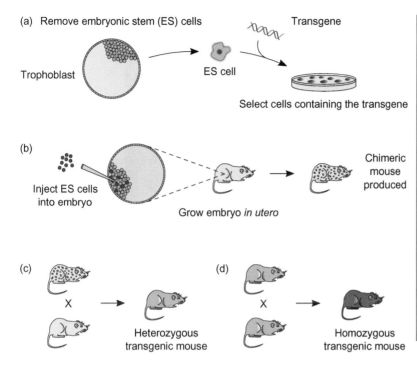

(a) Remove embryonic stem (ES) cells

Trophoblast

ES cell

Transgene

Select cells containing the transgene

(b)

Inject ES cells into embryo

Grow embryo *in utero*

Chimeric mouse produced

(c)

X

Heterozygous transgenic mouse

(d)

X

Homozygous transgenic mouse

Fig. 16.10 Production of transgenic mice using embryonic stem cell technology. (a) Embryonic stem (ES) cells are removed from an early embryo and cultured. The target transgene is inserted into the ES cells, which are grown on selective media. (b) The ES cells containing the transgene are injected into the ES cells of another embryo, where they are incorporated into the cell mass. The embryo is implanted into a pregnant mother, and a chimeric transgenic mouse is produced. (c) By crossing the chimera with a normal mouse, some heterozygous transgenics will be produced. If they are then self-crossed, homozygous transgenics will be produced in about 25 per cent of the offspring, as shown in (d).

into the **Atlantic salmon** (*Salmo salar*) under the control of a promoter from the **Ocean Pout** (*Zoarces americanus*). The modification enhances the early-stage growth rate and means that the salmon reach their final size in around 18 months instead of 3 years. Commercial development of the fish, named **AquAdvantage**, was by the company **AquaBounty**. In a now familiar trend, regulatory approval was slowed by environmental and health concerns, but the US Food and Drug Administration (FDA) approved the fish as safe to eat in 2010 and for sale in 2015. The AquAdvantage salmon was the first transgenic animal to be sold as human food, with the first fish sold in Canada in 2017 and in the USA in 2021. As with all of these approvals, it is interesting to note how the different views are portrayed during the process; antagonistically is often the appropriate word.

The study of development is one area of transgenic research that is currently yielding much useful information. By implanting genes into embryos, features of development such as tissue-specific gene expression can be investigated. The cloning of genes from the fruit fly *Drosophila melanogaster*, coupled with the isolation and characterisation of **transposable elements** (P elements) that can be used as vectors, has enabled the production of stable transgenic *Drosophila* lines. Thus, the fruit fly, which has been a major contributor to the field of classical genetic analysis, is now being studied at the molecular level by employing the full range of gene manipulation techniques.

In mammals, the mouse is the most useful model system for investigating embryological development, and the expression of many

A modified salmon was the first GM animal to be sold for human consumption.

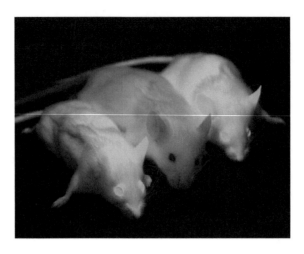

Fig. 16.11 Expression of a GFP transgene in mice. The gene for enhanced green fluorescent protein (eGFP) was inserted into NOD/SCID mice. When illuminated with ultraviolet light, the GFP transgenic mice show green fluorescence. In this photograph, two GFP mice are shown, with a non-GFP transgenic from the parental strain in the middle. *Source:* From Moen, I. *et al.* (2012). *BMC Cancer*, 12, 21. DOI [htpps://doi:10.1186/1471-2407-12-21]. Used under Licence CC-BY-2.0 [https://creativecommons.org/licenses/by/2.0/].

The mouse has proven to be one of the most useful model organisms for transgenesis research, with many different variants having been produced for a variety of purposes.

transgenes has been studied in this organism. One often-seen photograph of transgenic mice is shown in Fig. 16.11. Mice have also been used widely as animal models for disease states. One celebrated example is the oncomouse, generated by **Philip Leder** and his colleagues at Harvard University. Mice were produced in which the *c-myc* **oncogene** and sections of the **mouse mammary tumour (MMT) virus** gave rise to breast cancer. The oncomouse has a place in history as the first complex animal to be granted a patent in the USA. Other transgenic mice with disease characteristics include the prostate mouse (prostate cancer), mice with severe combined immunodeficiency syndrome (SCIDS), and mice that show symptoms of Alzheimer disease.

Many new variants of transgenic mice have been produced, and have become an essential part of research into human disease. Increased knowledge of molecular genetics and the continued development of the techniques of transgenic animal production have enabled mice to be generated in which specific genes can be either activated or inactivated. As outlined in Section 12.1, where a gene is inactivated or replaced with a mutated version, a knockout mouse is produced. If an additional gene function is established, this is sometimes called a knockin mouse. The use of knockout mice in cystic fibrosis (CF) research is one example of the technology being used both in basic research and in developing gene therapy procedures. Mice have been engineered to express mutant CF alleles, including the prevalent **ΔF508 mutation** that is responsible for most serious CF presentations. Having a mouse model enables researchers to carry out experiments that would not be possible in humans, although (as with developmental studies) results may not be exactly the same as would be the case in a human subject.

An early example of protein production in transgenic animals was the expression of human **tissue plasminogen activator (TPA)** in transgenic mice in 1987. The transgene protein product was secreted into the milk of lactating mice. Producing a therapeutic protein in milk provides an ideal way of ensuring a reliable supply from lactating animals, and downstream processing to obtain purified protein is relatively straightforward. This approach was used by scientists from the Roslin Institute near Edinburgh, working in conjunction with the biotechnology company **PPL Therapeutics**. In 1990, a transgenic sheep called *Tracy* was born, with the gene for the protease inhibitor α_1-antitrypsin. Yields of human proteins of around $40\,\mathrm{g\,L^{-1}}$ were produced from milk, demonstrating the great potential for this technology, and in 1997, the team was able to produce human **blood coagulation** factor IX (FIX) in transgenic sheep. However, PPL Therapeutics was unable to develop the technology on a commercial basis and was wound up in 2003. The first drug derived from transgenic goat milk to be approved for therapeutic use was an antithrombin anticoagulant named ATryn, approved by the European Medicines Agency (EMA) in 2006, and by the FDA in 2009. The product was voluntarily withdrawn by the European producer in 2018 for commercial reasons – again demonstrating that the scientific challenges are not the only ones that affect the success of a particular product.

> The use of transgenic animals as producers of high-value products is more fully developed than is the case for transgenic plants, although there are often scientific and commercial problems in making a success of the technology.

Despite slow progress to approval and market, using transgenic animals as bioreactors continues to be developed and on a technical level is now well established, with many biotechnology companies involved. If approval and commercial viability issues can be overcome, the technology will undoubtedly deliver significant benefits on a much broader scale than has been the case to date.

Our final word on transgenic animals brings the technology closer to humans. In January 2001, the birth of the first transgenic **non-human primate** (NHP), a rhesus monkey, was announced. This was developed by scientists in Portland, Oregon. He was named ANDi, which stands for inserted DNA (written backwards!). A marker gene from jellyfish, which produces green fluorescent protein (GFP) (Fig. 16.11) was used to confirm the transgenic status of a variety of ANDi's cells (although he did not glow green like the transgenic mice in Fig. 16.11). ANDi was significant in that he paved the way for the development of NHP models for diseases like Huntington disease (HD) and other complex neurological conditions that are difficult to study in animals such as transgenic mice.

> Transgenic non-human primates are being developed as models for complex human diseases; as might be expected, this area raises some difficult ethical questions.

16.4 | Future Trends

Transgenic technology is an area with significant challenges in scientific, commercial and societal contexts. As we have seen, the science is in fact often the simplest part of the process of delivering the benefits of a transgenic plant or animal to the intended end-user. Trying to predict what the next 20 years will bring is a bit of a guess, but there are some pointers that we can use. Let's look at two of them.

16.4.1 Transgenesis or Genome Editing?

As we discussed in Chapter 12, the emergence of **genome editing** and its widespread adoption has for many applications changed the way that gene manipulation is achieved. The use of RNA interference (RNAi) has also provided an option for some experiments. As making a transgenic is a complex process, if gene editing or RNAi can produce the same outcome, these are likely to be the method of choice. For knockout or knockdown procedures, the CRISPR-based editing systems may make traditional mouse transgenesis obsolete in the near future. Genome editing is less likely to replace transgenesis fully when the aim is to insert an entire intact gene from a different organism, as editing large sections of DNA is still a tricky procedure, although technical advances are being made at an impressive rate.

As we noted at the start of this chapter, there are other variants of transgenesis. We have just considered the *intragenic* modifications that genome editing achieves, but there is also the *cisgenic* approach, using target genes from closely related species that could be crossed to generate the same effect. In these cases, the advantage in using a gene manipulation procedure is that it can greatly reduce, or even remove the need for, the crossing and back-crossing that is required to get the intended change into a near-isogenic background. Where the generation time of the target is lengthy, this can be a significant advantage.

One current area of debate is how gene-edited organisms should be classified and regulated. Many scientists argue that because the majority of changes are intragenic, these are not transgenic organisms and therefore should not be classed as such, with less stringent regulation. Others, particularly in the environmental and health-focused camp, are adamant that any method that causes deliberate changes to the genome needs to be regulated at least as closely as transgenics. This area will probably take several years to settle, and may involve different approaches from different countries.

A final variant of the 'genic' approach is what is sometimes called a **xenogenic** change, where the target gene is synthesised in the laboratory. With the availability of genome sequence databases and the tools of bioinformatics and synthetic biology, this approach is likely to become a significant addition to the field over the next few years.

16.4.2 Gene Drives

Our final look into the future again takes us into the realms of a controversial subject, that of the **gene drive**. The concept is not new, as gene drives occur naturally as a function of **allelic fitness** in the range of genotypes within a population, and the potential of gene drives for altering allele frequencies was considered in the early-1960s. Although the topic is complex, with population genetics,

Gene editing based on the CRISPR-Cas9 system will have a major impact on how genome modifications are achieved.

mathematical modelling and ecological considerations all contributing to the investigation of the process, we can summarise the key points in fairly simple terms.

On balance, a normal Mendelian inheritance pattern for a gene with two alleles will segregate and transmit each at close to 50 per cent frequency. Small variations in this tend to even out over time, unless there is a 'drive' that pushes the transmission away from the 50:50 balance. Adopting this concept to cause a deliberate skewing of the allele transmission pattern can change the allelic frequency significantly over a relatively short period (depending on the generation time – insects breed much faster than elephants). This can be a positive change, where the allele is transmitted preferentially or can result from alternative genotypes being less fit than the version containing the target allele. In essence, this can enable an engineered allele to be 'driven' through a target population much faster than would normally be the case, if the allelic fitness can be adjusted to achieve this.

The main focus for artificial gene drives at present is on the possibility of controlling malaria by targeting the mosquitoes (*Anopheles* genus) that carry the trypanosome (usually one of six species from the *Plasmodium* genus, with **P. falciparum** being the most common). The aim is to either **suppress** or **modify** the mosquito population to prevent transmission of the disease agent. Suppression can be achieved in a number of ways, with a 2021 study showing that a trait to spread female infertility in a population could achieve this effectively in a large-cage laboratory situation. Currently the active technology has only been laboratory-tested, but initial field trials have been carried out and the possibility of releasing active control traits is within sight.

The gene drive represents a significant departure from the use of transgenic technology to date (although it is not itself a transgenic-based procedure, as the target allelic variants are mostly generated by gene editing). In most transgenic applications, the aim is to ensure that the transgene does *not* cause unintended negative effects; in a suppression gene drive, the aim is to deliberately cause an intended negative effect. For this reason, a precautionary approach is even more important than for standard transgenic applications. Gene drives have great potential, but there is also some caution at present, which is a sensible way to proceed.

> Altering the allele frequency of a target gene offers a way to control disease transmission by insect populations, with mosquito-borne malaria as the current focus. Deliberate release of gene-drive mosquitoes with a negative intended impact is, however, a contentious issue and requires careful evaluation.

16.5 | Conclusion: Changing Genomes and Attitudes

In this chapter, we have looked at ways to change the genome by the incorporation of genes from different sources. Although *transgenic* is the term that most people would recognise (along with *GMO*), there are also subtleties in that *intragenic*, *cisgenic* and *xenogenic* approaches can be used to alter genomes. Transgenic plants and animals can be constructed by a variety of methods, including techniques such as *vector-mediated transfer*, *biolistics*, *microinjection* and **embryonic stem cell technology**. The aim of transgenesis is to change the characteristics of a crop or farm animal, investigate gene expression and the mechanisms of underlying disease and produce useful proteins for use in therapy or biotechnology processes.

The science of transgenesis is difficult, but great success has been achieved across a number of applications. The generation of **transgenic mouse models** for a range of diseases is a critical part of medical research, and production of therapeutic *monoclonal antibodies* in mice has been a particularly successful use of the technology. The generation of *transgenic non-human primates* holds great promise for modelling complex neurological diseases, but also raises additional ethical concerns about using animals in this way.

In plants, GM crops have now been grown for over 25 years, with *herbicide-* and *insect-resistant* varieties being the most common. **Nutritional biofortification** has been achieved with *Golden Rice*, which could have a significant impact on alleviating *vitamin A deficiency (VAD)* – in particular, the prevention of blindness in children which often leads to death.

The paradox that GMOs illustrate is that although the science has been achieved, the deployment has been hampered by anti-GMO organisations and pressure groups such as *Greenpeace*. Antagonists often cite 'big biotech' as solely profit-motivated and responsible for causing potentially irreversible damage to human health and the environment. However, many anti-GMO groups have a multinational profile and significant levels of funding, and use exactly the same tactics that they claim the biotech industry uses – arguably with much more disinformation and anti-science propaganda, as well as direct illegal action in many cases. The fact that over 150 Nobel Laureates finally lost patience and countered the disinformation used to stir up anti-GMO sentiment is in some ways a backhanded compliment to those groups who have been successful in their campaigns.

No responsible scientist disputes the need for careful evaluation of any potential risks associated with transgenic or gene-edited GMOs, with appropriate control measures put in place where required. Whilst some biotech companies have undoubtedly been irresponsible (and in some cases negligent) in aspects of their business, the same applies to some anti-GMO groups. However, there is hope; many on both sides of the debate are not so polarised. A sensible and considered middle way is needed if we are to benefit from the technology, which will be essential if we are to address population and climate challenges in the future. Changing attitudes as well as genomes will be required.

Further Reading

Abrahamian, P. *et al.* (2020). Plant virus-derived vectors: applications in agricultural and medical biotechnology. *Ann. Rev. Virol.*, 7, 513–35. URL [www.annualreviews.org/doi/10.1146/annurev-virology-010720-054958#]. DOI [https://doi.org/10.1146/annurev-virology-010720-054958].

Bier, E. (2022). Gene drives gaining speed. *Nat. Rev. Genet.*, 23, 5–22. URL [www.nature.com/articles/s41576-021-00386-0#citeas]. DOI [https://doi.org/10.1038/s41576-021-00386-0].

Burgio, G. (2018). Redefining mouse transgenesis with CRISPR/Cas9 genome editing technology. *Genome Biol.*, 19, 27. URL [https://genomebiology.biomedcentral.com/articles/10.1186/s13059-018-1409-1]. DOI [https://doi.org/10.1186/s13059-018-1409-1].

Hammond, A. *et al.* (2021). Gene-drive suppression of mosquito populations in large cages as a bridge between lab and field. *Nat. Commun.*, 12, 4589. URL [www.nature.com/articles/s41467-021-24790-6#citeas]. DOI [https://doi.org/10.1038/s41467-021-24790-6].

Montagu, M. V. (2019). The future of plant biotechnology in a globalized and environmentally endangered world. *Genet. Mol. Biol.*, 43 (1 suppl. 2). URL [www.ncbi.nlm.nih.gov/pmc/articles/PMC7216575/]. DOI [https://doi.org/10.1590/1678-4685-GMB-2019-0040].

Roberts, R. J. (2018). The Nobel Laureates' campaign supporting GMOs. *Journal of Innovation & Knowledge*, 3 (2), 61–5. URL [www.sciencedirect.com/science/article/pii/S2444569X18300064]. DOI [https://doi.org/10.1016/j.jik.2017.12.006].

Snyder, B. R. and Chan, A. W. S. (2018). Progress in developing transgenic monkey model for Huntington's disease. *J. Neural Transm.*, 125 (3), 401–17. URL [www.ncbi.nlm.nih.gov/pmc/articles/PMC5826848/]. DOI [https://doi.org/10.1007/s00702-017-1803-y].

Wu, F. *et al.* (2021). Allow Golden Rice to save lives. *Proc. Natl. Acad. Sci. U. S. A.*, 118 (51). URL [www.pnas.org/doi/full/10.1073/pnas.2120901118]. DOI [https://doi.org/10.1073/pnas.2120901118].

Websearch

People

Two for this chapter, with different challenges in their respective careers. Investigate the contribution that *Ingo Potrykus* has made to the development of Golden Rice, and contrast this with the controversy that arose from some of the work of *Arpad Pusztai*.

Places

With headquarters in *Los Baños*, in the *Philippines*, the *International Rice Research Institute* coordinates a number of offices and projects across Asia and Africa, and collaborates with other organisations to further its aims. Have a look at how the organisation is structured, its aims and projects, and the impact that the work has had.

Processes

Find some information about *gene protection technology*, and the controversy around its proposed use. You will find information under different terms such as *suicide seeds*, traitor technology and genetic use restriction technology (GURT).

Reflections

What topics in this chapter have you found most challenging? Look for resources that help to illustrate the key points.

Transgenic plants and animals

Transgenic plants and animals — involves → insertion of a transgene into the genome

generation of → transgenic plants

for:
- research
- resistance traits
- tolerance traits
- nitrogen fixation
- improved nutrition
- biofortification
- vaccines

typical procedure: insert the transgene ▸ select transformed cells ▸ grow callus tissue and embryo ▸ grow to a mature plant

e.g.:
- ice-minus bacteria — for — frost protection
- glyphosate resistance — to — control weeds
- Bt plants — to — control insect pests

are examples of modification of crop plants for yield improvement by decreasing losses

Golden Rice — made by — adding two genes to synthesise β-carotene in the endosperm — to — improve nutritional quality by biofortification — to help — alleviate the symptoms of vitamin A deficiency — but still generates — strong opposition from anti-GMO groups

regulatory approval for:
- food safety
- environment
- field trials

if successful → approval for intended use

insertion of a transgene into the genome — using:

vector-mediated methods e.g. Ti plasmids, plant viruses, adenovirus, AAV, HSV, lentivirus

physical or chemical methods e.g. transfection, transformation, microinjection, electroporation, biolistics, liposomes

to produce a permanent and transmissible change in the genome

can also be achieved by gene editing or silencing techniques

plants / animals — generates very strong reactions

polarised perspectives

pro-GMO cite	anti-GMO cite
benefits for food security, new medical treatments, and sustainable agriculture and aquaculture	risks to human health and the environment, and biotech greed, question need for GMOs cf. organics

generation of → transgenic animals

for:
- research
- disease models
- mAb production
- therapeutics
- improved nutrition
- sustainability

typical procedure: isolate embryonic stem cells ▸ insert transgene ▸ select transformed cells ▸ inject into embryo ▸ embryo develops to term ▸ chimeric mouse ▸ cross to get heterozygote ▸ cross to get homozygote

→ making a transgenic animal is more complex than making a plant

- mice — supermouse, oncomouse; mAbs using hybridoma method
- sheep — potential as bioreactors for, e.g. mAb production
- goats
- pigs — immuno-suppressed to enable xenograft transplants
- non-human primates (NHPs) — first = ANDi; possible models for complex diseases in humans

→ use of NHPs raises additional ethical issues

Chapter 17 Summary

Learning Objectives

When you have completed this chapter, you will be able to:

- Define the terms associated with organismal cloning, including reproductive and therapeutic cloning
- Outline the history, development, range and scope of organismal cloning technology
- Describe the methods used to produce cloned plants and animals, including nuclear transfer and embryo splitting techniques
- Identify the current and potential uses of reproductive and therapeutic cloning
- Discuss the ethical issues involved in organismal cloning

Key Words

Clone, Dolly, molecular cloning, organismal cloning, reproductive cloning, stem cell technology, therapeutic cloning, preformationism, homunculus, epigenesis, proteome, interactome, temporal, spatial, Aristotle, August Weismann, Hans Spemann, nuclear transfer, somatic cell nuclear transfer (SNCT), embryo splitting, nuclear totipotency, totipotent, pluripotent, multipotent, irreversibly differentiated, F. C. Steward, *Rana pipiens*, Robert Briggs, Thomas King, Sir John Gurdon, *Xenopus laevis*, Steen Willadsen, Sir Ian Wilmut, Keith Campbell, cell cycle, G_0 phase, Megan, Morag, Dolly Parton, Ron James, Bonnie, telomeres, large offspring syndrome (LOS), Polly, human cloning, Richard Seed, black-footed ferret, de-extinction, Christmas Island rat, *Rattus macleari*, brown rat, *Rattus norvegicus*.

Chapter 17

The Other Sort of Cloning

On 5 July 1996, a lamb was born at the Roslin Research Institute near Edinburgh. It was an apparently normal event, yet it marked the achievement of a milestone in biological science. The lamb was a clone, and was named Dolly. She was the first organism to be cloned using adult differentiated cells as the donor nuclei, which is what makes the achievement such a groundbreaking event. In this chapter, we will look briefly at this area of genetic technology.

In this book so far, we have been considering the topic of molecular cloning, where the aim of an experimental process is to isolate a gene sequence for further analysis and use. In organismal cloning, the aim is to generate an organism from a cell that carries a complete set of genetic instructions. We have looked at the methods for generating transgenic organisms in Chapter 16, and a discussion of organismal cloning is a natural extension to this, although transgenic organisms are not necessarily clones. In a similar way, a clone need not necessarily be transgenic. Thus, although not strictly part of gene manipulation technology, organismal cloning has become a major part of genetics in a broader sense. From a wider public perspective, organismal cloning is seen as an issue for concern, and thus a discussion of the topic is essential, even in a book where the primary aim is to illustrate the techniques of gene manipulation.

Organismal cloning can be further subdivided according to the purpose of the procedure. An intention to generate a 'copy' of the original organism is termed **reproductive cloning**. Advances in **stem cell technology** open up the use of cloning embryos to enable production of matched tissue types for use in research and potentially in the treatment of disease. This aspect of cloning is called therapeutic cloning.

Organismal cloning can be divided into reproductive cloning and therapeutic cloning, each of which has a different purpose.

17.1 Early Thoughts and Experiments

The announcement of the birth of Dolly in a paper in the journal *Nature* in February 1997 rocked the scientific community. However, the basic scientific principles that underpin organismal cloning have had a long history, and can be traced back to the early days of experimental embryology in the late-1800s and early-1900s. The early embryologists were seeking answers to the central question of development – how does a complex multicellular organism develop from a single fertilised egg?

Historically two contrasting theories have been proposed to explain development. One is that everything is preformed in some way, and development is simply the unfolding of this already existing pattern. This theory is called preformationism, and at its most extreme was thought of as a fully formed organism called a **homunculus**, or 'little man', sitting inside the sperm, ready to 'grow' into a new individual during development. The alternative view is epigenesis, in which development is seen as an iterative process in which cells communicate with each other, and with their environment, as development proceeds. As with many opposing theories, there are aspects of each that can be considered valuable, even today. Certainly the genetic information is already 'preformed'; thus, the genome *could* be considered as a set of instructions that enable the unfolding of all structures in the embryo. However, with the growing appreciation of the importance of the proteins in the cell (the **proteome**), and the many and varied interactions between cell components (the **interactome**), it became clear that development is an interactive process that involves many factors, and that differential gene expression is the mechanism by which complexity is generated from the genetic information in the embryo. In addition to this control of gene expression, which essentially directs the process of embryogenesis, movement of cells and the formation of defined patterns enable structural complexity to arise. Overall, the process of development therefore involves a series of complex interactive steps in both a **temporal** and **spatial** context. This is illustrated in Fig. 17.1.

> Development of the foetus from a single cell is a complex process that involves many interactions between cells and their environment.

> Differential gene expression and interactions of various components in both temporal and spatial contexts are required if the development of the embryo is to proceed normally.

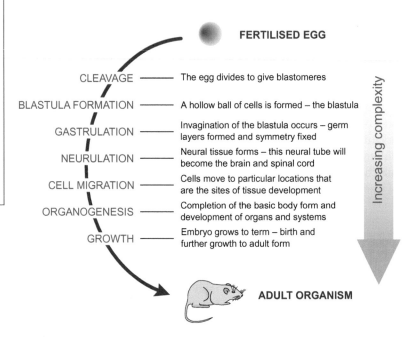

Fig. 17.1 Developmental sequence in the mouse. The journey from fertilised egg to adult organism involves cell proliferation by successive cell divisions, cellular differentiation and generation of complex patterns during embryogenesis. This results in a hierarchy of organisation from individual cells to tissues, organs and organ systems. All these events must be coordinated and regulated in time and space if the process is to proceed successfully to completion.

FERTILISED EGG

CLEAVAGE — The egg divides to give blastomeres

BLASTULA FORMATION — A hollow ball of cells is formed – the blastula

GASTRULATION — Invagination of the blastula occurs – germ layers formed and symmetry fixed

NEURULATION — Neural tissue forms – this neural tube will become the brain and spinal cord

CELL MIGRATION — Cells move to particular locations that are the sites of tissue development

ORGANOGENESIS — Completion of the basic body form and development of organs and systems

GROWTH — Embryo grows to term – birth and further growth to adult form

Increasing complexity

ADULT ORGANISM

17.1.1 First Steps towards Cloning

Our current knowledge of embryological development, as shown in Fig. 17.1, has been established over a long period. The first embryologist appears to have been **Aristotle**, who is credited with establishing an early version of the theory of epigenesis. More recently, **August Weismann** attempted to explain development as a unidirectional process. In 1885, he proposed that the genetic information of cells diminishes with each cell division as development proceeds. This set a number of scientists to work, trying to prove or disprove the theory. Results were somewhat contradictory, but in 1902, **Hans Spemann** managed to split a 2-cell salamander embryo into two parts, using a hair taken from his baby son's head! Each half developed into a normal organism. Further work confirmed this result, and also showed that in cases where a nucleus was separated from the embryo, and cytoplasm was retained to effectively give a complete cell, a normal organism developed. Thus, Weismann's idea of diminishing genetic resources was shown to be incorrect – all cells in developing embryos retained the ability to programme the entire course of development.

Early experiments in organismal cloning used the technique of embryo splitting to generate cloned organisms.

In 1938, Spemann published his book *Embryonic Development and Induction*, detailing his work. In this, he proposed what he called 'a fantastical experiment', in which the nucleus would be removed from a cell and implanted into an egg from which the nucleus had been removed. Spemann could not see any way of achieving this, hence his caution by using the word 'fantastical'. However, he had proposed the technique that would later become known as cloning by nuclear transfer (more fully known as **somatic cell nuclear transfer** or SCNT). Unfortunately, he did not live to see this attempted; the first cloning success was not achieved until 1952, 11 years after his death.

The idea of transplanting an intact nucleus into an egg cell to generate a clone had been proposed as early as 1938. The technique would not be developed fully until much later, culminating in 1996 with the birth of Dolly.

17.1.2 Nuclear Totipotency

The work of Spemann was an important part of the development of modern embryology – indeed, he is often called the 'father' of this discipline. He had proposed nuclear transfer and had demonstrated cloning by embryo splitting. The basis of these two techniques is shown in Fig. 17.2. The experiments with embryo splitting that had refuted Weismann's ideas showed that embryo cells retain the capacity to form all cell types. This became known as the concept of **nuclear totipotency**, which is now a fundamental part of developmental genetics. A cell is said to be totipotent if it can direct the formation of all cells in the organism. If it can direct a more limited number of cell types, it is said to be pluripotent or multipotent. Extending this along the developmental timeline, a cell that is not capable of being reprogrammed to direct development under appropriate conditions is said to be **irreversibly differentiated**.

The developmental status of cells is important, with the concept of nuclear totipotency an important part of the thinking that leads to experiments in SCNT-based cloning.

Nuclear totipotency is in many ways self-evident, as an adult organism has many different types of cell. The original zygote genome, passed on by successive mitotic divisions, must have the capacity to generate these different cells. However, the key in

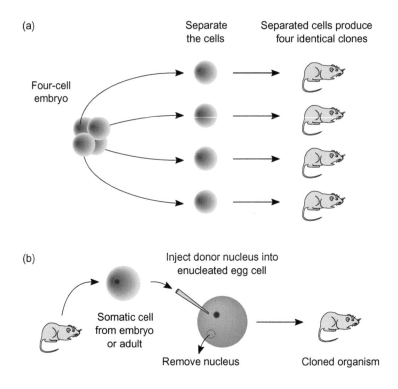

(a)

Four-cell embryo

Separate the cells

Separated cells produce four identical clones

(b)

Somatic cell from embryo or adult

Inject donor nucleus into enucleated egg cell

Remove nucleus

Cloned organism

Fig. 17.2 Two methods for animal cloning. (a) Embryo splitting. Cells from an early embryo (shown here as a 4-cell embryo) are separated and allowed to continue development. Each cell directs the process of development to produce a new individual. The four organisms are genetically identical clones. In (b), the technique of nuclear transfer is shown. A nucleus from a somatic cell (from an embryo or an adult) is transplanted into an enucleated egg cell. This is termed somatic cell nuclear transfer (SCNT). If development can be sustained, a clone develops. Note that in this case, the clone is not absolutely genetically identical to its 'parent'. Mitochondrial DNA is inherited from the cytoplasm as a separate genome, and will therefore have been derived from the egg cell. This is known as a maternal inheritance pattern.

developing cloning techniques was not that this idea was disputed, but rather that it was centred around attempts to see when embryo cells became irreversibly differentiated, and perhaps lost the *capacity* (but not the *genes*) to be totipotent. The next experiments to shed light on this area were carried out in the 1950s.

17.2 | Frogs and Toads and Carrots

Plant development is somewhat simpler than animal development, largely because there are fewer types of cell to arrange in the developing structure. However, the concept of nuclear totipotency is just as valid in plants as it is in animals. In fact, one of the early unequivocal experimental demonstrations of nuclear totipotency was provided by

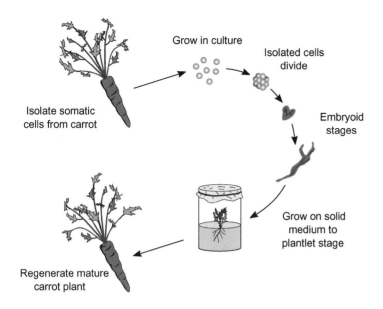

Grow in culture

Isolated cells divide

Isolate somatic cells from carrot

Embryoid stages

Grow on solid medium to plantlet stage

Regenerate mature carrot plant

Fig. 17.3 Cloning carrots. Somatic cells can be removed and grown in tissue culture. Under appropriate conditions, the cells begin to divide, then develop into somatic embryos known as embryoids (the 'heart' and 'torpedo' stages are shown here). These can be transferred to a solid growth medium for plantlet development. The final stage is regeneration of the complete organism.

the humble carrot in the late 1950s. Work by **F. C. Steward** and his colleagues at Cornell University showed that carrot plants could be regenerated from somatic (body) tissue, as shown in Fig. 17.3. This technique is now often used in the propagation of valuable plants in agriculture. The ability of plants to regenerate has of course been exploited for many years by taking cuttings and grafting – these are essentially asexual cloning procedures.

Amphibians provide useful systems for embryological research in that the eggs are relatively large and plentiful, and development proceeds under less stringent environmental conditions than would be required for mammalian embryos. The frog *Rana pipiens* was used by **Robert Briggs** and **Thomas King** around 1952 to carry out Spemann's 'fantastical experiment'. Using nuclei isolated from blastula cells, they were able to generate cloned embryos, some of which developed into tadpoles (Fig. 17.4). As more work was done, it became apparent that early embryo cells could direct development, but that cell nuclei isolated from cells of older embryos were much less likely to generate clones. It was becoming clear that there was a point at which the cell DNA could not easily be 'deprogrammed' and used to direct the development of a new organism, and this remained the limiting factor in cloning research for many years. The work was later extended by **Sir John Gurdon** at Oxford, using the toad *Xenopus laevis*. Some success was achieved, with fertile adult toads being generated from intestinal cell nuclei. In some of the experiments, serial transfers were used, with the implanted nuclei allowed to develop, then these cells used for the isolation of nuclei for further transfer. There was, however, a good deal of debate as to whether Gurdon had in fact used fully differentiated cells or if contamination with primordial reproductive cells had produced the results.

The frog *Rana pipiens* was used in animal nuclear transfer experiments in the early-1950s.

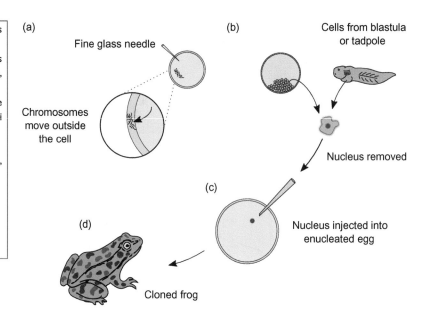

Fig. 17.4 Amphibian cloning. This is the type of procedure used by Briggs and King in 1952. (a) Egg cells are pricked with a fine glass needle, which activates the egg and causes the chromosomes to move outside the cytoplasm. (b) Donor cell nuclei are removed from early embryos or tadpoles. The donor nucleus is injected into the enucleated egg (c), and development can proceed, resulting in an adult cloned organism (d). Early embryo donor nuclei gave better results than nuclei taken from later stages of development.

17.3 | A Famous Sheep – The Breakthrough Achieved

The work with frogs and toads established the feasibility of SCNT-based cloning in animals, and opened the way for the work that would result in the cloning of mammals.

Despite the uncertainty surrounding Gurdon's experiments, the work with frogs and toads seemed to demonstrate that animal cloning by nuclear transfer was feasible if the right conditions could be established. Although the work seemed promising, cloning using a fully differentiated adult cell remained elusive. It was suspected that the manipulations used in the experiments could be damaging the donor nuclei, and by the 1970s, it was clear that further development of the techniques was required. Reports of mice cloned from early embryo nuclei appeared in 1977, but the work was again somewhat inconclusive and difficult to repeat.

Development of the techniques for nuclear transfer cloning continued, with sheep and cattle as the main targets due to their potential for biotechnological applications in agriculture and in the production of therapeutic proteins. By the mid-1980s, several groups had success with nuclear transfer from early embryos. A key figure at this time was **Steen Willadsen**, a Danish scientist who, in 1984, achieved the first cloning of a sheep, using nuclear transfer from early embryo cells. In 1985, Willadsen cloned a cow from embryo cells, and in 1986, he achieved the same feat using cells from older embryos. The work with the older embryos (at the 64- to 128-cell stage) was not published, but Willadsen had demonstrated that cloning from older cells might not be impossible, as most people thought. By this stage, several companies had become involved with cloning technology, particularly in cattle, and the future looked promising for the technology, the scientists who could do it and the industry. However, as with the *Flavr Savr* transgenic tomato, there were problems in establishing cloning on a commercial footing, and by the early-1990s, the promise had all but evaporated.

At around the same time that commercial interests were developing around cloning, a scientist called **Ian Wilmut** (now Sir Ian), working near Edinburgh in what would become the Roslin Institute, was busy generating transgenic sheep. He was keen to try to improve the efficiency of this somewhat hit-or-miss procedure, and cloning seemed an attractive way of doing this. If cells could be grown in culture, and the target transgenes added to these cells rather than being injected into fertilised eggs, the transgenic cells could be selected and used to clone the organism. This approach had been successful in mice, using embryonic stem cells, although it proved impossible to isolate the equivalent cells from sheep, cattle or pigs. However, older cells, derived from a foetus or an adult organism, could be grown easily in culture. Thus, a frustrating impasse existed – if only the older cells could be coaxed into directing development when used in a nuclear transfer experiment, then the process would work. This was the step that informed opinion said was impossible.

The use of older mature cells as donors in SCNT experiments was proving to be the problem area for reproductive cloning in the 1980s and early-1990s.

In 1986, Wilmut attended a scientific meeting in Dublin, where he heard about the 64- to 128-cell cattle cloning experiments from a vet who had worked with Willadsen. This encouraged him to continue with the cloning work. The key developments came when **Keith Campbell** joined Wilmut's group in 1990. Campbell was an expert on the cell cycle, and was able to develop techniques for growing cells in culture and then causing the cells to enter a quiescent stage of the cycle known as G_0 **phase**. Wilmut and Campbell thought that this might be a critical factor, perhaps the key to success. They were proved correct when, in 1995, **Megan** and **Morag** were born. These were lambs that had been produced by nuclear transfer using cultured cells derived from early embryos. Extensions to the work were planned, supported by PPL Therapeutics, a biotechnology firm set up in 1987 to commercialise transgenic sheep technology. Wilmut and Campbell devised a complex experiment in which they would use embryo cells, foetal cells and adult cells in a cloning procedure. A strange quirk of history appears at this point – the adult cells came from a vial that had been stored frozen at PPL Therapeutics for 3 years, and their source animal was long forgotten. However, it was known that the cells were from the udder of a 6-year old Finn Dorset ewe. The cloning process is summarised in Fig. 17.5.

From the adult cell work, 29 embryos were produced from 277 udder cells. These were implanted into Scottish Blackface surrogate mothers, and some 148 days later, on 5 July 1996, one lamb was born. She was named after the singer **Dolly Parton** (make the connection yourself!). Dolly is shown in Fig. 17.6. What had once seemed impossible had been achieved.

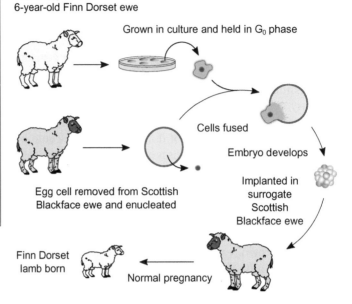

Fig. 17.5 The cloning method by which Dolly was produced. Mammary cells from an adult Finn Dorset ewe were isolated, cultured and held in the G_0 phase of the cell cycle. These were the donor cells. Egg cells taken from a different breed (Scottish Blackface) were enucleated to act as recipients. The donor and recipient cells were fused and cultured. The embryos were then implanted into surrogate Blackface mothers and pregnancies established. The Finn Dorset lamb that was named Dolly was born some 5 months later.

Mammary cells taken from 6-year-old Finn Dorset ewe

Grown in culture and held in G_0 phase

Cells fused

Embryo develops

Egg cell removed from Scottish Blackface ewe and enucleated

Implanted in surrogate Scottish Blackface ewe

Finn Dorset lamb born

Normal pregnancy

Fig. 17.6 Dolly. The first mammal to be cloned using a fully differentiated somatic adult cell as the source of the donor nucleus, Dolly represented a significant milestone in biology. She is shown here as a lamb, with her Scottish Blackface surrogate mother. *Source:* Photograph courtesy of the Roslin Institute, the University of Edinburgh, Roslin, Scotland, UK. Reproduced with permission under Licence CC-BY-NC-ND-4.0 [https://creativecommons.org/licenses/by-nc-nd/4.0/].

17.4 | Beyond Dolly

The birth of Dolly demonstrated that adult differentiated cells could, under the appropriate conditions, give rise to clones. Whilst the magnitude of this scientific achievement was appreciated by Wilmut and his colleagues at the time, the extent of the public reaction caught them a little by surprise. Suddenly Wilmut, Campbell and **Ron James** of PPL Therapeutics were in the spotlight, unfamiliar ground for scientists. Dolly herself became something of a celebrity in the media, with both the serious science press and the popular press maintaining an interest. She was mated with a Welsh mountain ram in 1997 and gave birth to a female lamb called **Bonnie** in April 1998, thus demonstrating that generation of an organism by reproductive cloning appeared to have no effect on normal reproductive processes in the clone.

Being a celebrity sheep, Dolly quite naturally was subjected to detailed scrutiny, and thus when signs of possible premature ageing were noted when she was around 5 years old, there was some cause for concern. One possible reason for this could be the progressive shortening of chromosomal telomeres that occurs during successive rounds of cell division, as this is thought to be associated with the ageing process. As the donor cell that created Dolly was 6 years old, she could be considered as effectively 6 when she was born (at least from a cellular genetics point of view); thus, perhaps this had an impact on her normal lifespan. There were also signs of arthritis, but there is doubt as to whether this was in any way associated with the fact that she was a clone.

The issue of telomere length and ageing is an interesting area of debate that has been highlighted by reproductive cloning.

Dolly died on 14 February 2003, from a progressive retroviral lung disease to which sheep are susceptible. Her death did not appear to be a consequence of her clone status, and her remains were preserved and are displayed in the Royal Museum in Edinburgh (part of the National Museum of Scotland). Some commentators have made much of the fact that she attained approximately half a normal lifespan, and that this was offset by having 'lived the other half already' as the cell from the donor organism. This is probably much too simplistic, and more research is needed to establish the physiological and genetic features of cloned organisms that may arise as a consequence of the cloning process. As work has progressed, it has become apparent that there is much complexity associated with this type of procedure, and that donor cell DNA may not in fact be completely reprogrammed during the SCNT process. This can lead to difficulties in that gene expression may not fully reflect the requirements of embryological development in the clone. It also seems that in around a third of cases, cloned organisms are larger than normal at birth, which is termed **large offspring syndrome (LOS)**.

Although organismal cloning technology is now well established, it still presents significant challenges in terms of expertise, technical difficulty and funding.

Having achieved success with Dolly, Wilmut and his colleagues went on to produce the first transgenic cloned sheep, a Poll Dorset clone (named **Polly**) carrying the gene for factor IX. Thus, the goal of producing transgenics using nuclear transfer technology had been achieved by the late-1990s, and at that time seemed to offer great potential.

17.4.1 Potential Unfulfilled?

Over 20 mammalian species have now been cloned, using either the embryo splitting technique or nuclear transfer. They include cats, dogs, mice, goats, cattle, rabbits, horses, donkeys and monkeys.

However, it is **human cloning** that tends to generate interest in the wider arena, generating both sensible technical and ethical debate, and also more extreme views and claims. The perceived stakes among some scientists can tend to push boundaries into areas that are either fantasy, fraud or simply nonsensical, and (as with all scientific work) careful evaluation, peer review and vigilance are required.

Most scientists consider that nuclear transfer cloning of humans should not be attempted, and many countries (around 50 or so) have either banned all cloning of humans or have banned reproductive, but not therapeutic, cloning. However, proposals to clone humans have been put forward, notably by **Richard Seed** in 1997, and by an Italian reproductive expert who stated that he was setting out to establish cloning of humans. Cloning of human embryos by embryo splitting was reported as far back as 1993, and in 2004, a team in South Korea published a paper in the journal *Science*, stating that they had cloned a human embryo in a test tube. However, an independent investigation into this paper (and a second related one published in 2005) found no evidence to support the claims and uncovered fabrication of data. Both papers were retracted in 2006.

> The possibility of reproductive cloning of humans is often presented too simplistically and sensationally. It generates significant debate and is banned, but that does not necessarily stop scientists from attempting it or claiming to have achieved it.

Although used for many specialised applications, the potential of cloning technology has not yet been exploited fully. The process is still highly technical and resource-intensive, and it is therefore difficult to see how this can be adopted as a routine technique that is accessible and cost-effective.

17.4.2 The Future of Organismal Cloning

Organismal cloning is undoubtedly an impressive technical feat that has made significant contributions to our understanding of genome function and embryological development. The fact that the end result is genetically (if not epigenetically and developmentally) identical to the donor genome means, somewhat paradoxically, that its use is relatively narrow in scope. More recent advances such as gene editing offer different ways to manipulate genomes, and in many laboratories, the emphasis has shifted away from SNCT-based cloning and transgenesis to using systems such as the CRISPR-Cas9 editing technology. As with organismal cloning, there are still many questions about the use of gene editing methods that, despite the precision of the technique, may still result in damage to the target cells and **off-target effects** that cannot be predicted or controlled easily. However, it seems likely that gene editing is the technology most likely to deliver advances that will eventually help to address many of the key questions and challenges in the treatment of disease conditions, and other important biomedical and biotechnological applications, across a range of organisms.

> The future of organismal cloning is likely to involve an amalgam of stem cell technology, therapeutic cloning and genome editing.

Human gene editing is generally confined to somatic cells for therapeutic research and development, with germ line gene editing currently prohibited in most countries. The 'CRISPR babies' issue that we looked at in Chapter 15, resulting in a prison sentence for the scientist involved, was a reminder that the regulation of research in this area requires not only a strong regulatory and legal framework, but also acceptance of, and compliance with, the constraints by individual scientists.

Other developments in non-human organismal cloning may be less controversial but still present technical and resource challenges. Cloning of endangered species is one area that may deliver benefits in some areas. The first success in this area came at the end of 2020 when a **black-footed ferret** was cloned using cells from an animal that had been frozen when it died some 30 years ago. Scientists hope that this can both save endangered (or already extinct) species and also have the potential to reintroduce genetic diversity into wild populations where this may be an advantage to the remaining populations. However, cloning long-extinct animals is an unlikely scenario; a recent attempt at what is called de-extinction, with the **Christmas Island rat** (*Rattus macleari*, extinct in 1908), showed how difficult this is likely to be. As no viable cells are available from long-dead animals, their genomes have to be sequenced to enable reconstruction. Despite sequencing the genome with 60 times coverage, about 5 per cent (some 2 600 genes) was still incomplete, using the **brown rat** (*Rattus norvegicus*) as a reference genome. Thus, even for a relatively recent extinction, not all of the genome sequence was recoverable. If critical species-defining genes are among the missing ones, this is a problem – if the 'modern' gene sequences are used to fill in the missing genes, this is not the original genome. As the evolutionary divergence between the two species is some 2.6 million years, there is likely to have been significant genetic drift across this period, with no way of determining what the original sequences were.

17.5 | Conclusion: From Genome to Organism

The key theme of this chapter has been 'how do we get from a fertilized egg to an adult organism?', in terms of both the fundamental science of how the genome works and the various stages of embryological development. The early concept of *preformationism* was replaced by *epigenesis*, with development being an iterative process that is a complex interaction between genetic and non-genetic (epigenetic) factors, in both temporal and spatial dimensions.

Hans Spemann was a major contributor to the field, having demonstrated *embryo splitting* in 1902 and mentioned cloning by *nuclear transplantation* in 1938, when he proposed his 'fantastical experiment' using the technique that would later become known as *somatic cell nuclear transfer* or *SCNT*. The concepts of *nuclear totipotency*, *pluripotency* and *irreversible differentiation* are important in that reprogramming of the developmental status of a donor cell's nucleus is required if it is to direct all stages of development when transferred into a recipient egg cell.

Organismal cloning had a golden spell in the 1950s, when elegant experiments with organisms as diverse as frogs and carrots demonstrated that cloning of plants and animals could be achieved. Cloning in mammals proved more difficult, but by the mid-1980s, elements of the technology were in place to enable cloning using donor nuclei from early embryos. The major breakthrough was the use of donor nuclei from adult mammalian cells, resulting in the birth of *Dolly* in 1996.

Despite significant achievements in organismal cloning, the technical difficulties and level of resource (both personnel and infrastructure) needed to clone animals has meant that the applications have been relatively restricted. In humans, there is the added complexity of major ethical issues to consider, with all attempts at *reproductive cloning* banned in most countries where the issue has been considered. Some countries permit research on *therapeutic cloning* using embryos, which is the area where most promise lies. When coupled with techniques such as *gene/genome editing*, the potential benefits in a range of diverse applications are almost unlimited, provided that the *off-target effects* and damage caused by manipulations can be controlled to an acceptable level.

As with so many aspects of modern genetics, the future will no doubt hold many contentious and interesting developments in the field of organismal cloning. Regardless of how the application of cloning technology develops, when the histories are written, the central character will be a sheep named after a country and Western singer – bizarre, but strangely appropriate in a field that stirs the imagination like few others in science.

Further Reading

Borsos, M. and Torres-Padilla, M.-E. (2016). Building up the nucleus: nuclear organization in the establishment of totipotency and pluripotency during mammalian development. *Genes Dev.*, 30, 611–21. URL [http://genesdev.cshlp.org/content/30/6/611.full]. DOI [https://doi.org/10.1101/gad.273805.115].

Gurdon, J. B. (2013). The cloning of a frog. *Development*, 140 (12), 2446–8. URL [www.researchgate.net/publication/236957730_The_cloning_of_a_frog]. DOI [https://doi.org/10.1242/dev.097899].

Häyry, M. (2018). Ethics and cloning. *Br. Med. Bull.*, 128 (1), 15–21. URL [https://academic.oup.com/bmb/article/128/1/15/5094025]. DOI [https://doi.org/10.1093/bmb/ldy031].

Kolata, G. (1997). *Clone: The Road to Dolly and the Path Ahead*. Allen Lane, The Penguin Press, London.

Liu, Z. *et al.* (2018). Cloning of macaque monkeys by somatic cell nuclear transfer. *Cell*, 172 (4), 881–7. URL [www.cell.com/fulltext/S0092-8674(18)30057-6]. DOI [https://doi.org/10.1016/j.cell.2018.01.020].

Websearch

People

The German embryologist *Hans Spemann* is often called 'the father of developmental biology', and is one of the most important figures in early twentieth-century biology. Find out what his main contributions were to the field, and try to get a sense of his pivotal role in establishing an experimental approach to embryology.

Places

The *Roslin Institute* will always (rightly) be associated with the cloning breakthrough that produced Dolly in 1996. What current research is carried out at the Institute, and how does this reflect the changes that have occurred in biology since Dolly was born?

Processes

Have a look in a little more detail at the topic of *epigenesis*, and *epigenetics* in particular. Find out what techniques are now available for studying epigenetic factors.

Reflections

What topics in this chapter have you found most challenging? Look for resources that help to illustrate the key points.

Organismal cloning

involves → **generating identical organisms by non-sexual processes**

used to clone → animals

and also → animals

via

nuclear transfer
requires →
- **enucleated recipient** — from → **an egg cell**
- **donor cell nucleus** — from → **embryo cell** / **foetal cell** / **adult cell**

to produce → **individual cloned organisms**

embryo splitting
to → **separate embryo cells** → to produce → **individual cloned organisms**

achieved with, *e.g.*
mouse
sheep
goat
cow
monkey
pig
deer
horse
dog
cat

but → **technical and financial constraints limit the use of reproductive cloning**

→ **animal cloning raises a number of significant ethical issues**

arising from → nuclear totipotency

where → **all somatic cells of an organism have the same genome**

but can be →
- **totipotent** → **can direct development of all types of cell in a cloned organism**
- **pluripotent or multipotent** → **can direct development of some types of cell in a cloned organism**
- **irreversibly differentiated**

used to clone → plants

by →

regeneration from somatic cells
need to → **isolate plant cells** — that are → totipotent

then → **grow in tissue culture** → to → **generate embryoids** → then → **grow into a cloned plant**

methods for grafting and cutting
for → **direct asexual propagation** — has been → **used in horticulture for centuries**

→ **few direct ethical issues around plant cloning, but some questions about the use of genetically modified plants**

Concept Map 17

Glossary

Abundance class (1) Used to describe the composition of a genome in terms of the type of DNA sequences present, *e.g.* repetitive and single copy. (2) Used to describe the relative abundance of different mRNA molecules in a cell at any given time.

Accuracy Refers to how close a measured value is to the true value.

A-chain One of the polypeptide chains in the insulin protein.

Acquired immune deficiency syndrome (AIDS) Disease of the immune system, causative agent the HIV virus (*q.v.*), characterised by deficiency in T-lymphocytes rendering sufferers susceptible to infection and other clinical conditions.

Activase Recombinant-derived tissue plasminogen activator, the first rDNA therapeutic to be produced from mammalian cells in the early-1980s.

ADA See *adenosine deaminase.*

Adapter (*or adaptor*) A synthetic single-stranded, non-self-complementary oligonucleotide used in conjunction with a linker to add cohesive ends to DNA molecules.

Adaptive regulation Regulation of gene expression in response to changing environmental conditions such as the availability of nutrients.

ADAR See *adenosine deaminase acting on RNA.*

Adenine (A) Nitrogenous base found in DNA and RNA.

Adenosine Nucleoside composed of ribose and adenine.

Adenosine deaminase (ADA) Enzyme that converts adenosine to inosine. Deficiency causes toxic nucleosides to build up, which causes immune system deficiency and results in SCIDS (*q.v.*).

Adenosine deaminase acting on RNA (ADAR) A deaminase that acts on double-stranded RNA, changing adenosine to inosine and destabilising RNA structure. Involved in gene regulation and disease.

Adsorption (1) Adhesion of molecules or macromolecules to a substrate or matrix. (2) Also used to describe the attachment of a bacteriophage to the bacterial cell surface.

Affinity chromatography Technique used to separate cellular components by selective binding and release using an appropriate medium to which the component of interest will bind. An example is the selection of mRNA from total RNA using oligo(dT)-cellulose.

Affinity tag A tag (additional moiety or sequence) added to a recombinant protein to enable purification by affinity chromatography (*q.v.*).

Agarose Jelly-like matrix, extracted from seaweed, used as a support in the separation of nucleic acids by gel electrophoresis.

Agrobacterium tumefaciens Bacterium that infects plants and causes crown gall disease. Carries a plasmid (the Ti plasmid, *q.v.*) used for gene manipulation in plants.

AIDS See *acquired immune deficiency syndrome.*

Alcohol oxidase (AOX) Enzyme of alcohol metabolism. The gene promoter can be used for the construction of expression vectors in yeast.

Algorithm A series of rules or instructions used in computer programming.

Alkaline phosphatase An enzyme that removes 5′ phosphate groups from the ends of DNA molecules, leaving 5′ hydroxyl groups.

Allele One of two or more variants of a particular gene.

Allele-specific oligonucleotide (ASO) Oligonucleotide with a sequence that can be matched precisely to a particular allele by using stringent hybridisation conditions.

Allogenic (or *allogeneic*) In transplant procedures, taken from a different member of the same species.

Allorejection Rejection, usually immunorejection (*q.v.*), of an allograft transplant.

Alpha-1-antitrypsin Protease inhibitor used in the treatment of cystic fibrosis.

Alpha peptide Part of the β-galactosidase protein, encoded by the *lac Z*′ gene fragment.

Alzheimer disease Neurodegenerative disease, characterised by progressive deterioration in cognitive function.

Ampicillin (Ap) A semisynthetic β-lactam antibiotic.

Amplicon Refers to the product of a polymerase chain reaction.

ANDi First genetically modified rhesus monkey.

Aneuploidy Variation in chromosome number where single chromosomes are affected; thus, the chromosome complement is not an exact multiple of the haploid chromosome number.

Annealing Two complementary strands of nucleic acid joining together by base-pairing to generate a double-stranded molecule.

Annotate Add information to a gene or genome sequence database to enrich the data set and identify key features.

Anopheles Genus of the mosquito, carries the infectious agent of malaria (trypanosome of the *Plasmodium* genus, *q.v.*).

Antibody An immunoglobulin that specifically recognises and binds to an antigenic determinant on an antigen.

Anticodon The three bases on a tRNA molecule that are complementary to the codon on the mRNA.

Antigen A molecule that is bound by an antibody. Also used to describe molecules that can induce an immune response, although these are more properly described as immunogens.

Antiparallel The arrangement of complementary DNA strands, which run in different directions with respect to their 5′ → 3′ polarity.

Antisense mRNA Produced from a gene sequence inserted in the opposite orientation, so that the transcript is complementary to the normal mRNA and can therefore bind to it and prevent translation.

Antisense oligonucleotide (ASO) An oligonucleotide that is complementary to part of an mRNA sequence to which it can bind and prevent translation.

AOX See *alcohol oxidase*.

AP-PCR See *arbitrarily primed PCR.*

Arabidopsis thaliana Small plant favoured as a research organism for plant molecular biologists.

Arbitrarily primed PCR (AP-PCR) PCR using low-stringency random primers, useful in the technique of RAPD analysis (*q.v.*) for genomic fingerprinting.

Arctic Apple Apple with levels of polyphenol oxidase reduced by gene silencing using RNAi (*q.v.*). Reduces browning when the apple is cut or bitten.

ARTs See *assisted reproductive technologies.*

ASO See *allele-specific oligonucleotide* and *antisense oligonucleotide.*

Aspergillus nidulans Ascomycete fungus useful for genetics research and biotechnology.

Assisted reproductive technologies (ARTs) A range of procedures and treatments to help overcome problems in reproduction.

ATryn The first therapeutic anticoagulant made in transgenic goat milk approved for use.

Autologous Cells or tissues from the same individual (*e.g.* blood cells taken from, engineered and replaced in the patient).

Autoradiograph Image produced on X-ray film in response to the emission of radioactive particles.

Auxotroph A cell that requires nutritional supplements for growth.

Auxotrophic marker A mutation in a biosynthetic pathway gene that produces a requirement for a nutritional supplement to enable cell growth.

BAC Bacterial artificial chromosome, used for cloning large fragments of genomic DNA.

Bacillus thuringiensis Bacterium used in crop protection, and in the generation of Bt plants that are resistant to insect attack. The bacterium produces a toxin that affects the insect.

Bacterial alkaline phosphatase (BAP) See *alkaline phosphatase.*

Baculovirus A particular type of virus that infects insect cells, producing large inclusions in the infected cells.

Bait protein A protein used in the yeast two-hybrid system to identify protein~protein interaction.

BAL 31 nuclease An exonuclease that degrades both strands of a DNA molecule at the same time.

Barcoding (1) Addition of DNA sequences at the ends of target fragments to enable identification, often used in multiplex reactions such as PCR and DNA sequencing. (2) Use of a short unique DNA sequence in the genome of an organism as an identifier for taxonomic classification.

Base calling In DNA sequencing, base calling interprets the information at a particular position in the sequence and 'calls' the most likely base.

Basic local alignment search tool (BLAST) A computer-based alignment tool to compare DNA sequences and align areas of sequence similarity.

B-chain One of the polypeptide chains in the insulin protein.

Beta carotene Provitamin A, a precursor required for synthesis of vitamin A. Engineered into Golden Rice, an example of biofortification of a foodstuff.

Beta galactosidase An enzyme encoded by the *lac Z* gene. Splits lactose into glucose and galactose.

Beta thalassaemia A blood disorder that affects haemoglobin function. Target for the first use of gene editing as a therapy.

Binary vector A vector system in which two plasmids are used. Each provides specific functions that complement each other to provide full functionality.

Bioconjugate A molecule formed by joining at least two other molecules together, one of which is of biological origin. Adding chemically reactive groups to a protein would be an example of a bioconjugate; often used as reporter tag for bioassays.

Bioethics Ethics in the context of biological issues.

Bioinformatics The discipline of collating and analysing biological information, especially genome sequence information.

Biolistic Refers to a method of introducing DNA into cells by bombarding them with microprojectiles which carry the DNA.

Biological containment Refers to the biological properties of a cell or vector system that makes it unlikely that the strain will survive outside the laboratory environment. *Cf. physical containment.*

Biological data set Refers to any data set of biological origin, but in context is taken to mean the large databases of sequence and interaction data that have arisen as a consequence of modern molecular biology investigations.

Biome A biogeographical unit that is distinct from others with respect to geophysical and ecological characteristics. Includes all the organisms that inhabit the area. A broader (and usually much larger in terms of area) concept than habitat.

Bioreactor (1) A vessel used for the cultivation of biological materials such as cells, ranging from simple single-container batch culture vessels to large complex tank systems with multiple monitoring devices. (2) More recently, the term has become associated with the use of transgenic plants to produce recombinant proteins or vaccines.

Biosimilar A biological therapeutic (*e.g.* a monoclonal antibody) that is similar to another variant. Not the same as a generic drug, in which the active chemical constituent is exactly the same as the preceding branded variant.

Biotechnology Generic word to describe the application of bioscience for the benefit of humankind. Encompasses a wide range of disciplines and procedures, but is often mistakenly thought to refer exclusively to the industrial-scale use of genetically modified microorganisms.

BLAST See *basic local alignment search tool.*

Blunt ends DNA termini without overhanging 3′ or 5′ ends. Also known as *flush ends.*

Bonnie The first lamb born to the cloned sheep Dolly (*q.v.*).

Bovine somatotropin (BST) Bovine growth hormone, produced as rBST for use in dairy cattle to increase milk production.

Bridge amplification Method used for the production of PCR-amplified DNA fragment libraries for the Illumina next-generation DNA sequencing method.

BST See *bovine somatotropin.*

Bt plants Transgenic plants carrying the toxin-producing gene from *Bacillus thuringiensis* as a means to protect the plant from insect attack.

CAAT box A sequence located approximately 75 base-pairs upstream from eukaryotic transcription start sites. This sequence is one of those that enhance binding of RNA polymerase.

Caenorhabditis elegans Nematode worm used as a model organism in developmental and molecular studies.

Calf intestinal phosphatase (CIP or CIAP) See *alkaline phosphatase.*

cAMP See *cyclic AMP.*

CAR See *chimeric antigen receptor.*

CAR T-cell A T-cell engineered to express chimeric antigen receptors (CAR, *q.v.*) on the cell surface. Used in targeting cancer cells in CAR T-cell therapy.

Carbohydrate Molecule that contains carbon, hydrogen and oxygen, empirical formula CH_2O. Important in energy storage and conversion reactions in the cell; examples include glucose, fructose and lactose. Polymers of carbohydrates are known as polysaccharides and are used as energy storage compounds.

Cas CRISPR-associated protein, used to describe endonucleases such as Cas9 and Cas13 that are associated with the CRISPR gene editing system.

Cassette A modular vector construct that has a set of standard components and a cloning site that can be used to generate recombinants in a consistent format.

Catabolic Reactions of the cell that break down molecules into smaller components, such as in energy-generating pathways.

Ccd B protein Encodes a toxin that kills bacterial cells containing non-desired constructs in Gateway cloning procedures.

C-chain (C-peptide) The polypeptide chain that links the A- and B-chains in the proinsulin molecule. Proteolytic removal of the C-chain generates the insulin protein.

CCS See *circular consensus sequencing.*

cDNA DNA that is made by copying mRNA using the enzyme reverse transcriptase.

cDNA library A collection of clones prepared from the mRNA of a given cell or tissue type, representing the genetic information expressed by such cells.

cdPCR Chip digital PCR, where the PCR is carried out on custom-designed 'chip'. Also refers to chamber digital PCR.

Cell membrane Lipid bilayer-based structure containing proteins embedded in or on the membrane. Acts as a selectively permeable barrier that separates the cell from its environment.

Cell wall Found in bacteria, fungi, algae and plants, the cell wall is a rigid structure that encloses the cell membrane and contents. Composed of a variety of polysaccharide-based components such as peptidoglycan (bacteria) chitin (fungi) and cellulose (plants, algae and fungi).

Central Dogma Statement regarding the unidirectional transfer of information from DNA to RNA to protein.

Centrifugation Method of separating components by spinning at high speed. The g-forces cause materials to pellet or move through the centrifugation medium. Uses include spinning down whole cells, cell debris, precipitated nucleic acids or other components. Also used in ultracentrifuges for separating macromolecules under gradient centrifugation.

CFTR gene (protein) The gene and protein involved in defective ion transport that causes cystic fibrosis (*q.v.*).

Chimera (also *chimaera*) An organism (usually transgenic) composed of cells with different genotypes.

Chimeric antigen receptor (CAR) Variable antigen-binding site on an antibody, used to describe the engineered chimeric antibodies on T-cells in CAR T-cell (*q.v.*) therapy.

ChIP assay See *chromatin immunoprecipitation (ChIP) assay.*

Chlamydomonas reinhardtii A unicellular green alga that is the model organism for this taxonomic level.

Chromatin immunoprecipitation (ChIP) assay Method of identifying regions of DNA that are associated with particular proteins by immunoprecipitation of the protein/DNA region.

Chromatography Method of separating various types of molecules based on their affinity or physical behaviour when passed through a matrix and eluted with a suitable solvent.

Chromogenic substrate A compound that generates a coloured product when processed, usually in an enzyme-catalysed reaction.

Chromosomal abnormalities (aberrations) Term used to describe genetic conditions that involve gain or loss of whole chromosomes or parts of chromosomes.

Chromosome A DNA molecule carrying a set of genes. There may be a single chromosome, as in bacteria, or multiple chromosomes, as in eukaryotic organisms.

Chromosome jumping Technique used to isolate non-contiguous regions of DNA by 'jumping' across gaps that may appear as a consequence of uncloned regions of DNA in a gene library.

Chromosome walking Technique used to isolate contiguous cloned DNA fragments by using each fragment as a probe to isolate adjacent cloned regions.

Chymosin (chymase) Enzyme used in cheese production, available as a recombinant product.

Circular consensus sequencing (CCS) In next-generation sequencing technology, the CCS method (also known as HiFi sequencing) processes each strand of the insert fragment multiple times and therefore builds a consensus sequence.

Cistron A sequence of bases in DNA that specifies one polypeptide.

Clone (1) A colony of identical organisms; often used to describe a cell carrying a recombinant DNA fragment. (2) Used as a verb to describe the generation of recombinants. (3) A complex organism (*e.g.* sheep) generated from a totipotent cell nucleus by nuclear transfer into an enucleated ovum.

Clone bank See *cDNA library, genomic library*.

Clone contig Refers to contiguous (*q.v.*) cloned sequences. A series of contiguous clones represents the arrangement of the DNA sequence as it would be found *in vivo*.

CLR See *continuous long-read*.

Clustered regularly interspaced short palindromic repeats (CRISPR) Short repeated sequences in prokaryotes, used to drive an adaptive 'immune system' that is able to attack viruses (phages) infecting the cell. The CRISPR system has been developed into a powerful gene editing technology.

CNV See *copy number variant*.

Coding strand The strand of DNA that carries the genetic information for specifying the amino acid sequence, *i.e.* the mRNA-like strand. *Cf. non-coding sequence*.

Codon The three bases in mRNA that specify a particular amino acid during translation.

Cohesive ends Those ends (termini) of DNA molecules that have short complementary sequences that can stick together to join two DNA molecules. Often generated by restriction enzymes.

Cointegration Formation of a single DNA molecule from two components, usually refers to the formation of a larger plasmid vector from two smaller component plasmids.

COL7A1 The *COL7A1* gene is a collagen gene that has recently been included in a gel formulation for topical delivery of gene therapy directly to the skin.

Combinatorial screening Method for screening clone banks or other populations of cells by pooling colonies to reduce the number of processes (*e.g.* PCR reactions).

Comparative genomics Use of genome analysis techniques to establish similarities and differences between genomes of organisms in a population or across different taxonomic levels. *Cf. structural genomics, functional genomics*.

Competent Refers to bacterial cells that are able to take up exogenous DNA.

Competitor RT-PCR Technique used to quantify the amount of PCR product by spiking samples with known amounts of a competitor sequence.

Complementary DNA See *cDNA*.

Complementation Process by which genes on different DNA molecules interact. Usually a protein product is involved, as this is a diffusible molecule that can exert its effect away from the DNA itself. For example, a *lac Z*$^+$ gene on a plasmid can complement a mutant (*lac Z*$^-$) gene on the chromosome by enabling the synthesis of β-galactosidase.

Computer hardware The computer machinery that provides the infrastructure for running the software programs.

Computer software The programs that run on computer hardware. Software packages enable manipulation and processing of various sorts of data, text and graphics.

Concatemer A DNA molecule composed of a number of individual pieces joined together *via* cohesive ends (*q.v.*).

Confirmation bias The tendency to accept or select information and evidence that supports a personally held view; care is needed to exclude confirmation bias from the scientific process.

Congenital Present at birth, usually used to describe genetically derived abnormalities.

Conjugation Plasmid-mediated transfer of genetic material from a 'male' donor bacterium to a 'female' recipient.

Consensus sequence A sequence that is found in most examples of a particular genetic element, and which shows a high degree of conservation. An example is the CAAT box (*q.v.*).

Containment See *biological containment, physical containment*.

Continuous long-read (CLR) NGS method that generates a single readthrough of a long DNA fragment.

Copy number (1) The number of plasmid molecules in a bacterial cell. (2) The number of copies of a gene in the genome of an organism.

Copy number variant (CNV) Variation in the number of copies of a particular DNA sequence between individuals, usually of significant length. Can be used in mapping studies and population genetics.

Cosmid A hybrid vector made up of plasmid sequences and the cohesive ends (*cos* sites) of bacteriophage lambda.

cos **site** The region generated when the cohesive ends of lambda DNA join together.

Covalent bond relatively strong molecular bond in which the electron configurations of the constituent atoms is satisfied by sharing electrons.

CRISPR See *clustered regularly interspaced short palindromic repeats*.

Crown gall disease Plant disease caused by the Ti plasmid of *Agrobacterium tumefaciens*, in which a 'crown gall' of tissue is produced after infection.

CRP See *cyclic AMP receptor protein*.

Crystallisation Method of concentrating a solute by forming crystalline structures, usually following on from the evaporation of solvent.

Ct See *cycle threshold*.

C-terminus Carboxyl terminus, defined by the –COOH group of an amino acid or protein.

Cycle threshold (Ct) In PCR, the cycle threshold in the number of amplification cycles needed to generate a distinct and reliable positive result.

Cyclic AMP (cAMP) Cyclised form of the nucleoside adenosine monophosphate, important in a range of cellular signalling mechanisms.

Cyclic AMP receptor protein (CRP) Also known as catabolite gene activator protein (CAP). Protein involved in regulating expression of catabolic operons, CRP binds cyclic AMP (*q.v.*) and acts as a positive regulator when operons are de-repressed.

Cystic fibrosis (CF) Disease affecting lungs and other tissues, caused by ion transport defects in the CFTR gene and protein (*q.v.*).

Cytosine (C) Nitrogenous base found in DNA and RNA.

Data curation The various techniques and processes that are used to source, assemble, check and annotate information in a database.

Data mining Refers to the technique of searching databases (of various types) for information.

Data warehouse Term to describe a data storage 'facility', usually storage space on a computer where the data are held in electronic formats of various types.

Deaminase An enzyme that removes amine groups from a molecule. See also *adenosine deaminase (ADA)* and *severe combined immunodeficiency syndrome (SCIDS)*.

Deductive A form of reasoning by which general inferences lead to specific conclusions, sometimes called the 'top–down' approach. Often used in science as a component of the hypothetico-deductive method, where a hypothesis is tested by experiment to see if the premises are supported by the results. *Cf. inductive.*

Dehydration synthesis A form of condensation reaction where two molecules are joined by the removal of water.

Deoxynucleoside triphosphate (dNTP) Triphosphorylated ('high-energy') precursor required for synthesis of DNA, where N refers to one of the four bases (A, G, T or C).

Deoxyribonuclease I (DNase I) A nuclease enzyme that hydrolyses (degrades) single- and double-stranded DNA.

Deoxyribonucleic acid (DNA) A condensation heteropolymer composed of nucleotides. DNA is the primary genetic material in all organisms, apart from some RNA viruses. Usually double-stranded.

Deoxyribose The sugar found in DNA.

Dependent variable The variable that is measured in an experiment. *Cf. independent variable.*

Deproteinisation Removal of protein contaminants from a preparation of nucleic acid.

Diabetes mellitus (DM) Condition where high levels of blood glucose exist due to problems in regulation of glucose levels. May be caused by insulin deficiency or other defects in the glucose regulation system. See also *insulin-dependent diabetes mellitus* and *non-insulin-dependent diabetes mellitus*.

Dicer Enzyme involved in RNA interference (RNAi, *q.v.*).

Dideoxynucleoside triphosphate (ddNTP) A modified form of dNTP used as a chain terminator in DNA sequencing.

Digital object identifier (DOI) A unique code for the web location(s) of a digital object. In science, mostly used to reference literature or data sources.

Digital PCR (dPCR) A development of the PCR method used to quantify nucleic acids. See also *droplet digital PCR*.

Diploid Having two sets of chromosomes. *Cf. haploid.*

Directed evolution Method of engineering changes in protein structure by producing random changes and selecting the most useful, analogous to an evolutionary process. *Cf. rational design.*

Disarmed vector A vector in which some characteristic (*e.g.* conjugation) has been disabled.

DMD See *Duchenne muscular dystrophy.*

DNA chip A DNA microarray used in the analysis of gene structure and expression. Consists of oligonucleotide sequences immobilised on a 'chip' array.

DNA fingerprinting See *genetic fingerprinting.*

DNA footprinting Method of identifying regions of DNA to which regulatory proteins will bind.

DNA ligase Enzyme used for joining DNA molecules by the formation of a phosphodiester bond between a 5′ phosphate and a 3′ OH group.

DNA microarray See *DNA chip.*

DNA polymerase An enzyme that synthesises a copy of a DNA template.

DNA profiling Term used to describe the various methods for analysing DNA to establish the identity of an individual.

DNA shuffling Refers to the generation of different combinations of gene sequence components by using the exon sequences in different combinations. Can be used in protein engineering, although also occurs *in vivo*, where it is usually referred to as exon shuffling.

DNase protection Method used to determine protein-binding regions on DNA sequences where the protein protects the DNA from nuclease digestion.

DOI See *digital object identifier.*

Dolly The first organism to be cloned using the somatic cell nuclear transfer (SCNT) method with the donor nucleus taken from a fully differentiated adult cell from a 6-year-old Dorset Finn sheep.

Domain (1) In protein structure, a domain is a distinct region of the protein that often folds independently from the rest of the polypeptide chain. (2) In website terminology, a domain name is used to identify a website address. Usually prefixed by *http://* and/or *www.* Often ends in a generic term (such as *.co*, *.com*, *.ac*, *.gov*) and may have a country identifier also (thus *.ac.uk* is an academic institution in the United Kingdom). *Cf. uniform resource locator.*

Dot-blotting Technique in which small spots, or 'dots', of nucleic acid are immobilised on a nitrocellulose or nylon membrane for hybridisation. *Cf. slot-blot.*

Double-strand break (DSB) A break involving both strands of a DNA molecule. In isolation, causes serious mutation. Also generated as part of the CRISPR-Cas9 gene editing system (*q.v.*).

Down-regulation Reduction in the level of expression of a gene in response to a regulatory signal or process.

Downstream processing (DSP) Refers to the procedures used to purify products (usually proteins) after they have been expressed in bacterial, fungal or mammalian cells.

Down syndrome Clinical condition resulting from trisomy-21 (three copies of chromosome 21), a consequence of non-disjunction during meiosis.

Droplet digital PCR (ddPCR) A development of digital PCR (*q.v.*) using water-in-oil emulsion droplets.

Drosophila melanogaster Fruit fly used as a model organism in genetic, developmental and molecular studies.

DSB See *double-strand break*.

Duchenne muscular dystrophy (DMD) X-linked (*q.v.*) muscle-wasting disease caused by defects in the gene for the protein dystrophin (*q.v.*).

Dystrophin Large protein linking the cytoskeleton to the muscle cell membrane, defects in which cause muscular dystrophy.

Eastern blotting Method for determining post-translational modifications in proteins. *Cf. Northern, Southern* and *Western blotting*.

Electrophoretic mobility shift assay (EMSA) Method of determining protein-binding sites on DNA fragments on the basis of their reduced mobility, relative to unbound DNA, in gel electrophoresis experiments.

Electroporation Technique for introducing DNA into cells by giving a transient electric pulse.

ELISA See *enzyme-linked immunosorbent assay*.

ELSI Sometimes used as shorthand to describe the ethical, legal and social implications of genetic engineering.

Embryo splitting Technique used to clone organisms by separating cells in the early embryo, which then go on to direct development and produce identical copies of the organism.

Embryonic stem (ES) cells Derived from cells in the late blastocyst stage of embryogenesis, ES cells are pluripotent and have the capacity for unlimited self-renewal.

Emergent properties Refers to the appearance of new characteristics as complexity increases, often not predictable from knowledge of the component parts that make up the next level of complexity. Sometimes described as 'the whole is greater than the sum of the parts'.

emPCR See *emulsion PCR*.

EMSA See *electrophoretic mobility shift assay*.

Emulsion PCR (emPCR or ePCR) A PCR reaction within a water-in-oil emulsion using picolitre-volume droplets.

End labelling Adding a radioactive molecule onto the end(s) of a polynucleotide.

Endonuclease An enzyme that cuts within a nucleic acid molecule, as opposed to an exonuclease (*q.v.*), which digests DNA from one or both ends.

End-point PCR Standard PCR technique where the reaction is run for a specific number of cycles.

Enhancer A sequence that enhances transcription from the promoter of a eukaryotic gene. May be several thousand base-pairs away from the promoter.

Ensembl Browser for looking at DNA sequence data from genomic sequencing projects. Ensembl was developed by EMBL and the Wellcome Trust Sanger Institute.

Enzyme A protein that catalyses a specific reaction.

Enzyme-linked immunosorbent assay (ELISA) Technique for detection of specific antigens by using an antibody linked to an enzyme that generates a coloured product. The antigens are fixed onto a surface (usually a 96-well plastic plate), and thus large numbers of samples can be screened at the same time.

Enzyme replacement therapy (ERT) Therapeutic procedure in which a defective enzyme function is restored by replacing the enzyme itself. *Cf. gene therapy*.

Epigenesis Theory of development that regards the process as an iterative series of steps, in which the various signals and control events interact to regulate development.

epPCR See *error-prone PCR*.

EPSP synthase An enzyme (5-enolpyruvylshikimate-3-phosphate synthase) of amino acid biosynthesis that is inhibited by the herbicide glyphosate.

Error-prone PCR (epPCR) PCR carried out under low stringency conditions, thus generating variants of PCR products. Can be used in directed mutagenesis techniques.

Escherichia coli The most commonly used bacterium in molecular biology.

Ethidium bromide A molecule that binds to DNA and fluoresces when viewed under ultraviolet light. Used as a stain for DNA.

Eukaryotic The property of having a membrane-bound nucleus.

Exome The part of the genome composed of exons, *i.e.* the expressed sequences that generate mature mRNA molecules after post-transcriptional processing.

Exome sequencing Sequencing the exome region of a genome only; avoids non-expressed regions. Also known as whole exome sequencing (WES).

Exon Region of a eukaryotic gene that is expressed *via* mRNA.

Exon shuffling See *DNA shuffling*.

Exonuclease An enzyme that digests a nucleic acid molecule from one or both ends.

Expressivity The degree to which a particular genotype generates its effect in the phenotype. *Cf. penetrance*.

Expressome (1) The entire complement of expressed components in any given cell – thus includes primary transcripts, mature mRNAs and proteins. (2) A complex of RNA polymerase, mRNA and a ribosome formed during coupled transcription and translation in bacteria.

Extrachromosomal element A DNA molecule that is not part of the host cell chromosome.

F508del (delta-F508, ΔF508) Single amino acid deletion and most common mutation in the CFTR gene (*q.v.*). Causes cystic fibrosis by misfolding of the CFTR protein.

Factor IX (FIX) One of a family of blood clotting factors; can be produced in transgenic sheep.

FEN1 Flap-editing nuclease, in gene manipulation used as part of the prime editing method to trim single-stranded flaps of DNA.

Fibrin Insoluble protein involved in blood clot formation.

Filtration Method of separating solid and liquid components of a suspension by passing through a filter.

Finished sequence data Refers to a completed DNA or protein sequence in which anomalies and missing regions have been resolved.

Flavr Savr (*sic*) Transgenic tomato in which polygalacturonase (*q.v.*) synthesis is restricted using antisense technology. Despite the novel science, the Flavr Savr was not a commercial success.

Flush ends See *blunt ends*.

Fok I Type IIS restriction enzyme from *Flavobacterium okeanokites*, used in engineered nucleases such as ZFNs and TALENS (*q.v.*).

Foldback DNA Class of DNA which has palindromic or inverted repeat regions that re-anneal rapidly when duplex DNA is denatured.

Formulation Used to describe the 'recipe' used for the production of a pharmaceutical or other product.

Freeze-drying Technique for concentrating solutes by removal of water under vacuum at low temperature.

Functional genomics Field of study that is used to investigate how genes are expressed in the context of their role in the genome, as opposed to in isolation. *Cf. structural genomics, comparative genomics*.

Fusion protein A hybrid recombinant protein that contains vector-encoded amino acid residues at the N-terminus.

GAL 4 promoter Yeast gene promoter used upstream of a reporter gene in the yeast hybrid systems. See *Y1H, Y2H* and *Y3H*.

Gamete Refers to the haploid male (sperm) and female (egg) cells that fuse to produce the diploid zygote (*q.v.*) during sexual reproduction.

Gateway cloning System developed by Invitrogen that enables simplified transfer of DNA sequences between plasmid cloning vectors.

Gel electrophoresis Technique for separating nucleic acid molecules on the basis of their movement through a gel matrix under the influence of an electric field. See *agarose* and *polyacrylamide*.

Gel retardation See *electrophoretic mobility shift assay*.

GenBank One of the original DNA sequence databases, and one of the main repositories for genome sequence data.

Gene The unit of inheritance, located on a chromosome. In molecular terms, usually taken to mean a region of DNA that encodes one function. Broadly, therefore, one gene encodes one protein.

Gene bank See *genomic library*.

Gene cloning The isolation of individual genes by generating recombinant DNA molecules, which are then propagated in a host cell which produces a clone that contains a single fragment of the target DNA.

Gene drive Process, natural or engineered, to skew the transmission rate of an allele in a population (usually 50% for an allele pair in a Mendelian transmission pattern). Potential to use gene drives to control populations, *e.g. Anopheles* mosquitoes (*q.v.*) for the control of malaria.

Gene editing See *genome editing*.

Gene protection technology Range of techniques used to ensure that particular commercially derived recombinant constructs cannot be used without some sort of control or process, usually supplied by the company marketing the recombinant. Also known as genetic use restriction technology and genetic trait control technology.

Gene silencing Down-regulation of gene expression using methods such as antisense RNA or RNA interference (*q.v.*).

Gene therapy (GT) The use of cloned genes in the treatment of genetically derived malfunctions. May be delivered *in vivo* or *ex vivo*. May be offered as gene addition or gene replacement versions.

Genetic code The triplet codons that determine the types of amino acid that are inserted into a polypeptide during translation. There are 61 codons for 20 amino acids (plus three stop codons), and the code is therefore referred to as *degenerate*.

Genetic fingerprinting A method which uses radioactive probes to identify bands derived from hypervariable regions of DNA (*q.v.*). The band pattern is unique for an individual, and can be used to establish identity or family relationships.

Genetic mapping Low-resolution method to assign gene locations (loci) to their position on the chromosome. *Cf. recombination frequency mapping, physical mapping*.

Genetic marker A phenotypic characteristic that can be ascribed to a particular gene.

Genetic use restriction technology (GURT) See *gene protection technology*.

Genetically modified organism (GMO) An organism in which a genetic change has been engineered. Usually used to describe transgenic or gene-edited plants and animals.

Genome Used to describe the complete genetic complement of a virus, cell or organism.

Genome editing Method for altering parts of an individual genome, most commonly using a technique based on the CRISPR-Cas9 system (*q.v.*) or similar.

Genome-wide association study (GWAS) Statistical approach to analysing genome variation and association with disease traits.

Genomic library A collection of clones which together represent the entire genome of an organism.

Genomics The study of genomes, particularly genome sequencing.

GFP See *green fluorescent protein*.

GIGO From computer science, meaning 'garbage in, garbage out'. Refers to the need to ensure that input data are robust and accurate if valid results are to be obtained when processed by various computer applications.

Glyphosate Herbicide that inhibits EPSP synthase (*q.v.*).

GMO See *genetically modified organism*.

Golden Gate cloning System for assembling fragments of DNA into contiguous sequences using type IIS restriction enzymes.

Golden Rice (GR) Rice engineered to contain β-carotene in the endosperm. Potential to address vitamin A deficiency in a significant way, but progress to full deployment hampered by anti-GMO groups.

Green fluorescent protein (GFP) Protein useful as a marker for various processes in cell biology and rDNA techniques. As the name suggests, GFP-expressing tissues fluoresce with a green colour.

Guanine (G) Nitrogenous base found in DNA and RNA.

Guide RNA (gRNA) RNA sequence involved in specifying the target sequence for the CRISPR-Cas9 genome editing process.

GURT See *genetic use restriction technology*, *gene protection technology*.

HAC Human artificial chromosome.

Half-life Time taken for half of something to decay or become non-functional; usually applies to radioactive decay and reduction of enzyme activity.

HAMA See *human anti-mouse antibody response*.

Haploid Having one set of chromosomes. *Cf. diploid*.

Hapten A small molecule used to link a biological molecule to, for example, a fluorophore to produce a bioconjugate (*q.v.*) as part of what is sometimes called 'click chemistry'.

HBsAg The surface antigen of hepatitis B virus, also known as the Australia antigen. Can be assayed to detect viral infection.

HDR See *homology-directed repair*.

Hemi-methylation Addition of a methyl (CH_3) group to one DNA strand only, involved in regulation of cellular development. *Cf. methylation*.

Hepatitis Inflammatory disease of the liver, most often caused by infection with the hepatitis virus (types A–E). The likelihood of infection and impact of the disease depend on the type and other factors such as general health and lifestyle.

Hereditary transthyretin amyloidosis polyneuropathy Rare disease resulting from extracellular amyloid deposition. Caused by mutations in the transthyretin gene. Notable as the first approval for mRNA-based therapy using small interfering RNA (siRNA) was granted in 2018 for a therapeutic agent (Onpattro) to treat the condition.

Heterologous Refers to gene sequences that are not identical, but show variable degrees of similarity.

Heteropolymer A polymer composed of different types of monomer. Most protein and nucleic acid molecules are heteropolymers.

Heterozygous Refers to a diploid organism (cell or nucleus) which has two different alleles at a particular locus.

HGP Human Genome Project (*q.v.*).

High-throughput sequencing (HTS) Large-scale DNA sequencing using next-generation sequencing techniques.

HIV See *human immunodeficiency virus*.

Hogness box See *TATA box*.

Homologous (1) Refers to paired chromosomes in diploid organisms. (2) Used to strictly describe DNA sequences that are identical; however, the percentage homology between related sequences is sometimes quoted.

Homologous recombination (HR) See *homology-directed repair*.

Homology-directed repair (HDR) Cellular DNA repair mechanism that uses a region of sequence homology to repair a break in the DNA helix. Most common form is homologous recombination (*q.v.*).

Homopolymer A polymer composed of only one type of monomer, such as polyphenylalanine (protein) or polyadenine (nucleic acid).

Homozygous Refers to a diploid organism (cell or nucleus) which has identical alleles at a particular locus.

Host A cell used to propagate recombinant DNA molecules.

Hot-start PCR A method of delaying the start of a PCR reaction to reduce non-specific amplification during the warm-up phase.

Human anti-mouse antibody (HAMA) response Immune reaction against the murine components of a therapeutic antibody when used in humans.

Human Genome Project (HGP) The multinational collaborative effort to determine the DNA sequence of the human genome.

Human immunodeficiency virus (HIV) Retrovirus, causative agent of acquired immune deficiency syndrome (AIDS, *q.v.*).

Humira The first fully human therapeutic monoclonal antibody, Humira is the trade name for an inhibitor of tumour necrosis factor alpha (anti-TNF-α). Has been the bestselling pharmaceutical worldwide from 2014 to 2020.

Humulin Recombinant-derived human insulin.

Huntington disease (HD) Middle age-onset autosomal dominant degenerative condition. Sometimes known as Huntington's chorea, so-called because of the characteristic physical movements of affected individuals.

Hybrid arrest translation (HART) Techniques used to identify the protein product of a cloned gene, in which translation of its mRNA is prevented by the formation of a DNA~mRNA hybrid.

Hybrid release translation Technique in which a particular mRNA is selected by hybridisation with its homologous, cloned DNA sequence, and is then translated to give a protein product that can be identified.

Hydrolysis Reaction where two covalently joined molecules are split apart by the addition of the elements of water – in effect, the reversal of a dehydration synthesis reaction.

Hyperchromic effect Change in absorbance of nucleic acids, depending on the relative amounts of single-stranded and double-stranded forms. Used as a measurement in denaturation/renaturation studies.

Hypothesis Possible explanation for an observed event, tested by experiment and data analysis.

Hypothesis-free Approach to scientific investigation where no hypothesis is being tested. An example is a genome-wide association study (*q.v.*) where statistical techniques are used to see if there are correlations that were previously not identified by traditional investigation.

Hypothetico-deductive Term used to describe a hypothesis-driven form of scientific method. See also *deductive, inductive*.

Ice-minus bacteria Bacteria engineered to disrupt the normal ice-forming process, used to protect plants from frost damage.

IDDM See *insulin-dependent diabetes mellitus*.

IFA See *indirect immunofluorescence assay*.

IGF-1 See *insulin-like growth factor*.

Illumina sequencing A next-generation sequencing method for DNA using a massively parallel sequencing-by-synthesis technique.

Immunorejection Rejection of transplanted tissues due to immune-based non-compatibility.

Indel Insertion/deletion; refers to either an insertion of additional bases or deletion of bases during error-prone repair. Indels generate mutations and are introduced deliberately during gene editing protocols.

Independent variable The variable that is set in an experiment (*e.g.* time, temperature). *Cf. dependent variable*.

Indirect immunofluorescence assay (IFA) Assay using secondary antibody detection methods to identify antigens.

Inductive Form of reasoning by which specific observations are used to draw general conclusions. *Cf. deductive*.

Information technology (IT) Commonly (and somewhat loosely) used to describe computer-based manipulation of data sets of various types.

Inosine Nucleoside found in tRNA, sometimes used in synthetic oligonucleotides at degenerate positions as it can pair with all the other DNA bases.

Insertion vector A bacteriophage vector that has a single cloning site into which DNA is inserted.

Insertional inactivation Insertion of sequence that causes inactivation of a gene. Often used to inactivate a marker gene in a vector to enable selection of clones using selective growth media, *e.g.* media containing an antibiotic.

In silico Used to describe virtual experiments carried out in a computer, often by manipulation of sequence data. One example is to predict the effect on a protein's structure of altering specific parts of the gene sequence.

Insulin Protein hormone involved in the regulation of blood glucose levels. Has been available in recombinant form since the 1980s.

Insulin-dependent diabetes mellitus (IDDM) An autoimmune disease caused by a number of factors. Characterised by the destruction of the insulin-producing β-cells in the pancreas. *Cf. Non-insulin-dependent diabetes mellitus*.

Insulin-like growth factor (IGF-1) Polypeptide hormone, synthesis of which is stimulated by growth hormone. Implicated in some concerns about the safety of using recombinant bovine growth hormone in cattle to increase milk yields.

Interactome Term used to describe the set of interactions between cellular components such as proteins and other metabolites. *Cf. genome, proteome, transcriptome, metabolome*.

Intermediate vector Plasmid vector used in conjunction with a second plasmid in, for example, the binary Ti system. The two plasmids complement each other to enable full functionality.

Intervening sequence Region in a eukaryotic gene that is not expressed *via* the processed mRNA.

Intragenic Within the genome, refers to gene editing procedures where the gene is not from an external source. *Cf. cisgenic, transgenic, xenogenic.*

Intron See *intervening sequence.*

Inverse PCR (IPCR) Method for using PCR to amplify DNA for which there are no sequence data available for primer design. Works by circularising and inverting the target sequence so that primers within a known sequence area can be used.

Inverted palindrome A palindrome (*q.v.*) sequence that is the same on each of the two strands of DNA when read in the same direction (*e.g.* 5′-GAATTC-3′ and its complement). *Cf. mirror palindrome.*

Inverted repeat A short sequence of DNA that is repeated, usually at the ends of a longer sequence, in a reverse orientation.

Ion Torrent NGS method that measures ion release in a massively parallel microchip-based method.

IPTG *iso*-propyl-thiogalactoside, a gratuitous inducer which de-represses transcription of the lac operon.

Isoschizomers Different restriction endonucleases (*q.v.*) that recognise the same DNA sequence and have the same cutting pattern.

IT See *information technology.*

Klenow fragment A fragment of DNA polymerase I that lacks the 5′ → 3′ exonuclease activity.

Knockdown Refers to a reduction in gene expression, often as the result of modification of the target gene sequence (or control sequences) using rDNA techniques.

Knockin mouse A transgenic mouse in which a gene function has been added or 'knocked in'. Used primarily to generate animal models for the study of human disease (*Cf. knockout mouse*).

Knockout (KO) mouse A transgenic mouse in which a gene function has been disrupted or 'knocked out'. Used primarily to generate animal models for the study of human disease, *e.g.* cystic fibrosis (*Cf. knockin mouse*).

Lambda (λ) Bacteriophage used in vector construction, commonly for the generation of insertion and replacement vectors. Replacement (substitution) vectors are often used for genomic library construction and can accept DNA fragments of up to some 23 kb in length.

Leber congenital amaurosis Eye disorder that affects the retina. Target disease for the first *in vivo* gene therapy trial in 2020.

LIC See *ligation-independent cloning.*

Ligation Joining two fragments of DNA together by sealing the break in the phosphodiester backbone using DNA ligase.

Ligation-independent cloning (LIC) Method of cloning DNA fragments using complementary overhanging sequences rather than cohesive ends generated by restriction enzyme digestion. Useful for the assembly of vector constructs.

Linkage mapping Genetic mapping (*q.v.*) technique used to establish the degree of linkage between genes. See also *recombination frequency mapping.*

Linker A synthetic self-complementary oligonucleotide that contains a restriction enzyme recognition site. Used to add cohesive ends (*q.v.*) to DNA molecules that have blunt ends (*q.v.*).

Lipase Enzyme that hydrolyses fats (lipids).

Liposome (lipoplex) Lipid-based method for delivering gene therapy.

Locus The site at which a gene is located on a chromosome.

Lysogenic Refers to bacteriophage infection that does not cause lysis of the host cell.

Lytic Refers to bacteriophage infection that causes lysis of the host cell.

mAb See *monoclonal antibody*.

Macromolecule Large polymeric molecule made up of monomeric units, commonly used to describe proteins (monomers are amino acids) and nucleic acids (monomers are nucleotides).

Massively parallel Refers to next-generation sequencing methods where millions of short fragments are sequenced concurrently. *Cf. single-molecule real-time (SMRT) sequencing.*

Melting temperature (T_m) The temperature at which a duplex DNA or DNA~RNA molecule 'melts' to separate the strands. Depends on base composition and various other factors. A high G·C content increases the T_m due to the three hydrogen bonds, instead of two, in an A·T or A·U base-pair.

Messenger RNA (mRNA) The ribonucleic acid molecule transcribed from DNA that carries the codons specifying the sequence of amino acids in a protein.

Metabolome Refers to the population of metabolites in a cell, which, together with proteomics and transcriptomics (*q.v.*), can give a snapshot of the cell's activity at any given point in time.

Methylation Epigenetic modification that adds a methyl (CH_3) group to DNA, involved in the regulation of cellular processes. In bacteria, DNA methylation can protect against the bacterium's own restriction enzyme; in mammalian cells, methylation is involved in cellular development. *Cf. hemi-methylation.*

Microinjection Introduction of DNA into the nucleus or cytoplasm of a cell by insertion of a microcapillary and direct injection.

Micro RNAs (miRNAs) Short RNA molecules synthesised as part of the RNA interference mechanism (*q.v.*).

Microsatellite DNA Type of sequence repeated many times in the genome. Based on dinucleotide repeats, microsatellites are highly variable and can be used in mapping and profiling studies.

MinION Small NGS DNA sequencer using nanopore technology to sequence DNA as a series of single molecules.

Minisatellite DNA Type of sequence based on variable number tandem repeats (VNTRs; *q.v.*). Used in genetic mapping and profiling studies.

Mini-Ti vector Vector for cloning in plant cells, based on part of the Ti plasmid of *Agrobacterium tumefaciens* (*q.v.*).

miRNA See *micro RNAs*.

Mirror palindrome A palindrome (*q.v.*) where the DNA sequence on a single strand is palindromic (*e.g.* 5'-TAGGAT-3'). *Cf. inverted palindrome.*

Molecular cloning Alternative term for gene cloning.

Molecular ecology Use of molecular biology and recombinant DNA techniques in studying ecological topics.

Molecular palaeontology Use of molecular techniques to investigate the past, as in DNA profiling from mummified or fossilised samples.

Monoclonal antibody (mAb) Antibody that recognises a single antigenic determinant. Plays a major role in the treatment of disease by antibody therapy.

Monogenic Trait caused by a single gene. *Cf. polygenic trait.*

Monosomic Diploid cells in which one of a homologous pair of chromosomes has been lost. *Cf. trisomy.*

Monozygotic Refers to identical twins, generated from the splitting of a single embryo at an early stage.

Mosaic An embryo or organism in which not all the cells carry identical genomes.

mRNA See *messenger RNA.*

Multifactorial Caused by many factors, *e.g.* genetic trait in which many genes and environmental influences may be involved.

Multi-locus probe DNA probe used to identify several bands in a DNA fingerprint or profile. Generates the 'barcode' pattern in a genetic fingerprint.

Multiple cloning site (MCS) A short region of DNA in a vector that has recognition sites for several restriction enzymes.

Multiplex PCR A PCR reaction in which multiple target sequences are amplified at the same time. *Cf. singleplex PCR.*

Multipotent Cell which can give rise to a range of differentiated cells. *Cf. totipotent, pluripotent.*

Mus musculus The mouse, a major model organism for molecular genetic studies. Many different types of transgenic mice are available.

Mutagenesis *in vitro* Introduction of defined mutations in a cloned sequence by manipulation in the test tube. See also *oligonucleotide-directed mutagenesis.*

Mutant An organism (or gene) carrying a genetic mutation.

Mutation An alteration to the sequence of bases in DNA. May be caused by insertion, deletion or modification of bases.

Muteins Refers to proteins that have been engineered by the incorporation of mutational changes.

Native protein A recombinant protein that is synthesised from its own N-terminus, rather than from an N-terminus supplied by the cloning vector.

Negative control (1) Control of gene expression by switching off or inhibiting a gene, mostly at the transcriptional level. (2) A control sample in an experiment without the test agent or target, used to check for any reagent contamination or undesired reactivity. *Cf. positive control.*

Nested fragments A series of nucleic acid fragments that differ from each other (in terms of length) by one or only a few nucleotides.

Nested PCR Form of PCR where two sets of primers are used, one pair being internal (nested) with respect to the other pair.

Neurospora crassa An ascomycete fungus, commonly called red bread mould. Used as a model organism for research and in biotechnology applications.

Neutral molecular polymorphism A molecular polymorphism that has no adverse effect and can be used to tag a gene sequence with which it co-segregates.

Next-generation sequencing (NGS) Range of advanced automated methods for high-throughput sequencing of DNA.

NHEJ See *non-homologous end joining*.

Nick translation Method for labelling DNA with radioactive dNTPs.

Nickase In gene editing, inactivation of one of the nuclease domains reduces the activity of the enzyme and results in single-strand breaks ('nicks', hence the term) rather than double-strand breaks.

NIDDM See *non-insulin-dependent diabetes mellitus*.

Non-Hodgkin lymphoma (NHL) Cancer of the lymphatic system. In litigation against Monsanto (now part of Bayer), it is claimed that glyphosate increases the risk of NHL.

Non-homologous end joining (NHEJ) DNA repair mechanism that functions in the absence of any additional donor DNA sequence and generates error-prone repairs that tend to introduce insertions or deletions (indels, *q.v.*) into the sequence. *Cf. homology-directed repair.*

Non-insulin-dependent diabetes mellitus (NIDDM) Form of diabetes, also known as type II diabetes. Often linked with obesity and inactivity, it is an increasing problem, particularly in developed countries where overnutrition (*q.v.*) is an issue.

Northern blotting Transfer of RNA molecules onto membranes for the detection of specific sequences by hybridisation. *Cf. Southern, Eastern* and *Western blotting.*

N-terminus Amino terminus, defined by the $-NH_2$ group of an amino acid or protein.

Nuclear transfer Method for cloning organisms in which a donor nucleus is taken from a somatic cell and transferred to the recipient ovum. Also known as somatic cell nuclear transfer (SCNT).

Nuclease An enzyme that hydrolyses phosphodiester bonds.

Nucleotide A nucleoside bound to a phosphate group.

Nucleoid Region of a bacterial cell in which the genetic material is located.

Nucleus Membrane-bound region in a eukaryotic cell that contains the genetic material.

Nutritional biofortification Improvement of the nutritional quality of a foodstuff by increasing the amount of a specific nutrient. The best example currently is Golden Rice (*q.v.*), which has been engineered to produce β-carotene in the endosperm. Potential for alleviating the symptoms of vitamin A deficiency.

Occam's razor Philosophical and scientific principle that encourages consideration of the simplest explanation for any event or phenomenon. Also called the law of economy or law of parsimony.

Oligo(dT)-cellulose Short sequence of deoxythymidine residues linked to a cellulose matrix, used in the purification of eukaryotic mRNA.

Oligolabelling See *primer extension*.

Oligomer General term for a short sequence of monomers.

Oligonucleotide A short sequence of nucleotides.

Oligonucleotide-directed mutagenesis Process by which a defined alteration is made to DNA using a synthetic oligonucleotide. *Cf. site-directed mutagenesis*.

Oncomouse Transgenic mouse engineered to be susceptible to cancer.

One-hybrid system (Y1H) Refers to the yeast one-hybrid system. See *Y1H, Y2H* and *Y3H*.

Online Mendelian Inheritance in Man (OMIM) A database of genetically based conditions.

Onpattro The first approved RNAi-based drug for treatment of hereditary transthyretin amyloidosis polyneuropathy (*q.v.*).

Open reading frame (ORF) A sequence of DNA that is a potential gene coding sequence, where the reading frame is 'open'. Further investigation can define the regulatory elements and thus confirm ORFs as genes. Considered by some to be an unhelpful term, but widely used in genome sequence analysis.

Operator Region of an operon, close to the promoter, to which a repressor protein binds.

Operon A cluster of bacterial genes under the control of a single regulatory region.

Opines Group of modified amino acids encoded by the Ti plasmid of *Agrobacterium tumefaciens* (*q.v.*).

ORF See *open reading frame*.

Organismal cloning The production of an identical copy of an individual organism by techniques such as embryo splitting or nuclear transfer. Used to distinguish the process from molecular cloning (*q.v.*).

Ornithine transcarbamylase deficiency (OTCD) Enzyme deficiency condition targeted for gene therapy.

Over-medicalisation The practice of using pharmaceutical or medical intervention when this is not essential, usually only a problem found in developed countries' healthcare systems.

Overnutrition Consumption of more food than is required to maintain a healthy lifestyle. Largely a condition found in developed countries, overnutrition can lead to lifestyle-choice diseases such as obesity, diabetes and cancers. *Cf. undernutrition*.

PacBio sequencing Method for NGS sequencing that uses the single-molecule real-time approach in a zero-mode waveguide (ZMW, *q.v.*).

Packaging *in vitro* Method to produce viable recombinant phage particles by self-assembly of the components in a test tube.

Palindrome Usually refers to a DNA sequence that reads the same on both strands when read in the same (*e.g.* 5′ → 3′) direction (See *inverted palindrome*). Examples include many restriction enzyme recognition sites. *Cf. mirror palindrome*.

PAM See *protospacer adjacent motif.*

PCR See *polymerase chain reaction.*

Pedigree analysis Determination of the transmission characteristics of a particular gene by examination of family histories.

pegRNA In the prime gene editing method, the guide RNA is known as prime editing guide RNA.

Penetrance The proportion of individuals with a particular genotype that show the genotypic characteristic in the phenotype. *Cf. expressivity.*

Phage See *bacteriophage.*

Phage display Technique used for the production of chimeric antibodies by expressing them on the surface of a bacteriophage such as M13. Also used to study protein~protein interactions.

Phagemid A vector containing plasmid and phage sequences.

Pharm animal Transgenic animal used for the production of pharmaceuticals.

Phosphodiester bond A bond formed between the 5′ phosphate and the 3′ hydroxyl groups of two nucleotides.

Physical containment Refers to engineering protective measures to contain GM experiments and ensure that no host/vector systems can escape into the environment. Includes rooms under negative pressure, filtration and secure access and egress in laboratory design. *Cf. biological containment.*

Physical mapping Mapping genes with reference to their physical location on the chromosome. Generates the next level of detail, compared to genetic mapping (*q.v.*).

Physical marker A sequence-based tag that labels a region of the genome. There are several such tags that can be used in mapping studies. *Cf. RFLP, STS.*

Plant-made pharmaceuticals (PMPs) Therapeutics made in plants by using the plant as a bioreactor. The production of vaccines in virus-like particles is one example.

Plaque A cleared area on a bacterial lawn caused by infection by a lytic bacteriophage.

Plasmid A circular extrachromosomal element found naturally in bacteria and some other organisms. Engineered plasmids are used extensively as vectors for cloning.

Plasmin Active form of the protease derived from plasminogen that acts to hydrolyse fibrin, the constituent of blood clots.

Plasminogen Precursor of plasmin, converted to the active form by tissue plasminogen activator (*q.v.*).

Plasmodium Genus of trypanosome, the causative agent of malaria. The most common species involved in malarial transfer is *P. falciparum*, carried by mosquitoes of the genus *Anopheles* (*q.v.*).

Ploidy number Refers to the number of sets of chromosomes, *e.g.* haploid, diploid, triploid, tetraploid, hexaploid.

Pluripotent Cell which can give rise to a range of differentiated cells. *Cf. multipotent, totipotent.*

Polyacrylamide A cross-linked matrix for gel electrophoresis (*q.v.*) of small fragments of nucleic acids, primarily used for electrophoresis of DNA. Also used for electrophoresis of proteins.

Polycistronic Refers to an RNA molecule encoding more than one function. Many bacterial operons are expressed *via* polycistronic mRNAs.

Polygalacturonase (PG) Enzyme involved in pectin degradation. Target for antisense control in the Flavr Savr tomato (*q.v.*).

Polygenic trait A trait determined by the interaction of more than one gene, *e.g.* eye colour in humans.

Polyhedra Capsid structures in baculoviruses, composed of the protein polyhedrin.

Polylinker See *multiple cloning site*.

Polymerase chain reaction (PCR) A method for the selective amplification of DNA sequences. Several variants exist for different applications.

Polymorphism Refers to the occurrence of many allelic variants of a particular gene or DNA sequence motif. Can be used to identify individuals by genetic mapping and DNA profiling techniques.

Polynucleotide A polymer made up of nucleotide monomers.

Polynucleotide kinase (PNK) An enzyme that catalyses the transfer of a phosphate group onto a 5' hydroxyl group.

Polypeptide A chain of amino acid residues. *Cf. protein.*

Polyphenol oxidases (PPOs) A group of enzymes that cause browning when surfaces of fruit and vegetables are exposed to air, *e.g.* after cutting or harvesting. Down-regulated to produce the Artic Apple (*q.v.*).

Posilac Commercial form of recombinant bovine somatotropin (rBST).

Positional cloning Cloning genes for which little information is available, apart from their location on the chromosome.

Positive control (1) Control of gene expression by switching on genes in response to a cellular signal. (2) A control in an experiment that has a sample of the test agent or target, used to check that the reaction is working as anticipated.

Post-transcriptional gene silencing (PTGS) Alternative name for RNA interference (RNAi, *q.v.*), often used to describe the use of RNAi in plants.

Post-translational modification (PTM) Modification of a protein after it has been synthesised. An example would be the addition of sugar residues to form a glycoprotein.

PPIs Protein~protein interactions.

Preformationism An early concept that all development is pre-coded in the zygote, and that embryological development is simply the unfolding of this information. *Cf. epigenesis.*

Pribnow box Sequence found in prokaryotic promoters that is required for transcription initiation. The consensus sequence (*q.v.*) is TATAAT.

Primary database Database in which data are deposited without, or with minimal, manipulation. May have sophisticated search and process tools.

Primary transcript The initial, and often very large, product of transcription of a eukaryotic gene. Subjected to processing to produce the mature mRNA molecule.

Primer extension Synthesis of a copy of a nucleic acid from a primer. Used in labelling DNA and in determining the start site of transcription.

Probe A labelled molecule used in hybridisation procedures.

Processivity Refers to the sequential addition of components by an enzyme; for example, DNA and RNA polymerases are processive enzymes.

Proinsulin Precursor of insulin that includes an extra polypeptide sequence that is cleaved to generate the active insulin molecule.

Prokaryotic The property of lacking a membrane-bound nucleus, *e.g.* bacteria such as *Escherichia coli*.

Promoter (P) DNA sequence(s) lying upstream from a gene, to which RNA polymerase binds.

Pronucleus One of the nuclei in a fertilised egg prior to fusion of the gametes.

Prophage A bacteriophage maintained in the lysogenic state in a cell.

Prostate mouse Transgenic mouse model for prostate cancer.

Protease Enzyme that hydrolyses polypeptides.

Protein A condensation (dehydration) heteropolymer composed of amino acid residues linked together by peptide bonds to give a polypeptide.

Proteome Refers to the population of proteins produced by a cell. *Cf. genome, transcriptome.*

Protoplast A cell from which the cell wall has been removed.

Protospacer adjacent motif (PAM) In the CRISPR system, the PAM is a short 2- to 6-bp sequence present in the target DNA, but absent from the CRISPR sequence. This prevents the system from cleaving the bacterial CRISPR region.

Pseudomonas syringae Bacterium that is a plant pathogen. Ice-forming properties used to develop ice-minus bacteria (*q.v.*).

PTGS See *post-transcriptional gene silencing*.

Pull-down assay Method to selectively purify a protein~DNA complex from a reaction mixture, often using an affinity tag (*q.v.*).

Purine A double-ring nitrogenous base such as adenine and guanine.

Pyrimidine A single-ring nitrogenous base such as cytosine, thymine and uracil.

Pyrococcus furiosus Hyperthermophilic bacterium from which a thermostable polymerase can be purified for use in PCR and other polymerase-dependent protocols.

Quadrivalent recombinant influenza vaccine A vaccine designed to recognise four different strains of the influenza virus.

Qualitative Information based on non-numerical formats, such as observations, descriptions, interviews, *etc. Cf. quantitative.*

Quantification cycle (Cq) The cycle number in real-time qPCR that indicates the detection threshold. The Cq is usually set well below the maximum cycle number in a standard end-point PCR, as the most accurate results are obtained as soon as a clearly defined signal is generated.

Quantitative Data based on numerical formats, such as measurement of a variable or collation of a set of measurements. *Cf. qualitative.*

Quantitative PCR (qPCR) A form of PCR reaction designed to quantify the amount of DNA in the original sample.

Radiolabelling Short for radioactive labelling; method used to incorporate radioactive isotopes into biological molecules. An example is labelling nucleic acids with ^{32}P-dNTPs to prepare high-specific activity probes for use in hybridisation experiments.

Random amplified polymorphic DNA (RAPD) PCR-based method of DNA profiling that involves amplification of sequences using random primers. Generates a type of genetic fingerprint that can be used to identify individuals.

RAPD See *random amplified polymorphic DNA.*

Rational design Process of engineering changes in protein structure by using existing knowledge to design such changes. *Cf. directed evolution.*

Reading frame The pattern of triplet codon sequences in a gene. There are three reading frames for any given strand of DNA, depending on which nucleotide is the start point. Insertion and deletion mutations can disrupt the reading frame and have serious consequences, as often the entire coding sequence becomes nonsense after the point of mutation.

Real-time PCR Alternative name for quantitative PCR (qPCR, *q.v.*).

Recessive dystrophic epidermolysis bullosa (RDEB) Distressing condition where a dysfunctional collagen gene (*COL7A1*) results in skin blistering easily. In 2022, a novel gene therapy with topical application in a gel formulation was developed to treat the condition.

Recombinant DNA (rDNA) A DNA molecule made up of sequences that are not normally joined together.

Recombination frequency mapping Method of genetic mapping that uses the number of crossover events that occur during meiosis to estimate the distance between genes. *Cf. physical mapping.*

Redundancy In molecular biology, refers to the fact that some amino acids can be specified by multiple codons; thus, the third base (the 'wobble' position in the codon) is effectively redundant from an informatics point of view.

Reference genome A genome that is used as a comparator for other genomes and gene sequences. The reference genome is usually a representative 'average' genome that is continually updated and annotated with regard to gene sequence identification, filling in any sequence gaps, anomalies, *etc.*

Renaturation kinetics Method of analysing the complexity of genomes by studying the patterns obtained when DNA is denatured and allowed to renature.

Repeat variable di-residue (RVD) A key feature of TALEs (transcription activator-like effectors, used to produce TALENs). The TALE has a highly conserved repeat region of 33–35 amino acids, within which is a variable region of two amino acids known as the repeat variable di-residue (RVD) region.

Replacement vector A bacteriophage vector in which the cloning sites are arranged in pairs, so that the section of the genome between these sites can be replaced with insert DNA.

Replication Copying the genetic material during the cell cycle. Also refers to the synthesis of new phage DNA during phage multiplication.

Replicon A piece of DNA carrying an origin of replication.

Restriction enzyme An endonuclease that cuts DNA at sites defined by its recognition sequence.

Restriction fragment length polymorphism (RFLP) A variation in the locations of restriction sites bounding a particular region of DNA, such that the fragment defined by the restriction sites may be of different lengths in different individuals.

Restriction mapping Technique used to determine the location of restriction sites in a DNA molecule.

Reverse transcriptase (RTase) An RNA-dependent DNA polymerase found in retroviruses, used *in vitro* for the synthesis of cDNA.

RFLP See *restriction fragment length polymorphism*.

Ribonuclease (RNase) An enzyme that hydrolyses RNA.

Ribonucleic acid (RNA) A condensation heteropolymer composed of ribonucleotides.

Ribosomal RNA (rRNA) RNA that is part of the structure of ribosomes.

Ribosome The 'jig' that is the site of protein synthesis. Composed of rRNA and proteins.

Ribosome binding site A region on an mRNA molecule that is involved in the binding of ribosomes during translation.

RISC See *RNA-induced silencing complex*.

RNA See *ribonucleic acid*.

RNAi See *RNA interference*.

RNA-induced silencing complex (RISC) Complex of protein and the nuclease slicer, involved in RNA interference (*q.v.*).

RNA interference (RNAi) Complex *in vivo* process involved in regulating gene expression post-transcriptionally by 'interfering' with transcript availability.

RNA processing The formation of functional RNA from a primary transcript (*q.v.*). In mRNA production, this involves removal of introns, addition of a 5′ cap and polyadenylation.

RNA-specific endonuclease An endonuclease activity in Cas13 protein that cuts RNA, proving to be useful in the development of RNA editing methods.

RT-PCR Reverse transcriptase (transcription) PCR, where a cDNA copy of mRNA is made and then amplified using PCR.

RT-qPCR Quantitative version of reverse transcriptase PCR, used to measure transcript levels. *Cf. RT-PCR*.

S1 mapping Technique for determining the start point of transcription using S1 nuclease.

S1 nuclease An enzyme that hydrolyses (degrades) single-stranded DNA. Used for a variety of procedures in gene manipulation.

Saccharomyces cerevisiae Unicellular yeast (baker's yeast, also known as budding yeast) that is extensively used as a model microbial eukaryote in molecular studies. Also used in the biotechnology industry for a range of applications, as well as in brewing and bread-making.

SCD See *sickle-cell disease*.

SCIDS See *severe combined immunodeficiency syndrome*.

Scintillation spectrometer (counter) A machine for determining the amount of radioactivity in a sample. Detects the emission of light from a fluor that is excited by radioactive disintegrations. Varies in efficiency according to the energy of the isotope, thus produces an estimate as counts per minute rather than actual disintegrations per minute.

SCNT See *nuclear transfer*.

Screening Identification of a clone in a genomic or cDNA library (*q.v.*) by using a method that discriminates between different clones.

Secondary database A database in which the information has been derived by manipulation of an existing data set.

Seedcorn funding Early-stage capital funding to enable a project or new company to become established. Often provided by public or private bodies who are willing to take a risk on the capital, or by charity foundations.

Selectable marker A gene that confers a selectable characteristic on an organism when expressed. Examples include resistance to antibiotics or other growth-limiting molecules, or nutritional pathway genes used as auxotrophic markers (*q.v.*).

Selection Exploitation of the genetics of a recombinant organism to enable desirable recombinant genomes to be selected over non-recombinants during growth.

Selective breeding Usually applies to breeding crop plants or farm animals by selecting organisms with desirable characteristics for mating. A long-standing part of the development of agriculture.

Sequence-tagged site (STS) Refers to a DNA sequence that is unique in the genome and which can be used in mapping studies. Usually identified by PCR amplification.

Severe combined immunodeficiency syndrome (SCIDS) A condition that results from a defective enzyme (adenosine deaminase, ADA, *q.v.*).

Shine–Dalgarno sequence See *ribosome binding site*.

Short tandem repeat (STR) Short (2–5 base-pairs) repeat elements in the genome, also known as microsatellites. Useful for mapping and tagging genomes. *Cf. VNTRs.*

Sickle-cell disease (SCD) Disease of red blood cell morphology, caused by defective haemoglobin that makes the erythrocytes hard and 'sticky', and shaped like a sickle. When presenting in the heterozygous form, can confer a level of resistance to malaria in countries where this is prevalent, which acts as a selective advantage for the sickle-cell allele. Heterozygotes are said to have the sickle-cell trait. In 2019, Victoria Gray became the first person to be treated with gene editing for SCD.

Single-locus probe Probe used in DNA fingerprinting that identifies a single sequence in the genome. Diploid organisms therefore usually show two bands in a fingerprint, one allelic variant from each parent.

Single-molecule real-time (SMRT) sequencing Method of sequencing a single DNA molecule in a specialised nanometre-scale reaction chamber. Examples are the PacBio SMRT sequencing method and the nanopore method (Oxford Nanopore Technologies).

Single nucleotide polymorphism (SNP) Polymorphic pattern at a single base, essentially the smallest polymorphic unit that can be identified.

Single-pass sequence data Sequence data that have been produced by a single read; often contain anomalies or errors. *Cf. draft sequence, finished sequence data.*

Singleplex PCR A standard PCR reaction in which a single sequence is amplified. *Cf. multiplex PCR.*

siRNAs See *small interfering RNAs.*

Site-directed mutagenesis See *oligonucleotide-directed mutagenesis.*

Slicer Nuclease involved in RNA interference (RNAi, *q.v.*) as part of the RISC complex (*q.v.*).

Slot-blotting Similar to dot-blotting (*q.v.*), but uses slots in a template apparatus rather than circular dots.

Small interfering RNAs (siRNAs) RNAs involved in gene regulation via the process of RNA interference (RNAi, *q.v.*).

SME Term used to describe a company with few to many employees, but which has not yet reached 'large' size. Many successful SMEs grow into larger companies as they expand their range of products or services.

SNP See *single nucleotide polymorphism.*

Southern blotting Method for transferring DNA fragments onto a membrane for detection of specific sequences by hybridisation. *Cf. Northern, Eastern* and *Western blotting.*

Specific activity The amount of radioactivity per unit material; for example, a labelled probe might have a specific activity of 10^6 counts/minute per microgram. Also used to quantify the activity of an enzyme.

STR See *short tandem repeat.*

Structural gene A gene that encodes a protein product.

Structural genomics Field of study that investigates how genomes are structured and organised. *Cf. comparative genomics, functional genomics.*

STS See *sequence-tagged site.*

Stuffer fragment The section in a replacement vector (*q.v.*) that is removed and replaced with insert DNA.

Substitution vector See *replacement vector.*

TA cloning Sub-cloning procedure that avoids restriction enzyme digestion. Uses single base A and T overhangs to act as what are essentially very short cohesive ends.

TALENs See *transcription activator-like effector nucleases.*

Taq polymerase Thermostable DNA polymerase from the thermophilic bacterium *Thermus aquaticus.* Used in the polymerase chain reaction (*q.v.*).

TATA box Sequence found in eukaryotic promoters. Also known as the Hogness box, it is similar to the Pribnow box (*q.v.*) found in prokaryotes, and has the consensus sequence TATAAAT.

T-cell One of the key cell types of the immune system, targeted for engineering with chimeric antibodies in cancer therapy. See *CAR T-cell*.

T-DNA Region of Ti plasmid of *Agrobacterium tumefaciens* that can be used to deliver recombinant DNA into the plant cell genome.

Technology S-curve Progression in the development of a new technology. Tends to follow a broadly similar pattern with recognisable stages. Timescale for each of the stages can vary according to the speed of uptake of the technology.

Technology transfer In biotechnology, refers to the process of moving from laboratory-scale science to commercial production. Often a complex and risky stage in product development.

Telomere Region of repetitive DNA at the ends of a eukaryotic chromosome. Acts as a protection during cell division, but shortens with successive divisions until the cell can no longer divide.

Temperate Refers to bacteriophages that can undergo lysogenic infection of the host cell.

Terminal transferase An enzyme that adds nucleotide residues to the 3' terminus of an oligo- or polynucleotide.

Tetracycline (Tc) A commonly used antibiotic.

Text mining Using bibliographic databases to search for information, analogous to searching sequence databases. *Cf. data mining.*

Therapeutic cloning Organismal cloning to enable the isolation of embryonic stem cells for potential therapeutic purposes.

Thermal cycler Heating/cooling system for PCR applications. Enables denaturation, primer binding and extension cycles to be programmed and automated.

Thermus aquaticus Thermophilic bacterium from which *Taq* polymerase (*q.v.*) is purified. Other bacteria from this genus include *Thermus flavus* and *Thermus thermophilus*.

Three-hybrid system (Y3H) Refers to the yeast three-hybrid system. See *Y1H*, *Y2H* and *Y3H*.

Thymine (T) Nitrogenous base found in DNA only.

Ti plasmid Plasmid of *Agrobacterium tumefaciens* (*q.v.*) that causes crown gall disease.

Tissue plasminogen activator (TPA) A protease that occurs naturally, and functions in breaking down blood clots. Acts on an inactive precursor (plasminogen), which is converted to the active form (plasmin). This attacks the clot by breaking up fibrin, the protein involved in clot formation.

TOPO cloning Restriction-free cloning that can use the TA cloning mechanism (*q.v.*) or blunt-end cloning. Used for simplified sub-cloning into vectors. TOPO refers to the enzyme topoisomerase I used in the procedure.

Totipotent A cell that can give rise to all cell types in an organism. Totipotency has been demonstrated by cloning carrots from somatic cells, and by nuclear transfer experiments in animals.

TPA See *tissue plasminogen activator*.

Traitor technology Version of gene protection technology (*q.v.*), sometimes called 'genetic trait control technology'.

Trans-acting element A genetic element that can exert its effect without having to be on the same molecule as a target sequence. Usually such an element encodes a protein product (perhaps an enzyme or a regulatory protein) that can diffuse to the site of action.

Transcription (T_C) The synthesis of RNA from a DNA template.

Transcription activator-like effector nucleases (TALENs) Engineered nuclease used for gene editing, incorporates the *Fok* I restriction enzyme (*q.v.*).

Transcriptional unit The DNA sequence that encodes the RNA molecule, *i.e.* from the transcription start site to the stop site.

Transcriptome The population of RNA molecules that is expressed by a particular cell type. *Cf. genome, proteome*.

Transfection Introduction of purified phage or virus DNA into cells.

Transfer RNA (tRNA) A small RNA (approximately 75–85 bases) that carries the anticodon and the amino acid residue required for protein synthesis.

Transformant A cell that has been transformed by exogenous DNA.

Transformation The process of introducing DNA (usually plasmid DNA) into cells. Also used to describe the change in growth characteristics when a cell becomes cancerous.

Transgene The target gene involved in the generation of a transgenic (*q.v.*) organism.

Transgenic An organism that carries DNA sequences that it would not normally have in its genome, from a source that is reproductively incompatible. *Cf. cisgenic, intragenic, xenogenic*.

Translation (T_L) The synthesis of protein from an mRNA template.

Transposable element A genetic element that carries the information that allows it to integrate at various sites in the genome. Transposable elements are sometimes called 'jumping genes'.

TrEMBL Stands for 'Translated EMBL' and is a tool for deriving protein sequence data from nucleotide sequence databases.

Trisomy Aneuploid (*q.v.*) condition where an extra chromosome is present. A common example is the trisomy-21 condition that causes Down syndrome.

Tumour necrosis factor alpha (TNF-α) Protein involved in a number of cell signalling and immune functions, with mutated proteins causing a range of inflammatory diseases. The anti-TNF drug Humira (*q.v.*) was the biggest selling pharmaceutical from 2014 until COVID-19 vaccines in 2021.

Two-hybrid system (Y2H) Refers to the yeast two-hybrid system. See *Y1H, Y2H* and *Y3H*.

Undernutrition Not having enough food to sustain a healthy lifestyle; largely a symptom of agricultural and societal inequality in developing countries. *Cf. overnutrition.*

Uniform resource locator (URL) A sub-descriptor of a website address or domain. A URL locates a particular web page within a website. *Cf. domain (2).*

Unique sequence A DNA sequence present only once in the haploid genome. *Cf. abundance class.*

Uracil (U) Nitrogenous base found in RNA only.

URL See *uniform resource locator.*

Variable number tandem repeat (VNTR) Repetitive DNA composed of a number of copies of a short sequence, involved in the generation of polymorphic loci that are useful in genetic fingerprinting. Also known as hypervariable regions. See also *minisatellite DNA* and *microsatellite DNA.*

Vector A DNA molecule that is capable of replication in a host organism, and can act as a carrier molecule for the construction of recombinant DNA.

Venture capital (VC) Funding used to capitalise young companies to enable development to a more secure stage. Often following seedcorn funding, venture capital is often high risk for the investor.

Virulent Refers to bacteriophages that cause lysis of the host cell.

Vitamin A deficiency (VAD) Most common cause of childhood blindness, often leading to premature death. Addressed by supplements, but also potentially alleviated by dietary ingestion of β-carotene in Golden Rice *(q.v.).*

VNTR See *variable number tandem repeat.*

WES See *exome sequencing.*

Western blotting Transfer of electrophoretically separated proteins onto a membrane for probing with antibody. *Cf. Northern, Southern* and *Eastern blotting.*

WGS See *whole genome shotgun* and *whole genome sequencing.*

Whole genome sequencing Sequencing of entire genomes; often conflated with whole genome shotgun sequencing *(q.v.).*

Whole genome shotgun Method for large-scale sequencing of genomes.

Wobble Refers to the degenerate *(q.v.)* nature of the third base in many codons.

World Wide Web (www) The information resource hosted on the internet that enables almost instantaneous access to a vast amount of different types of information. Has become an essential part of everyday life, and is extensively used in all areas of science and technology.

Xenogenic Refers to a synthetic gene for insertion into a target organism. *Cf. cisgenic, intragenic, transgenic.*

Xenopus laevis African clawed toad, used by Sir John Gurdon in the first organismal cloning experiments in the 1960s and 1970s.

Xenotransplantation The use of tissues or organs from a non-human source for transplantation.

X-gal 5-bromo-4-chloro-3-indolyl-β-D-galactopyranoside – a chromogenic substrate for β-galactosidase; on cleavage, it yields a blue-coloured product.

X-linked SCIDS Form of severe combined immunodeficiency disease (SCIDS, *q.v.*) inherited as an X-linked transmission pattern rather than the autosomal version.

Y1H Yeast one-hybrid; system for assaying DNA~protein interactions.

Y2H Yeast two-hybrid; system for assaying protein~protein interactions.

Y3H Yeast three-hybrid; system for assaying protein interactions with small molecules.

YAC Yeast artificial chromosome, a vector for cloning very large pieces of DNA in yeast.

YCp Yeast centromere plasmid.

YEp Yeast episomal plasmid.

YIp Yeast integrative plasmid.

YRp Yeast replicative plasmid.

Zero-mode waveguide (ZMW) Small-volume compartment used for next-generation sequencing technology. The ZMW enables single-molecule real-time (SMRT) sequencing.

Zinc-finger nuclease (ZFN) Engineered nuclease used for gene editing, incorporates the *Fok* I restriction enzyme (*q.v.*).

Zygote Single-celled product of the fusion of a male and a female gamete (*q.v.*). Develops into an embryo by successive mitotic divisions.

Index

'F' entries refer to figures, and 'T' entries to tables.

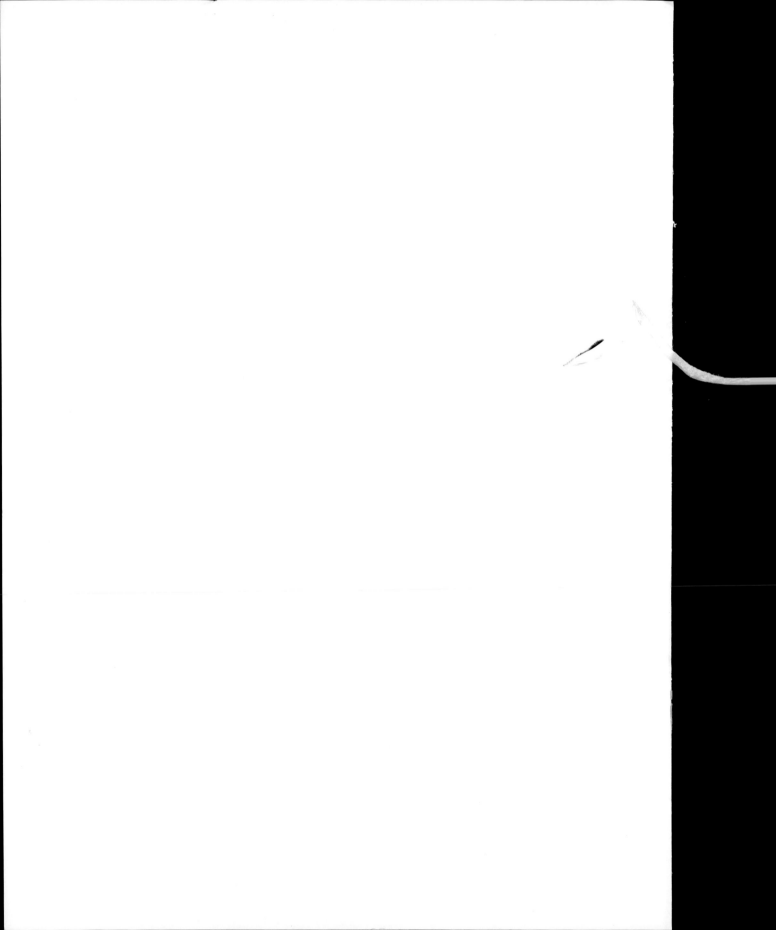